CAMBRIDGE STUDIES IN ADVANCED MATHEMATICS

Lectures in Logic and Set Theory Volume 1

This two-volume work bridges the gap between introductory expositions of logic or set theory on one hand, and the research literature on the other. It can be used as a text in an advanced undergraduate or beginning graduate course in mathematics, computer science, or philosophy. The volumes are written in a user-friendly conversational lecture style that makes them equally effective for self-study or class use.

Volume 1 includes formal proof techniques, a section on applications of compactness (including non-standard analysis), a generous dose of computability and its relation to the incompleteness phenomenon, and the first presentation of a complete proof of Gödel's second incompleteness theorem since Hilbert and Bernay's *Grundlagen*.

T0296054

LECTURES IN LOGIC
AND SET THEORY

Volume 1: Mathematical Logic

GEORGE TOURLAKIS
York University

CAMBRIDGE
UNIVERSITY PRESS

CAMBRIDGE UNIVERSITY PRESS
Cambridge, New York, Melbourne, Madrid, Cape Town, Singapore,
São Paulo, Delhi, Dubai, Tokyo, Mexico City

Cambridge University Press
The Edinburgh Building, Cambridge CB2 8RU, UK

Published in the United States of America by Cambridge University Press, New York

www.cambridge.org
Information on this title: www.cambridge.org/9780521168465

First published 2003
First paperback edition 2010

A catalogue record for this publication is available from the British Library

Library of Congress Cataloguing in Publication Data

Tourlakis, George J.
Lectures in logic and set theory / George Tourlakis.
p.   cm. – (Cambridge studies in advanced mathematics)
Includes bibliographical references and index.
Contents: v. 1. Mathematical logic – v. 2. Set theory.
ISBN 0-521-75373-2 (v. 1) – ISBN 0-521-75374-0 (v. 2)
1. Logic, Symbolic and mathematical. 2. Set theory. I. Title. II. Series.
QA9.2 .T68   2003
511.3 – dc21   2002073308

ISBN 978-0-521-75373-9 Hardback
ISBN 978-0-521-16846-5 Paperback

για την δεσποινα, την μαρινα και τον γιαννη

Contents

Preface

Both volumes in this series are about what mathematicians, especially logicians, call the "foundations" (of mathematics) – that is, the tools of the axiomatic method, an assessment of their effectiveness, and two major examples of application of these tools, namely, in the development of number theory and set theory.

There have been, in hindsight, two main reasons for writing this volume. One was the existence of notes I wrote for my lectures in mathematical logic and computability that had been accumulating over the span of several years and badly needed sorting out. The other was the need to write a small section on logic, "A Bit of Logic" as I originally called it, that would bootstrap my volume on set theory[†] on which I had been labouring for a while. Well, one thing led to another, and a 30 or so page section that I initially wrote for the latter purpose grew to become a self-standing volume of some 300 pages. You see, this material on logic is a good story and, as with all good stories, one does get carried away wanting to tell more.

I decided to include what many people will consider, I should hope, as being the absolutely essential topics in *proof*, *model*, and *recursion* theory – "absolutely essential" in the context of courses taught near the upper end of undergraduate, and at the lower end of graduate curricula in mathematics, computer science, or philosophy. But no more.[‡] This is the substance of Chapter I; hence its title "Basic Logic".

[†] A chapter by that name now carries out these bootstrapping duties – the proverbial "Chapter 0" (actually Chapter I) of volume 2.

[‡] These topics include the foundation and development of non-standard analysis up to the extreme value theorem, elementary equivalence, diagrams, and Löwenheim-Skolem theorems, and Gödel's first incompleteness theorem (along with Rosser's sharpening).

But then it occurred to me to also say something about one of the most remarkable theorems of logic – arguably *the* most remarkable – about the limitations of formalized theories: Gödel's second incompleteness theorem. Now, like most reasonable people, I never doubted that this theorem is true, but, as the devil is in the details, I decided to learn its proof – right from Peano's axioms. What better way to do this than writing down the proof, gory details and all? This is what Chapter II is about.[†]

As a side effect, the chapter includes many theorems and techniques of one of the two most important – from the point of view of foundations – "applied" logics (formalized theories), namely, Peano arithmetic (the other one, set theory, taking all of volume 2).

I have hinted above that this (and the second) volume are aimed at a fairly advanced reader: The level of exposition is designed to fit a spectrum of mathematical sophistication from third year undergraduate to junior graduate level (each group will find here its favourite sections that serve its interests and level of preparation – and should not hesitate to judiciously omit topics).

There are no specific prerequisites beyond some immersion in the "proof culture", as this is attainable through junior level courses in calculus, linear algebra, or discrete mathematics. However, some familiarity with concepts from elementary naïve set theory such as finiteness, infinity, countability, and uncountability will be an asset.[‡]

A word on approach. I have tried to make these lectures user-friendly, and thus accessible to readers who do not have the benefit of an instructor's guidance. Devices to that end include anticipation of questions, frequent promptings for the reader to rethink an issue that might be misunderstood if glossed over ("Pauses"), and the marking of important passages, by ☖, as well as those that can be skipped at first reading, by ☖ ☖.

Moreover, I give (mostly) *very* detailed proofs, as I know from experience that omitting details normally annoys students.

[†] It is strongly conjectured here that this is the only complete proof in print other than the one that was given in Hilbert and Bernays (1968). It is fair to clarify that I use the term "complete proof" with a strong assumption in mind: That the axiom system we start with is *just* Peano arithmetic. Proofs based on a stronger – thus technically more convenient – system, namely, *primitive recursive arithmetic*, have already appeared in print (Diller (1976), Smoryński (1985)). The difficulty with using Peano arithmetic as the starting point is that the only primitive recursive functions initially available are the successor, identity, plus, and times. An awful amount of work is needed – a preliminary "coding trick" – to prove that all the rest of the primitive recursive functions "exist". By then are we already midway in Chapter II, and only then are we ready to build Gödel numbers of terms, formulas, and proofs and to prove the theorem.

[‡] I have included a short paragraph nicknamed "a crash course on countable sets" (Section I.5, p. 62), which certainly helps. But having seen these topics before helps even more.

The first chapter has a lot of exercises (the second having proportionally fewer). Many of these have hints, but none are marked as "hard" vs. "just about right", a subjective distinction I prefer to avoid. In this connection here is some good advice I received when I was a graduate student at the University of Toronto: "Attempt all the problems. Those you can do, don't do. Do the ones you cannot".

What to read. Consistently with the advice above, I suggest that you read this volume from cover to cover – including footnotes! – skipping only what you already know. Now, in a class environment this advice may be impossible to take, due to scope and time constraints. An undergraduate (one semester) course in logic at the third year level will probably cover Sections I.1–I.5, making light of Section I.2, and will introduce the student to the elements of computability along with a hand-waving "proof" of Gödel's first incompleteness theorem (the "semantic version" ought to suffice). A fourth year class will probably attempt to cover the entire Chapter I. A first year graduate class has no more time than the others at its disposal, but it usually goes much faster, skipping over familiar ground, thus it will probably additionally cover Peano arithmetic and will get to see how Gödel's second theorem follows from Löb's derivability conditions.

Acknowledgments. I wish to offer my gratitude to all those who taught me, a group led by my parents and too large to enumerate. I certainly include my students here. I also include Raymond Wilder's book on the foundations of mathematics, which introduced me, long long ago, to this very exciting field and whetted my appetite for more (Wilder (1963)).

I should like to thank the staff at Cambridge University Press for their professionalism, support, and cooperation, with special appreciation due to Lauren Cowles and Caitlin Doggart, who made all the steps of this process, from refereeing to production, totally painless.

This volume is the last installment of a long project that would have not been successful without the support and warmth of an understanding family (thank you).

I finally wish to record my appreciation to Donald Knuth and Leslie Lamport for the typesetting tools $\TeX$ and $\LaTeX$ that they have made available to the technical writing community, making the writing of books such as this one almost easy.

George Tourlakis
Toronto, March 2002

I

Basic Logic

Logic is the science of reasoning. Mathematical logic applies to mathematical reasoning – the art and science of writing down *deductions*. This volume is about the *form*, *meaning*, *use*, and *limitations* of logical deductions, also called *proofs*. While the user of mathematical logic will practise the various proof techniques with a view of applying them in everyday mathematical practice, the student of the subject will also want to know about the power and limitations of the deductive apparatus. We will find that there are some inherent limitations in the quest to discover truth by purely *formal* – that is, *syntactic* – techniques. In the process we will also discover a close affinity between formal proofs and *computations* that persists all the way up to and including issues of limitations: Not only is there a remarkable similarity between the types of respective limitations (computations vs. uncomputable functions, and proofs vs. unprovable, but "true", sentences), but, in a way, you cannot have one type of limitation without having the other.

The modern use of the term mathematical logic encompasses (at least) the areas of proof theory (it studies the structure, properties, and limitations of proofs), model theory (it studies the interplay between syntax and meaning – or semantics – by looking at the algebraic structures where formal languages are interpreted), recursion theory (or computability, which studies the properties and limitations of algorithmic processes), and set theory. The fact that the last-mentioned will totally occupy our attention in volume 2 is reflected in the prominence of the term in the title of these lectures. It also reflects a tendency, even today, to think of set theory as a branch in its own right, rather than as an "area" under a wider umbrella.

1

Volume 1 is a brief study of the other three areas of logic[†] mentioned above. This is the point where an author usually apologizes for what has been omitted, blaming space or scope (or competence) limitations. Let me start by outlining what is included: "Standard" phenomena such as *completeness, compactness* and its startling application to analysis, *incompleteness* or *unprovability* (including a *complete* proof of the *second incompleteness theorem*), and a fair amount of *recursion theory* are thoroughly discussed. Recursion theory, or *computability*, is of interest to a wide range of audiences, including students with main areas of study such as computer science, philosophy, and, of course, mathematical logic. It studies among other things the phenomenon of uncomputability, which is closely related to that of unprovability, as we see in Section I.9.

Among the topics that I have deliberately left out are certain algebraic techniques in model theory (such as the method of *ultrapowers*), formal interpretations of one theory into another,[‡] the introduction of "other" logics (*modal, higher order, intuitionistic,* etc.), and several topics in recursion theory (oracle computability, Turing reducibility, recursive operators, degrees, Post's theorem in the arithmetic hierarchy, the analytic hierarchy, etc.) – but then, the decision to stop writing within 300 or so pages was firm. On the other hand, the topics included here form a synergistic whole in that I have (largely) included at every stage material that is prerequisite to what follows. The absence of a section on *propositional calculus* is deliberate, as it does not in my opinion further the understanding of logic in any substantial way, while it delays one's plunging into what really matters. To compensate, I include all tautologies as "propositional" (or Boolean) logical axioms and present a mini-course on propositional calculus in the exercises of this chapter (I.26–I.41, pp. 193–195), including the completeness and compactness of the calculus.

It is inevitable that the language of sets intrudes in this chapter (as it indeed does in all mathematics) and, more importantly, some of the results of (informal) set theory are needed here (especially in our proofs of the completeness and compactness *meta*theorems). Conversely, *formal* set theory of volume 2 needs some of the results developed here. This "chicken or egg" phenomenon is often called "bootstrapping" (not to be confused with "circularity" – which it is not[§]), the term suggesting one pulling oneself up by one's bootstraps.[¶]

[†] I trust that the reader will not object to my dropping the qualifier "mathematical" from now on.
[‡] Although this topic is included in volume 2 (Chapter I), since it is employed in the relative consistency techniques applied there.
[§] Only informal, or naïve, set theory notation and results are needed in Chapter I at the meta-level, i.e, outside the formal system that logic is.
[¶] I am told that Baron Münchhausen was the first one to apply this technique, with success.

⟨☇⟩ This is a good place to outline how our story will unfold: First, our objective is to *formalize* the rules of reasoning in general – as these apply to all mathematics – and develop their properties. In particular, we will study the interaction between formalized rules and their "intended meaning" (semantics), as well as the limitations of these formalized rules: That is, how good (= potent) are they for capturing the informal notions of truth?

Secondly, once we have acquired these tools of formalized reasoning, we start behaving (mostly[†]) as *users* of formal logic so that we can discover important theorems of two important mathematical theories: *Peano arithmetic* (Chapter II) and *set theory* (volume 2).

By *formalization* (of logic) we understand the faithful representation or simulation of the "reasoning processes" of mathematics *in general* (*pure* logic), or of a particular mathematical theory (*applied* logic: e.g., Peano arithmetic), within an activity that – in principle – is driven exclusively by the *form* or syntax of mathematical statements, totally ignoring their meaning.

We build, describe, and study the properties of this *artificial replica* of the reasoning processes – the formal theory – within "everyday mathematics" (also called "informal" or "real" mathematics), using the usual abundance of mathematical symbolism, notions, and techniques available to us, augmented by the descriptive power of English (or Greek, or French, or German, or Russian, or . . . , as particular circumstances or geography might dictate). This milieu within which we build, pursue, and study our theories is often called the *metatheory*, or more generally, *metamathematics*. The language we speak while at it, this mélange of mathematics and "natural language", is the *metalanguage*.

Formalization turns mathematical theories into mathematical objects that we can study. For example, such study may include interesting questions such as "is the continuum hypothesis provable from the axioms of set theory?" or "can we prove the consistency of (axiomatic) Peano arithmetic within Peano arithmetic?"[‡] This is analogous to building a "model airplane", a replica of the real thing, with a view of studying *through the replica* the properties, power, and limitations of the real thing.

But one can also use the formal theory to *generate theorems*, i.e., discover "truths" in the real domain by simply "running" the simulation that this theory-replica is.[§] Running the simulation "by hand" (rather than using the program

---

[†] Some tasks in Chapter II of this volume, and some others in volume 2, will be to treat the "theory" at hand as an object of study rather than using it, as a machine, to crank out theorems.

[‡] By the way, the answer to both these questions is "no" (Cohen (1963) for the first, Gödel (1938) for the second).

[§] The analogy implied in the terminology "running the simulation" is apt. For formal theories such as set theory and Peano arithmetic we can build within real mathematics a so-called "provability

of the previous footnote) means that you are acting as a "user" of the formal system, a formalist, proving theorems through it. It turns out that once you get the hang of it, it is easier and safer to reason formally than to do so informally. The latter mode often mixes syntax and semantics (meaning), and there is always the danger that the "user" may assign incorrect (i.e., convenient, but not *general*) meanings to the symbols that he[†] manipulates, a phenomenon that has distressed many a mathematics or computer science instructor.

"Formalism for the user" is hardly a revolutionary slogan. It was advocated by Hilbert, the founder of formalism, partly as a means of – as he believed[‡] – formulating mathematical theories in a manner that allows one to check them (i.e., run "diagnostic tests" on them) for freedom from contradiction,[§] but also as *the right way to "do" mathematics*. By this proposal he hoped to salvage mathematics itself, which, Hilbert felt, was about to be destroyed by the Brouwer school of intuitionist thought. In a way, his program could bridge the gap between the classical and the intuitionist camps, and there is some evidence that Heyting (an influential intuitionist and contemporary of Hilbert) thought that such a *rapprochement* was possible. After all, since meaning is irrelevant to a formalist, then all that he is doing (in a proof) is shuffling finite sequences of symbols, never having to handle or argue about infinite objects – a good thing, as far as an intuitionist is concerned.[¶]

predicate", that is, a relation $P(y, x)$ which is true of two natural numbers y and x just in case y *codes* a proof of the formula *coded* by x. It turns out that $P(y, x)$ has so simple a structure that it is programmable, say in the C programming language. But then we can write a program (also in C) as follows: "Systematically generate all the pairs of numbers (y, x). For each pair generated, if $P(y, x)$ holds, then print the formula coded by x". Letting this process run for ever, we obtain a listing of *all* the theorems of Peano arithmetic or set theory! This fact does not induce any insomnia in mathematicians, since this is an extremely impractical way to obtain theorems. By the way, we will see in Chapter II that either set theory or Peano arithmetic is sufficiently strong to *formally* express a provability predicate, and this leads to the incompletableness phenomenon.
[†] In this volume, the terms "he", "his", "him", and their derivatives are by definition gender-neutral.
[‡] This belief was unfounded, as Gödel's incompleteness theorems showed.
[§] Hilbert's *metatheory* – that is, the "world" or "lab" *outside the theory*, where the replica is actually manufactured – was *finitary*. Thus – Hilbert advocated – all this theory building and theory checking ought to be effected by *finitary means*. This ingredient of his "program" was consistent with peaceful coexistence with the intuitionists. And, alas, this ingredient was the one that – as some writers put it – destroyed Hilbert's program to found mathematics on his version of formalism. Gödel's incompleteness theorems showed that a finitary metatheory is not up to the task.
[¶] True, a formalist applies classical logic, while an intuitionist applies a different logic where, for example, double negation is not removable. Yet, unlike a Platonist, a Hilbert-style formalist does not believe – or he does not have to disclose to his intuitionist friends that he might believe – that infinite sets exist *in the metatheory*, as his tools are just finite symbol sequences. To appreciate the tension here, consider this anecdote: It is said that when Kronecker – the father of intuitionism – was informed of Lindemann's proof (1882) that π is transcendental, while he granted that this was an interesting result, he also dismissed it, suggesting that "π" – whose decimal expansion is, of

In support of the "formalism for the user" position we must definitely mention the premier paradigm, Bourbaki's monumental work (1966a), which is a formalization of a huge chunk of mathematics, including set theory, algebra, topology, and theory of integration. This work is strictly for the *user* of mathematics, not for the metamathematician who *studies* formal theories. Yet, it is fully formalized, true to the spirit of Hilbert, and it comes in a self-contained package, including a "Chapter 0" on formal logic.

More recently, the proposal to employ formal reasoning as a tool has been gaining support in a number of computer science undergraduate curricula, where logic and discrete mathematics are taught in a formalized setting, starting with a rigorous course in the two logical calculi (propositional and predicate), emphasizing the point of view of the *user* of logic (and mathematics) – hence with an attendant emphasis on "calculating" (i.e., writing and annotating formal) proofs. Pioneering works in this domain are the undergraduate text (1994) and the paper (1995) of Gries and Schneider.

I.1. First Order Languages

In the most abstract (therefore simplest) manner of describing it, a *formalized mathematical theory* consists of the following sets of things: A set of basic or primitive symbols, $\mathcal{V}$, used to build *symbol sequences* (also called strings, or expressions, or words) "over $\mathcal{V}$". A set of strings, **Wff**, over $\mathcal{V}$, called the *formulas* of the theory. Finally, a *subset* of **Wff**, called **Thm**, the set of *theorems* of the theory.[†]

Well, this is the *extension* of a theory, that is, the explicit set of objects in it. How is a theory "given"?

In most cases of interest to the mathematician it is given by $\mathcal{V}$ and two sets of simple *rules*: formula-building rules and theorem-building rules. Rules from the first set allow us to build, or *generate*, **Wff** from $\mathcal{V}$. The rules of the second set generate **Thm** from **Wff**. In short (e.g., Bourbaki (1966b)), *a theory consists of an alphabet of primitive symbols, some* rules *used to generate the "language of the theory"* (*meaning, essentially,* **Wff**) *from these symbols, and some* additional *rules used to generate the theorems.* We expand on this below:

---

course, infinite but not periodic – "does not exist" (see Wilder (1963, p. 193)). We are not to propound the tenets of intuitionism here, but it is fair to state that infinite sets *are* possible in intuitionistic mathematics as this has later evolved in the hands of Brouwer and his Amsterdam "school". However, such sets must be (like all sets of intuitionistic mathematics) *finitely generated* – just as our formal languages and the set of theorems are (the latter *provided* our axioms are too) – in a sense that may be familiar to some readers who have had a course in "automata and language theory". See Wilder (1963, p. 234)

[†] For a less abstract, but more detailed view of theories see p. 38.

I.1.1 Remark. What is a "rule"? We run the danger of becoming circular or too pedantic if we overdefine this notion. Intuitively, the rules we have in mind are string manipulation rules, that is, "black boxes" (or functions) that receive string inputs and respond with string outputs. For example, a well-known theorem-building rule receives as input a formula and a variable, and returns (essentially) the string composed of the symbol ∀, immediately followed by the variable and, in turn, immediately followed by the formula.[†]   □

(1) First off, the (*first order*) *formal language*, *L*, where the theory is "spoken",[‡] is a triple ($\mathscr{V}$, **Term**, **Wff**), that is, it has three important components, each of them a set.

> $\mathscr{V}$ is the *alphabet* or vocabulary of the language. It is the collection of the *basic* syntactic "bricks" (symbols) that we use to form *expressions* that are *terms* (members of **Term**) or *formulas* (members of **Wff**). We will ensure that the processes that build terms or formulas, using the basic building blocks in $\mathscr{V}$, are intuitively *algorithmic* or "mechanical".

> Terms will formally codify "objects", while formulas will formally codify "statements" about objects.

(2) *Reasoning* in the theory will be the process of discovering *true statements* about objects – that is, *theorems*. This discovery journey begins with certain formulas which codify statements that we take for granted (i.e., we accept without "proof" as "basic truths"). Such formulas are the *axioms*. There are two types of axioms:

> *Special* or *nonlogical* axioms are to describe specific aspects of any specific theory that we might be building. For example, "$x + 1 \neq 0$" is a special axiom that contributes towards the characterization of number theory over the natural numbers, $\mathbb{N}$.

> The other kind of axiom will be found in *all* theories. It is the kind that is "universally valid", that is, *not* theory-specific (for example, "$x = x$" is such a "universal truth"). For that reason this type of axiom will be called *logical*.

(3) Finally, we will need *rules* for reasoning, actually called *rules of inference*. These are rules that allow us to deduce, or derive, a true statement from other statements that we have already established as being true.[§] These rules will be chosen to be oblivious to meaning, being only concerned with

---

[†] This rule is usually called "generalization".

[‡] We will soon say what makes a language "first order".

[§] The generous use of the term "true" here is only meant for motivation. "Provable" or "deducible" (formula), or "theorem", will be the technically precise terminology that we will soon define to replace the term "true statement".

form. They will apply to statement "configurations" of certain *recognizable forms* and will produce (derive) new statements of some *corresponding recognizable forms* (See Remark I.1.1).

I.1.2 Remark. We may think of axioms of either logical or nonlogical type as special cases of rules, that is, rules that receive *no* input in order to produce an output. In this manner item (2) above is subsumed by item (3), and thus we are faithful to our abstract definition of theory where axioms were not mentioned.

An example, outside mathematics, of an inputless rule is the rule invoked when you type **date** on your computer keyboard. This rule receives no input, and outputs on your screen the current date. □

We next look carefully into (first order) formal languages.

There are two parts in each first order alphabet. The first, the collection of the *logical symbols*, is *common to all first order languages* regardless of which theory is "spoken" in them. We describe this part immediately below.

Logical Symbols

LS.1. *Object or individual variables.* An *object variable* is any one *symbol* out of the non-ending sequence $v_0, v_1, v_2, \ldots$. *In practice* – whether we are using logic as a tool or as an object of study – we agree to be sloppy with notation and use, *generically*, x, y, z, u, v, w with or without subscripts or primes as *names* of object variables.[†] This is just a matter of notational convenience. We allow ourselves to write, say, z instead of, say, $v_{1200000000560000009}$. Object variables (intuitively) "vary over" (i.e., are allowed to take values that are) the objects that the theory studies (numbers, sets, atoms, lines, points, etc., as the case may be).

LS.2. *The Boolean or propositional connectives.* These are the symbols "¬" and "∨".[‡] They are pronounced *not* and *or* respectively.

LS.3. *The existential quantifier*, that is, the symbol "∃", pronounced *exists* or *for some*.

LS.4. *Brackets*, that is, "(" and ")".

LS.5. *The equality predicate.* This is the symbol "=", which we use to indicate that objects are "equal". It is pronounced *equals*.

[†] Conventions such as this one are essentially agreements – effected in the metatheory – on how to be sloppy and get away with it. They are offered in the interest of user-friendliness.

[‡] The quotes are *not* part of the symbol. They serve to indicate clearly here, in particular in the case of "∨", what is part of the symbol and what is not (the following period).

The logical symbols will have a fixed interpretation. In particular, "=" will always be expected to mean *equals*.

The theory-specific part of the alphabet is not fixed, but varies from theory to theory. For example, in set theory we just add the nonlogical (or special) symbols, $\in$ and U. The first is a special *predicate symbol* (or just predicate) of *arity* 2, the second is a predicate symbol of arity 1.[†]

In number theory we adopt instead the special symbols S (intended meaning: successor, or " $+ 1$" function), $+$, $\times$, 0, $<$, and (sometimes) a symbol for the exponentiation operation (function) a^b. The first three are *function symbols* of arities 1, 2, and 2 respectively. 0 is a *constant symbol*, $<$ a predicate of arity 2, and whatever symbol we might introduce to denote a^b would have arity 2.

The following list gives the general picture.

Nonlogical Symbols

NLS.1. A (possibly empty) set of symbols for *constants*. We normally use the metasymbols[‡] a, b, c, d, e, with or without subscripts or primes, to stand for constants unless we have in mind some alternative "standard" formal notation in specific theories (e.g., $\emptyset$, 0, ω).

NLS.2. A (possibly empty) set of symbols for *predicate symbols* or *relation symbols* for each possible "arity" $n > 0$. We normally use P, Q, R generically, with or without primes or subscripts, to stand for predicate symbols. Note that $=$ is in the logical camp. Also note that theory-specific formal symbols are possible for predicates, e.g., $<$, $\in$.

NLS.3. Finally, a (possibly empty) set of symbols for *functions* for each possible "arity" $n > 0$. We normally use f, g, h, generically, with or without primes or subscripts, to stand for function symbols. Note that theory-specific formal symbols are possible for functions, e.g., $+$, $\times$.

I.1.3 Remark. (1) We have the option of assuming that each of the *logical* symbols that we named in **LS.1–LS.5** have no further "structure" and that the symbols are, ontologically, *identical to their names*, that is, they are just these exact signs drawn on paper (or on any equivalent display medium).

In this case, changing the symbols, say, $\neg$ and $\exists$ to $\sim$ and $\mathbf{E}$ respectively results in a "different" logic, but one that is, trivially, "isomorphic" to the one

---

[†] "Arity" is a term mathematicians have made up. It is derived from "ary" of "un*ary*", "bin*ary*", etc. It denotes the number of arguments needed by a symbol according to the dictates of correct syntax. Function and predicate symbols need arguments.

[‡] *Meta*symbols are *informal* (i.e., outside the formal language) symbols that we use within "everyday" or "real" mathematics – the *meta*theory – in order to describe, as we are doing here, the formal language.

we are describing: Anything that we may do in, or say about, one logic trivially translates to an equivalent activity in, or utterance about, the other as long as we systematically carry out the translations of all occurrences of $\neg$ and $\exists$ to $\sim$ and $\mathbf{E}$ respectively (or vice versa).

An alternative point of view is that the symbol names are *not* the same as (identical with) the symbols they are naming. Thus, for example, "$\neg$" names the connective we pronounce **not**, but we do not know (or care) exactly what the nature of this connective is (we only care about how it behaves). Thus, the name "$\neg$" becomes just a typographical expedient and may be replaced by other names that name the same object, **not**.

This point of view gives one flexibility in, for example, deciding how the variable symbols are "implemented". It often is convenient to think that the entire sequence of variable symbols was built from just two symbols, say, "v" and "$|$".[†] One way to do this is by saying that v_i is a name for the symbol sequence[‡]

$$\text{``} v \underbrace{|\ldots|}_{i \,|\text{'s}} \text{''}$$

Or, preferably – see (2) below – v_i might be a name for the symbol sequence

$$\text{``} v \underbrace{|\ldots|}_{i \,|\text{'s}} v \text{''}$$

Regardless of option, v_i and v_j will name distinct objects if $i \neq j$.

This is *not* the case for the *meta*variables ("abbreviated informal names") x, y, z, u, v, w. *Unless we say so explicitly otherwise, x and y may name the same formal variable, say, v_{131}.*

We will mostly abuse language and deliberately confuse names with the symbols they name. For example, we will say, e.g., "let v_{1007} *be* an object variable . . . " rather than "let v_{1007} *name* an object variable . . . ", thus *appearing* to favour option one.

(2) *Any two symbols* included in the alphabet are *distinct*. Moreover, if any of them are built from simpler "sub-symbols" – e.g., $v_0, v_1, v_2, \ldots$ might *really name* the strings $vv, v|v, v||v, \ldots$ – then none of them is a *substring* (or *subexpression*) of any other.[§]

[†] We intend these two symbols to be identical to their names. No philosophical or other purpose will be served by allowing "more indirection" here (such as "v names u, which actually names w, which actually is . . . ").

[‡] Not including the quotes.

[§] What we have stated under (2) are *requirements*, not metatheorems! That is, they are nothing of the sort that we can *prove* about our formal language within everyday mathematics.

(3) A formal language, just like a "natural" language (such as English or Greek), is "alive" and evolving. The particular type of evolution we have in mind is the one effected by *formal definitions*. Such definitions continually *add* nonlogical symbols to the language.[†]

Thus, when we say that, e.g., "∈ and U are the only nonlogical symbols of set theory", we are telling a small white lie. More accurately, we ought to have said that "∈ and U are the only 'primitive' nonlogical symbols of set theory", for we will add loads of other symbols such as $\cup, \omega, \emptyset, \subset, \subseteq$.

This evolution affects the (formal) language of *any* theory, not just set theory.   □

Wait a minute! If formal set theory is "the foundation of all mathematics", and if, ostensibly, this chapter on logic assists us to found set theory itself, then how come we are employing *natural numbers* like 1200000000560000009 as *subscripts* in the *names* of object variables? How is it permissible to already talk about "*sets* of symbols" when we are about to *found* a theory of *sets* formally? Surely we do not "have"[‡] any of these "items" yet, do we?

First off, the presence of subscripts such as 1200000000560000009 in

$$v_{1200000000560000009}$$

is a non-issue. One way to interpret what has been said in the definition is to view the various v_i as abbreviated names of the real thing, the latter being strings that employ the symbols v and | as in Remark I.1.3. In this connection saying that v_i is "implemented" as

$$v \underbrace{|\ldots|}_{i\,|\text{'s}} v \tag{1}$$

especially the use of "i" above, is only illustrative, thus totally superfluous. We can say instead that strings of type (1) *are* the variables which we define as follows *without* the help of the "natural number i" (this is a variation of how this is done in Bourbaki (1966b) and Hermes (1973)):

An "|-calculation" forms a string like this: Write a "|".[§] This is the "current string". Repeat a finite number of times: Add (i.e., concatenate) *one* | immediately to the right of the current string. Write this new string (it is now *the* current string).

[†] This phenomenon will be studied in some detail in what follows. By the way, any additions are made to the nonlogical side of the alphabet. All the logical symbols have been given, once and for all.
[‡] "Do not have" in the sense of having not formally defined – or proved to exist – or both.
[§] Without the quotes. These were placed to exclude the punctuation following.

Let us call any string that figures in some |-calculation a "|-string". A variable either *is* the string vv, or is obtained as the concatenation from left to right of v followed by an |-string, followed by v.

All we now need is the ability to generate as many as necessary distinct variables (this is the "non-ending sequence" part of the definition, p. 7): For any two variables we get a new one that is different from either one by forming the string "v, followed by the concatenation of the two |-parts, followed by v". Similarly if we had three, four, . . . variables. By the way, two strings of | are distinct iff[†] both occur in the same |-calculation, one, but not both, as the last string.

Another, more direct way to interpret what was said about object variables on p. 7 is to take the definition literally, i.e., to suppose that it speaks about the ontology of the variables.[‡] Namely, the subscript is just a a string of meaningless symbols taken from the list below:

$$0, 1, 2, 3, 4, 5, 6, 7, 8, 9$$

Again we can pretend that we know nothing about natural numbers, and whenever, e.g., we want a variable other than either of v_{123} or v_{321}, we may offer either of v_{123321} or v_{321123} as such a *new* variable.

O.K., so we have *not* used natural numbers in the definition. But we did say "sets" and also "non-ending sequence", implying the presence of *infinite* sets!

As we have already noted, on one hand we have "real mathematics", and on the other hand we have syntactic *replicas* of theories – the formal theories – that we built *within real mathematics*. Having built a formal theory, we can then choose to *use* it (acting like formalists) to generate theorems, the latter being codified as symbol sequences (formulas). Thus, the assertion "axiomatic set theory is the foundation of all mathematics" is just a colloquialism proffered in the metatheory that means that "within axiomatic set theory we can construct the known sets of mathematics, such as the reals $\mathbb{R}$ and the complex numbers $\mathbb{C}$, and moreover we can simulate what we informally do whenever we are working in real or complex analysis, algebra, topology, theory of measure and integration, functional analysis, etc., etc."

There is no circularity here, but simply an empirical boastful observation *in the metatheory* of what our simulator can do. Moreover, our metatheory does

---

[†] If and only if.

[‡] Why not just say *exactly* what a definition is meant to say rather than leave it up to interpretation? One certainly could, as in Bourbaki (1966b), make the ontology of variables crystal-clear right in the definition. Instead, we have followed the custom of more recent writings and given the definition in a quasi-sloppy manner that leaves the ontology of variables as a matter for speculation. This gives one the excuse to write footnotes like this one and remarks like I.1.3.

have sets and all sorts of other mathematical objects. In principle we can use any among those towards building or discussing the simulator, the formal theory.

Thus, the question is not whether we can use sets, or natural numbers, in our definitions, but whether restrictions apply. For example, can we use infinite sets?

If we are Platonists, then we have available in the metatheory all sorts of sets, including infinite sets, in particular the set of all natural numbers. We can use any of these items, speak about them, etc., as we please, when we are describing or building the formal theory *within our metatheory*.

Now, if we are not Platonists, then our "real" mathematical world is much more restricted. In one extreme, we have *no* infinite sets.[†]

We can still manage to define our formal language! After all, the "non-ending" sequence of object variables $v_0, v_1, v_2, \ldots$ can be *finitely generated* in at least two different ways, as we have already seen. Thus we can explain (to a true formalist or finitist) that "non-ending sequence" was an unfortunate slip of the tongue, and that we really meant to give a *procedure* of how to generate on demand a *new* object variable, different from whatever ones we may already have.

Two parting comments are in order: One, we have been somewhat selective in the use of the term "metavariable". We have called x, x', y metavariables, but have implied that the v_i are formal variables, even if they are just *names* of formal objects such that we do not know or do not care what they look like. Well, strictly speaking the abbreviations v_i are also metavariables, but they are endowed with a property that the "generic" metavariables like x, y, z' do not have: Distinct v_i names denote distinct object variables (cf. I.1.3).

Two, we should clarify that a formal theory, when *used* (i.e., the simulator is being "run") is a *generator* of strings, *not* a decider or "parser". Thus, it can *generate* any of the following: variables (if these are given by procedures), formulas and terms (to be defined), or theorems (to be defined). *Decision* issues, no matter how trivial, the system is not built to handle. These belong to the metatheory. In particular, the theory does not see whatever numbers or strings (like 12005) may be hidden in a variable name (such as v_{12005}).

Examples of decision questions: Is this string a term or a formula or a variable (finitely generated as above)? All these questions are "easy". They are algorithmically decidable in the metatheory. Or, is this formula a theorem? This is

---

[†] A finitist – and don't forget that Hilbert-style metatheory was finitary, ostensibly for political reasons – will let you have as many integers as you like in one serving, as long as the serving is *finite*. If you ask for more, you can have more, but never the set of all integers or an infinite subset thereof.

algorithmically undecidable in the metatheory if it is a question about Peano arithmetic or set theory.

I.1.4 Definition (Terminology about Strings). A symbol sequence or *expression* (or *string*) that is formed by using symbols exclusively out of a given set[†] *M* is called *a string over the set, or alphabet, M*.

If *A* and *B* denote strings (say, over *M*), then the symbol $A * B$, or more simply AB, denotes the symbol sequence obtained by listing first the symbols of *A* in the given left to right sequence, immediately followed by the symbols of *B* in the given left to right sequence. We say that *AB* is (more properly, denotes or names) the *concatenation* of the strings *A* and *B* in that order.

We denote the fact that the strings (named) *C* and *D* are *identical sequences* (but we just say that they are *equal*) by writing $C \equiv D$. The symbol $\not\equiv$ denotes the negation of the string equality symbol $\equiv$. Thus, if # and ? are (we do mean "are") symbols from an alphabet, then

$$\#?? \equiv \#?? \quad \text{but} \quad \#? \not\equiv \#??$$

We can also employ $\equiv$ in contexts such as "let $A \equiv \#\#?$", where we give the name *A* to the string ##?.[‡]

 In this book the symbol $\equiv$ will be exclusively *used in the metatheory* for equality of strings over some set *M*.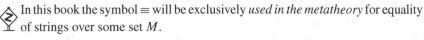

The symbol λ normally denotes the *empty* string, and we postulate for it the following behaviour:

$$A \equiv A\lambda \equiv \lambda A \qquad \text{for all strings } A$$

We say that *A occurs in B*, or is a *substring of B*, iff there are strings *C* and *D* such that $B \equiv CAD$.

For example, "(" occurs four times in the (explicit) string "¬(()∨)((", at *positions* 2, 3, 7, 8. Each time this happens we have an *occurrence* of "(" in "¬(()∨)((".

If $C \equiv \lambda$, we say that *A* is a *prefix* of *B*. If moreover $D \not\equiv \lambda$, then we say that *A* is a *proper prefix* of *B*. □

[†] A set that supplies symbols to be used in building strings is not special. It is just a set. However, it often has a special name: "alphabet".

[‡] Punctuation such as "." is not part of the string. One often avoids such footnotes by enclosing strings that are explicitly written as symbol sequences inside quotes. For example, if *A* stands for the string #, one writes $A \equiv$ "#". Note that we must not write "*A*", unless we mean a string whose only symbol *is A*.

I.1.5 Definition (Terms). The set of *terms*, **Term**, is the *smallest* set of strings over the alphabet $\mathscr{V}$ with the following two properties:

(1) All of the items in **LS.1** or **NLS.1** (x, y, z, a, b, c, etc.) are included.
(2) If f is a function[†] of arity n and $t_1, t_2, \ldots, t_n$ are included, then so is the string "$f t_1 t_2 \ldots t_n$".

The symbols t, s, and u, with or without subscripts or primes, will denote arbitrary terms. Since we are using them in the *metalanguage* to "vary over" terms, we naturally call them metavariables. They also serve – as variables – towards the definition (this one) of the *syntax* of terms. For this reason they are also called *syntactic variables*. □

I.1.6 Remark. (1) We often abuse notation and write $f(t_1, \ldots, t_n)$ instead of $f t_1 \ldots t_n$.

(2) Definition I.1.5 is an *inductive definition*.[‡] It defines a more or less "complicated" term by assuming that we already know what "simpler" terms look like. This is a standard technique employed in real mathematics. We will have the opportunity to say more about such inductive definitions – and their appropriateness – in a ☞☞-comment later on.

(3) We relate this particular manner of defining terms to our working definition of a theory (given on p. 6 immediately before Remark I.1.1 in terms of "rules" of formation). Item (2) in I.1.5 essentially says that we build new terms (from old ones) by applying the following *general rule*: Pick an arbitrary function symbol, say f. This has a specific formation rule associated with it that, for the appropriate number, n, of an already existing ordered list of terms, $t_1, \ldots, t_n$, will build the new term consisting of f, immediately followed by the ordered list of the given terms.

To be specific, suppose we are working in the language of number theory. There is a function symbol $+$ available there. The rule associated with $+$ builds the new term $+ts$ for any prior obtained terms t and s. For example, $+v_1 v_{13}$ and $+v_{121} + v_1 v_{13}$ are well-formed terms. We normally write terms of number theory in "infix" notation,[§] i.e., $t + s$, $v_1 + v_{13}$ and $v_{121} + (v_1 + v_{13})$ (note the intrusion of brackets, to indicate sequencing in the application of $+$).

[†] We will omit from now on the qualification "symbol" from terminology such as "function symbol", "constant symbol", "predicate symbol".
[‡] Some mathematicians will absolutely insist that we call this a *recursive* definition and reserve the term "induction" for "induction *proofs*". This is seen to be unwarranted hair splitting if we consider that Bourbaki (1966b) calls *induction proofs "démonstrations par récurrence"*. We will be less dogmatic: Either name is all right.
[§] Function symbol placed between the arguments.

A by-product of what we have just described is that *the arity of a function symbol f is whatever number of terms the associated rule will require as input.*

(4) A crucial word used in I.1.5 (which recurs in all inductive definitions) is "smallest". It means "least inclusive" (set). For example, we may easily think of a set of strings that satisfies both conditions of the above definition, but which is *not* "smallest" by virtue of having additional elements, such as the string "¬¬(".

Pause. Why is "¬¬(" *not* in the smallest set as defined above, and therefore not a term?

The reader may wish to ponder further on the import of the qualification "smallest" by considering the familiar (similar) example of $\mathbb{N}$, the set of natural numbers. The principle of induction in $\mathbb{N}$ ensures that this set is the *smallest* with the properties:

 (i) 0 is included, and
(ii) if n is included, then so is $n + 1$.

By contrast, all of $\mathbb{Z}$ (set of integers), $\mathbb{Q}$ (set of rational numbers), $\mathbb{R}$ (set of real numbers) satisfy (i) and (ii), but they are clearly *not* the "smallest" such.   □

I.1.7 Definition (Atomic Formulas). The set of *atomic formulas*, **Af**, contains precisely:

(1) The strings $t = s$ for every possible choice of terms t, s.
(2) The strings $Pt_1t_2 \ldots t_n$ for every possible choice of n-ary predicates P (for all choices of $n > 0$) and all possible choices of terms $t_1, t_2, \ldots, t_n$.   □

 We often abuse notation and write $P(t_1, \ldots, t_n)$ instead of $Pt_1 \ldots t_n$.

I.1.8 Definition (Well-Formed Formulas). The set of *well-formed formulas*, **Wff**, is the *smallest* set of strings or expressions over the alphabet $\mathscr{V}$ with the following properties:

(a) All the members of **Af** are included.
(b) If $\mathscr{A}$ and $\mathscr{B}$ denote strings (over $\mathscr{V}$) that are included, then $(\mathscr{A} \vee \mathscr{B})$ and $(\neg\mathscr{A})$ are also included.
(c) If $\mathscr{A}$ is[†] a string that is included and x is *any* object variable (*which may or may not occur (as a* substring*) in the string* $\mathscr{A}$), then the string $((\exists x)\mathscr{A})$ is also included. We say that $\mathscr{A}$ is the *scope* of $(\exists x)$.   □

[†] Denotes!

I.1.9 Remark.

(1) The above is yet another inductive definition. Its statement (*in the metalanguage*) is facilitated by the use of so-called *syntactic*, or meta, variables – $\mathcal{A}$ and $\mathcal{B}$ – used as *names* for *arbitrary* (indeterminate) formulas. In general, we will let calligraphic capital letters $\mathcal{A}$, $\mathcal{B}$, $\mathcal{C}$, $\mathcal{D}$, $\mathcal{E}$, $\mathcal{F}$, $\mathcal{G}$ (with or without primes or subscripts) be names for well-formed formulas, or just *formulas*, as we often say. The definition of **Wff** given above is standard. In particular, it permits well-formed formulas such as $((\exists x)((\exists x)x = 0))$ in the interest of making the formation rules "context-free".[†]

(2) The rules of syntax just given do not allow us to write things such as $\exists f$ or $\exists P$ where f and P are function and predicate symbols respectively. That quantification is deliberately restricted to act solely on object variables makes the language *first order*.

(3) We have already indicated in Remark I.1.6 where the arities (of function and predicate symbols) come from (Definitions I.1.5 and I.1.7 referred to them). These are numbers that are implicit ("hardwired") with the formation rules for terms and atomic formulas. Each function and each predicate symbol (e.g., $+$, $\times$, $\in$, $<$) has its own unique formation rule. This rule "knows" how many terms are needed (on the input side) in order to form a term or atomic formula. Therefore, since the theory, *in use*, applies rather than studies its formation rules, it is, in particular, ignorant of arities of symbols.

Now that this jurisdictional point has been made (cf. the concluding remarks about decision questions, on p. 12), we can consider an alternative way of making arities of symbols known (in the *metatheory*): Rather than embedding arities in the formation rules, we can hide them in the ontology of the symbols, not making them explicit in the name.

For example, a new symbol, say $*$, can be used to record arity. That is, we can think of a predicate (or function) symbol as consisting of two parts: an arity part and an "all the rest" part, the latter needed to render the symbol unique.[‡] For example, $\in$ may be actually the name for the symbol "$\in$**", where this latter name is identical to the symbol it denotes, or "what you see is what you get" – see Remark I.1.3(1) and (2), p. 8. The presence of the two asterisks declares the arity. Some people say this differently: They make available to the metatheory a "function", *ar*, from "the set of

---

[†] In some presentations, the formation rule in I.1.8(c) is "context-sensitive": It requires that x be *not* already quantified in $\mathcal{A}$.

[‡] The reader may want to glimpse ahead, on p. 166, to see a possible implementation in the case of number theory.

all predicate symbols and functions" (of a given language) to the natural numbers, so that for any function symbol f or predicate symbol P, $ar(f)$ and $ar(P)$ yield the arities of f and P respectively.[†]

(4) *Abbreviations*

Abr1. The string $((\forall x)\mathscr{A})$ abbreviates the string "$(\neg((\exists x)(\neg\mathscr{A})))$". Thus, for any explicitly written formula $\mathscr{A}$, the former notation is informal (metamathematical), while the latter is formal (within the formal language). In particular, $\forall$ is a metalinguistic symbol. "$\forall x$" is the *universal quantifier*. $\mathscr{A}$ is its scope. The symbol $\forall$ is pronounced *for all*.

We also introduce – in the metalanguage – a number of additional Boolean connectives in order to abbreviate certain strings:

Abr2. (*Conjunction*, $\wedge$) $(\mathscr{A}\wedge\mathscr{B})$ stands for $(\neg((\neg\mathscr{A})\vee(\neg\mathscr{B})))$. The symbol $\wedge$ is pronounced *and*.

Abr3. (*Classical or material implication*, $\rightarrow$) $(\mathscr{A}\rightarrow\mathscr{B})$ stands for $((\neg\mathscr{A})\vee\mathscr{B})$. $(\mathscr{A}\rightarrow\mathscr{B})$ is pronounced *if $\mathscr{A}$, then $\mathscr{B}$*.

Abr4. (*Equivalence*, $\leftrightarrow$) $(\mathscr{A}\leftrightarrow\mathscr{B})$ stands for $((\mathscr{A}\rightarrow\mathscr{B})\wedge(\mathscr{B}\rightarrow\mathscr{A}))$.

Abr5. To minimize the use of brackets in the metanotation we adopt standard *priorities* of connectives: $\forall$, $\exists$, and $\neg$ have the highest, and then we have (in decreasing order of priority) $\wedge$, $\vee$, $\rightarrow$, $\leftrightarrow$, and we agree not to use outermost brackets. All *associativities* are *right* – that is, if we write $\mathscr{A}\rightarrow\mathscr{B}\rightarrow\mathscr{C}$, then this is a (sloppy) counterpart for $(\mathscr{A}\rightarrow(\mathscr{B}\rightarrow\mathscr{C}))$.

(5) The language just defined, L, is *one-sorted*, that is, it has a single *sort* or *type* of object variable. Is this not inconvenient? After all, our set theory (volume 2 of these lectures) will have both *atoms* and *sets*. In other theories, e.g., geometry, one has points, lines, and planes. One would have hoped to have different "types" of variables, one for each.

Actually, to do this would amount to a totally unnecessary complication of syntax. We can (and will) get away with just *one* sort of object variable. For example, in set theory we will also introduce a 1-ary[‡] predicate, U, whose job is to "test" an object for "sethood".[§] Similar remedies are available to other theories. For example, geometry will manage with one sort of variable and unary predicates "Point", "Line", and "Plane".

[†] In mathematics we understand a function as a set of input–output pairs. One can "glue" the two parts of such pairs together, as in "$\in$**" – where "$\in$" is the input part and "**" is the output part, the latter denoting "2" – etc. Thus, the two approaches are equivalent.

[‡] More commonly called *unary*.

[§] People writing about, or teaching, set theory have made this word up. Of course, one means by it the property of being a set.

Apropos language, some authors emphasize the importance of the nonlogical symbols, taking at the same time the formation rules for granted; thus they say that we have a *language*, say, "$L = \{\in, U\}$" rather than "$L = (\mathcal{V}, \textbf{Term}, \textbf{Wff})$" where $\mathcal{V}$ has $\in$ and U as its only nonlogical symbols". That is, they use "language" for the nonlogical part of the alphabet.   □

A variable that is quantified is *bound in the scope of the quantifier*. Non-quantified variables are *free*. We also give below, by induction on formulas, precise (metamathematical) definitions of "free" and "bound".

I.1.10 Definition (Free and Bound Variables). An object variable x occurs *free* in a term t or atomic formula $\mathscr{A}$ iff it occurs in t or $\mathscr{A}$ as a substring (see I.1.4).

x occurs free in $(\neg \mathscr{A})$ iff it occurs free in $\mathscr{A}$.

x occurs free in $(\mathscr{A} \vee \mathscr{B})$ iff it occurs free in at least one of $\mathscr{A}$ or $\mathscr{B}$.

x occurs free in $((\exists y)\mathscr{A})$ iff x occurs free in $\mathscr{A}$, *and* y is not the same variable as x.[†]

The y in $((\exists y)\mathscr{A})$ is, of course, *not* free – even if it might be so in $\mathscr{A}$ – as we have just concluded in this inductive definition. We say that it is *bound* in $((\exists y)\mathscr{A})$. Trivially, terms and atomic formulas have no bound variables.   □

I.1.11 Remark. (1) Of course, Definition I.1.10 takes care of the defined connectives as well, via the obvious translation procedure.

(2) *Notation.* If $\mathscr{A}$ is a formula, then we often write $\mathscr{A}[y_1, \dots, y_k]$ to indicate our interest in the variables $y_1, \dots, y_k$, which may or may not be free in $\mathscr{A}$. Indeed, there may be other free variables in $\mathscr{A}$ that we may have chosen not to include in the list.

On the other hand, if we use *round* brackets, as in $\mathscr{A}(y_1, \dots, y_k)$, then we are implicitly asserting that $y_1, \dots, y_k$ is the *complete list* of free variables that occur in $\mathscr{A}$.   □

I.1.12 Definition. A term or formula is *closed* iff no free variables occur in it. A closed formula is called a *sentence*.

A formula is *open* iff it contains no quantifiers (thus, an open formula may also be closed).   □

[†] Recall that x and y are abbreviations of names such as $v_{1200098}$ and v_{11009} (which name distinct variables). However, it could be that both x and y name v_{101}. Therefore it is *not* redundant to say "*and* y is not the same variable as x". By the way, $x \not\equiv y$ says the same thing, by I.1.4.

I.2. A Digression into the Metatheory:
Informal Induction and Recursion

We have already seen a number of inductive or recursive definitions in Section I.1. The reader, most probably, has already seen or used such definitions elsewhere.

We will organize the common important features of inductive definitions in this section, for easy reference. We just want to ensure that our grasp of these notions and techniques, *at the metamathematical level*, is sufficient for the needs of this volume.

One builds a set S by *recursion*, or *inductively* (or by induction), out of two ingredients: a set of *initial objects*, $\mathcal{I}$, and a set of *rules* or *operations*, $\mathcal{R}$. A member of $\mathcal{R}$ – a rule – is a (possibly infinite) *table*, or *relation*, like

y_1	$\cdots$	y_n	z
a_1	$\cdots$	a_n	a_{n+1}
b_1	$\cdots$	b_n	b_{n+1}
$\vdots$		$\vdots$	$\vdots$

If the above rule (table) is called Q, then we use the notations[†]

$$Q(a_1, \ldots, a_n, a_{n+1}) \quad \text{and} \quad \langle a_1, \ldots, a_n, a_{n+1} \rangle \in Q$$

interchangeably to indicate that the *ordered sequence* or "row" $a_1, \ldots, a_n, a_{n+1}$ is present in the table.

We say that "$Q(a_1, \ldots, a_n, a_{n+1})$ holds" or "$Q(a_1, \ldots, a_n, a_{n+1})$ is true", but we often also say that "Q applied to $a_1, \ldots, a_n$ yields a_{n+1}", or that "a_{n+1} is a *result* or *output* of Q, when the latter receives *input* $a_1, \ldots, a_n$". We often abbreviate such inputs using *vector notation*, namely, $\vec{a}_n$ (or just $\vec{a}$, if n is understood). Thus, we may write $Q(\vec{a}_{n+1})$ for $Q(a_1, \ldots, a_n, a_{n+1})$.

A rule Q that has $n + 1$ columns is called $(n + 1)$-*ary*.

I.2.1 Definition. We say "a set T is *closed under an* $(n + 1)$-*ary rule* Q" to mean that whenever $c_1, \ldots, c_n$ are all in T, then $d \in T$ for *all* d satisfying $Q(c_1, \ldots, c_n, d)$. $\qquad \square$

With these preliminary understandings out of the way, we now state

---

[†] "$x \in A$" means that "x is a member of – or is in – A" in the informal set-theoretic sense.

I.2.2 Definition. S is *defined by recursion*, or by *induction*, from initial objects $\mathscr{I}$ and set of rules $\mathscr{R}$, provided it is the *smallest* (least inclusive) set with the properties

(1) $\mathscr{I} \subseteq S$,[†]
(2) S is closed under every Q in $\mathscr{R}$. In this case we say that S is $\mathscr{R}$-closed.

We write $S = \mathrm{Cl}(\mathscr{I}, \mathscr{R})$, and say that "$S$ is the *closure of $\mathscr{I}$ under $\mathscr{R}$*". □

We have at once:

I.2.3 Metatheorem (Induction on S). *If* $S = \mathrm{Cl}(\mathscr{I}, \mathscr{R})$ *and if some set T satisfies*

(1) $\mathscr{I} \subseteq T$, *and*
(2) T *is closed under every Q in $\mathscr{R}$,*

then $S \subseteq T$.

Pause. Why is the above a *meta*theorem?

The above principle of induction on S is often rephrased as follows: To prove that a property $P(x)$ holds for all members of $\mathrm{Cl}(\mathscr{I}, \mathscr{R})$, just prove that

(a) every member of $\mathscr{I}$ has the property, and
(b) the *property propagates with every rule in $\mathscr{R}$*, i.e., if $P(c_i)$ holds (is true) for $i = 1, \ldots, n$, and if $Q(c_1, \ldots, c_n, d)$ holds, then d too has the property $P(x)$ – that is, $P(d)$ holds.

Of course, this rephrased principle is valid, for if we let T be the set of all objects that have property $P(x)$ – for which set one employs the well-established symbol $\{x : P(x)\}$ – then this T satisfies (1) and (2) of the metatheorem.[‡]

I.2.4 Definition (Derivations and Parses). A $(\mathscr{I}, \mathscr{R})$-*derivation*, or simply *derivation* – if $\mathscr{I}$ and $\mathscr{R}$ are understood – is a finite sequence of objects $d_1, \ldots, d_n$

[†] From our knowledge of elementary informal set theory, we recall that $A \subseteq B$ means that every member of A is also a member of B.

[‡] We are sailing too close to the wind here! It turns out that not all properties $P(x)$ lead to *sets* $\{x : P(x)\}$. Our explanation was naïve. However, formal set theory, which is meant to save us from our naïveté, upholds the "principle" (a)–(b) using just a slightly more complicated explanation. The reader can see this explanation in our volume 2 in the chapter on cardinality.

$(n \geq 1)$ such that each d_i is

(1) a member of $\mathscr{T}$, or[†]
(2) for some $(r + 1)$-ary $Q \in \mathscr{R}$, $Q(d_{j_1}, \ldots, d_{j_r}, d_i)$ holds, and $j_l < i$ for $l = 1, \ldots, r$.

We say that d_i is *derivable within i steps*.
A derivation of an object A is also called a *parse* of *a*. □

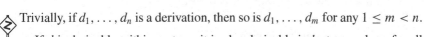 Trivially, if $d_1, \ldots, d_n$ is a derivation, then so is $d_1, \ldots, d_m$ for any $1 \leq m < n$.

If d is derivable within n steps, it is also derivable in k steps or less, for all $k > n$, since we can lengthen a derivation arbitrarily by adding $\mathscr{T}$-elements to it.

I.2.5 Remark. The following metatheorem shows that there is a way to "construct" $\mathrm{Cl}(\mathscr{T}, \mathscr{R})$ iteratively, i.e., one element at a time *by repeated application of the rules.*

This result shows definitively that our inductive definitions of terms (I.1.5) and well-formed formulas (I.1.8) fully conform with our working definition of theory, as an alphabet and a set of rules that are used to build formulas and theorems (p. 5). □

I.2.6 Metatheorem.

$\mathrm{Cl}(\mathscr{T}, \mathscr{R}) = \{x : x \text{ is } (\mathscr{T}, \mathscr{R})\text{-derivable within some number of steps, } n\}$

Proof. For notational convenience let us write

$$T = \{x : x \text{ is } (\mathscr{T}, \mathscr{R})\text{-derivable within some number of steps, } n\}.$$

As we know from elementary naïve set theory, we need to show here both $\mathrm{Cl}(\mathscr{T}, \mathscr{R}) \subseteq T$ and $\mathrm{Cl}(\mathscr{T}, \mathscr{R}) \supseteq T$ to settle the claim.

($\subseteq$) We do induction on $\mathrm{Cl}(\mathscr{T}, \mathscr{R})$ (using I.2.3). Now $\mathscr{T} \subseteq T$, since every member of $\mathscr{T}$ is derivable in $n = 1$ step. (Why?)

Also, T is closed under every Q in $\mathscr{R}$. Indeed, let such an $(r + 1)$-ary Q be chosen, and assume

$$Q(a_1, \ldots, a_r, b) \qquad\qquad (i)$$

[†] This "or" is inclusive: (1), or (2), or both.

and $\{a_1, \ldots, a_r\} \subseteq T$. Thus, each a_i has a $(\mathscr{I}, \mathscr{R})$-derivation. Concatenate all these derivations:

$$\ldots, a_1, \ldots, a_2, \ldots, \ldots, a_r$$

The above is a derivation (why?). But then, so is

$$\ldots, a_1, \ldots, a_2, \ldots, \ldots, a_r, b$$

by (i). Thus, $b \in T$.

($\supseteq$) We argue this – that is, "if $d \in T$, then $d \in \text{Cl}(\mathscr{I}, \mathscr{R})$" – by induction on the number of steps, n, in which d is derivable.

For $n = 1$ we have $d \in \mathscr{I}$ and we are done, since $\mathscr{I} \subseteq \text{Cl}(\mathscr{I}, \mathscr{R})$.

Let us make the induction hypothesis (I.H.) that for derivations of $\leq n$ steps the claim is true. Let then d be derivable within $n + 1$ steps. Thus, there is a derivation $a_1, \ldots, a_n, d$.

Now, if $d \in \mathscr{I}$, we are done as above (is this a "real case"?) If on the other hand $Q(a_{j_1}, \ldots, a_{j_r}, d)$, then for $i = 1, \ldots, r$ we have $a_{j_i} \in \text{Cl}(\mathscr{I}, \mathscr{R})$ by the I.H.; hence $d \in \text{Cl}(\mathscr{I}, \mathscr{R})$, since the closure is closed under all $Q \in \mathscr{R}$. □

I.2.7 Example. One can see now that $\mathbb{N} = \text{Cl}(\mathscr{I}, \mathscr{R})$, where $\mathscr{I} = \{0\}$ and $\mathscr{R}$ contains just the relation $y = x + 1$ (input x, output y). Similarly, $\mathbb{Z}$, the set of all integers, is $\text{Cl}(\mathscr{I}, \mathscr{R})$, where $\mathscr{I} = \{0\}$ and $\mathscr{R}$ contains just the relations $y = x + 1$ and $y = x - 1$ (input x, output y).

For the latter, the inclusion $\text{Cl}(\mathscr{I}, \mathscr{R}) \subseteq \mathbb{Z}$ is trivial (by I.2.3). For $\supseteq$ we easily see that any $n \in \mathbb{Z}$ has a $(\mathscr{I}, \mathscr{R})$-derivation (and then we are done by I.2.6). For example, if $n > 0$, then $0, 1, 2, \ldots, n$ is a derivation, while if $n < 0$, then $0, -1, -2, \ldots, n$ is one. If $n = 0$, then the one-term sequence 0 is a derivation.

Another interesting closure is obtained by $\mathscr{I} = \{3\}$ and the two relations $z = x + y$ and $z = x - y$. This is the set $\{3k : k \in \mathbb{Z}\}$ (see Exercise I.1). □

Pause. So, taking the first sentence of I.2.7 one step further, we note that we have just *proved* the *induction principle* for $\mathbb{N}$, for that is exactly what the "equation" $\mathbb{N} = \text{Cl}(\mathscr{I}, \mathscr{R})$ says (by I.2.3). Do you agree?

There is another way to view the iterative construction of $\text{Cl}(\mathscr{I}, \mathscr{R})$: The set is constructed *in stages*. Below we are using some more notation borrowed from informal set theory. For any sets A and B we write $A \cup B$ to indicate the set *union*, which consists of all the members found in A or B or in both. More generally, if we have a lot of sets, $X_0, X_1, X_2, \ldots$, that is, one X_i for every integer $i \geq 0$ – which we denote by the compact notation $(X_i)_{i \geq 0}$ – then we may wish to form a set that includes *all* the objects found as members all over the X_i, that is (using *inclusive*, or logical, "or"s below), form

$$\{x : x \in X_0 \text{ or } x \in X_1 \text{ or} \ldots\}$$

or, more elegantly and precisely,

$$\{x : \text{for some } i \geq 0, \ x \in X_i\}$$

The latter is called the union of the sequence $(X_i)_{i \geq 0}$ and is often denoted by

$$\bigcup_{i \geq 0} X_i \quad \text{or} \quad \bigcup_{i \geq 0} X_i$$

Correspondingly, we write

$$\bigcup_{i \leq n} X_i \quad \text{or} \quad \bigcup_{i \leq n} X_i$$

if we only want to take a finite union, also indicated clumsily as $X_0 \cup \ldots \cup X_n$.

I.2.8 Definition (Stages). In connection with $\text{Cl}(\mathscr{I}, \mathscr{R})$ we define the sequence of sets $(X_i)_{i \geq 0}$ by induction on n, as follows:

$$X_0 = \mathscr{I}$$

$$X_{n+1} = \left(\bigcup_{i \leq n} X_i \right)$$

$$\cup \left\{ b : \text{for some } Q \in \mathscr{R} \text{ and some } \vec{a}_n \text{ in } \bigcup_{i \leq n} X_i, \ Q(\vec{a}_n, b) \right\}$$

That is, to form X_{n+1} we append to $\bigcup_{i \leq n} X_i$ all the outputs of all the relations in $\mathscr{R}$ acting on all possible inputs, the latter taken from $\bigcup_{i \leq n} X_i$.

We say that X_i is built at *stage i*, from initial objects $\mathscr{I}$ and rule-set $\mathscr{R}$. ☐

In words, at stage 0 we are given the initial objects ($X_0 = \mathscr{I}$). At stage 1 we apply all possible relations to all possible objects *that we have so far* – they form the set X_0 – and build the 1st stage set, X_1, by appending the outputs to what we have so far. At stage 2 we apply all possible relations to all possible objects *that we have so far* – they form the set $X_0 \cup X_1$ – and build the 2nd stage set, X_2, by appending the outputs to what we have so far. And so on.

When we work in the metatheory, we take for granted that we can have simple inductive definitions on the natural numbers. The reader is familiar with several such definitions, e.g.,

$$a^0 = 1 \quad \text{(for } a \neq 0 \text{ throughout)}$$
$$a^{n+1} = a \cdot a^n$$

We will (meta)prove a general theorem on the feasibility of recursive definitions later on (I.2.13).

The following theorem connects stages and closures.

I.2.9 Metatheorem. *With the X_i as in I.2.8,*

$$\mathrm{Cl}(\mathscr{I}, \mathscr{R}) = \bigcup_{i \geq 0} X_i$$

Proof. ($\subseteq$) We do induction on $\mathrm{Cl}(\mathscr{I}, \mathscr{R})$. For the basis, $\mathscr{I} = X_0 \subseteq \bigcup_{i \geq 0} X_i$.

We show that $\bigcup_{i \geq 0} X_i$ is $\mathscr{R}$-closed. Let $Q \in \mathscr{R}$ and $Q(\vec{a}_n, b)$ hold, for some $\vec{a}_n$ in $\bigcup_{i \geq 0} X_i$. Thus, by definition of union, there are integers $j_1, j_2, \ldots, j_n$ such that $a_i \in X_{j_i}, i = 1, \ldots, n$. If $k = \max\{j_1, \ldots, j_n\}$, then $\vec{a}_n$ is in $\bigcup_{i \leq k} X_i$; hence $b \in X_{k+1} \subseteq \bigcup_{i \geq 0} X_i$.

($\supseteq$) It suffices to prove that $X_n \subseteq \mathrm{Cl}(\mathscr{I}, \mathscr{R})$, a fact we can prove by induction on n. For $n = 0$ it holds by I.2.2. As an I.H. we assume the claim for all $n \leq k$.

The case for $k + 1$: X_{k+1} is the union of two sets. One is $\bigcup_{i \leq k} X_i$. This is a subset of $\mathrm{Cl}(\mathscr{I}, \mathscr{R})$ by the I.H. The other is

$$\left\{ b : \text{ for some } Q \in \mathscr{R} \text{ and some } \vec{a} \text{ in } \bigcup_{i \leq k} X_i, \, Q(\vec{a}, b) \right\}$$

This too is a subset of $\mathrm{Cl}(\mathscr{I}, \mathscr{R})$, by the preceding observation and the fact that $\mathrm{Cl}(\mathscr{I}, \mathscr{R})$ is $\mathscr{R}$-closed. $\square$

 Worth Saying. An inductively defined set can be built by stages.

I.2.10 Definition (Immediate Predecessors, Ambiguity). If $d \in \mathrm{Cl}(\mathscr{I}, \mathscr{R})$ and for some Q and $a_1, \ldots, a_r$ it is the case that $Q(a_1, \ldots, a_r, d)$, then the $a_1, \ldots, a_r$ are *immediate Q-predecessors* of d, or just *immediate predecessors* if Q is understood; for short, i.p.

A pair $(\mathscr{I}, \mathscr{R})$ is called *ambiguous* if some $d \in \mathrm{Cl}(\mathscr{I}, \mathscr{R})$ satisfies any (or all) of the following conditions:

  (i) It has two (or more) distinct sets of immediate P-predecessors for some rule P.
  (ii) It has both immediate P-predecessors *and* immediate Q-predecessors, for $P \neq Q$.
  (iii) It is a member of $\mathscr{I}$, yet it has immediate predecessors.

If $(\mathscr{I}, \mathscr{R})$ is not ambiguous, then it is *unambiguous*. $\square$

I.2.11 Example. The pair $(\{00,0\}, \{Q\})$, where $Q(x, y, z)$ holds iff $z = xy$ (where "xy" denotes the concatenation of the strings x and y, in that order), is ambiguous. For example, 0000 has the two immediate predecessor sets $\{00,00\}$ and $\{0,000\}$. Moreover, while 00 is an initial object, it does have immediate predecessors, namely, the set $\{0,0\}$ (or, what amounts to the same thing, $\{0\}$). $\quad\square$

I.2.12 Example. The pair $(\mathscr{I}, \mathscr{R})$, where $\mathscr{I} = \{3\}$ and $\mathscr{R}$ consists of $z = x + y$ and $z = x - y$, is ambiguous. Even 3 has (infinitely many) distinct sets of i.p. (e.g., any $\{a, b\}$ such that $a + b = 3$, or $a - b = 3$).

The pairs that effect the definition of **Term** (I.1.5) and **Wff** (I.1.8) are unambiguous (see Exercises I.2 and I.3). $\quad\square$

I.2.13 Metatheorem (Definition by Recursion). *Let* $(\mathscr{I}, \mathscr{R})$ *be* unambiguous *and* $\mathrm{Cl}(\mathscr{I}, \mathscr{R}) \subseteq A$, *where A is some set. Let also Y be a set, and*[†] $h : \mathscr{I} \to Y$ *and g_Q, for each $Q \in \mathscr{R}$, be given functions. For any $(r+1)$-ary Q, an input for the function g_Q is a sequence $\langle a, b_1, \ldots, b_r \rangle$ where a is in A and the $b_1, \ldots, b_r$ are all in Y. All the g_Q yield outputs in Y.*

Under these assumptions, there is a unique *function* $f : \mathrm{Cl}(\mathscr{I}, \mathscr{R}) \to Y$ *such that*

$$y = f(x) \text{ iff } \begin{cases} y = h(x) \text{ and } x \in \mathscr{I} \\ \textbf{or}, \textit{for some } Q \in \mathscr{R}, \\ y = g_Q(x, o_1, \ldots, o_r) \text{ and } Q(a_1, \ldots, a_r, x) \textit{ holds}, \\ \textit{where } o_i = f(a_i), \textit{ for } i = 1, \ldots, r \end{cases} \quad (1)$$

The reader may wish to skip the proof on first reading.

Proof. Existence part. For each $(r + 1)$-ary $Q \in \mathscr{R}$, define $\widehat{Q}$ by[‡]

$$\widehat{Q}(\langle a_1, o_1\rangle, \ldots, \langle a_r, o_r\rangle, \langle b, g_Q(b, o_1, \ldots, o_r)\rangle) \text{ iff } Q(a_1, \ldots, a_r, b) \quad (2)$$

For any $a_1, \ldots, a_r, b$, the above definition of $\widehat{Q}$ is effected for all possible choices of $o_1, \ldots, o_r$ such that $g_Q(b, o_1, \ldots, o_r)$ is defined.

Collect now all the $\widehat{Q}$ to form a set of rules $\widehat{\mathscr{R}}$.
Let also $\widehat{\mathscr{I}} = \{\langle x, h(x)\rangle : x \in \mathscr{I}\}$.

[†] The notation $f : A \to B$ is common in informal (and formal) mathematics. It denotes a function f that receives "inputs" from the set A and yields "outputs" in the set B.

[‡] For a relation Q, writing just "$Q(a_1, \ldots, a_r, b)$" is equivalent to writing "$Q(a_1, \ldots, a_r, b)$ holds".

We will verify that the set $F = \mathrm{Cl}(\widehat{\mathscr{I}}, \widehat{\mathscr{R}})$ is a 2-ary relation that for every input yields *at most one* output, and therefore is a function. For such a relation it is customary to write, letting the context fend off the obvious ambiguity in the use of the letter F,

$$y = F(x) \quad \text{iff} \quad F(x, y) \tag{*}$$

We will further verify that replacing f in (1) above by F results in a valid equivalence (the "iff" holds). That is, F satisfies (1).

(a) We establish that F is a relation composed of pairs $\langle x, y \rangle$ (x is input, y is output), where $x \in \mathrm{Cl}(\mathscr{I}, \mathscr{R})$ and $y \in Y$. This follows easily by induction on F (I.2.3), since $\widehat{\mathscr{I}} \subseteq F$, and the property (of "containing such pairs") propagates with each $\widehat{Q}$ (recall that the g_Q yield outputs in Y).

(b) We next show that "if $\langle x, y \rangle \in F$ and $\langle x, z \rangle \in F$, then $y = z$", that is, F is "single-valued" or "well-defined", in short, it is a *function*.

 We again employ induction on F, *thinking of the quoted statement as a "property" of the pair $\langle x, y \rangle$*:

 Suppose that $\langle x, y \rangle \in \widehat{\mathscr{I}}$, and let also $\langle x, z \rangle \in F$.

 By I.2.6, $\langle x, z \rangle \in \widehat{\mathscr{I}}$, or $\widehat{Q}(\langle a_1, o_1 \rangle, \ldots, \langle a_r, o_r \rangle, \langle x, z \rangle)$, where $Q(a_1, \ldots, a_r, x)$ and $z = g_Q(x, o_1, \ldots, o_r)$, for some $(r+1)$-ary $\widehat{Q}$ and $\langle a_1, o_1 \rangle, \ldots, \langle a_r, o_r \rangle$ in F.

 The right hand side of the italicized "or" cannot hold for an *unambiguous* $(\mathscr{I}, \mathscr{R})$, since x cannot have i.p. Thus $\langle x, z \rangle \in \widehat{\mathscr{I}}$; hence $y = h(x) = z$.

 To prove that the property propagates with each $\widehat{Q}$, let

$$\widehat{Q}(\langle a_1, o_1 \rangle, \ldots, \langle a_r, o_r \rangle, \langle x, y \rangle)$$

 but also

$$\widehat{P}(\langle b_1, o_1' \rangle, \ldots, \langle b_l, o_l' \rangle, \langle x, z \rangle)$$

 where $Q(a_1, \ldots, a_r, x)$, $P(b_1, \ldots, b_l, x)$, and

$$y = g_Q(x, o_1, \ldots, o_r) \quad \text{and} \quad z = g_P(x, o_1', \ldots, o_l') \tag{3}$$

 Since $(\mathscr{I}, \mathscr{R})$ is unambiguous, we have $Q = P$ (hence also $\widehat{Q} = \widehat{P}$), $r = l$, and $a_i = b_i$ for $i = 1, \ldots, r$.

 By I.H., $o_i = o_i'$ for $i = 1, \ldots, r$; hence $y = z$ by (3).

(c) Finally, we show that F satisfies (1). We do induction on $\mathrm{Cl}(\widehat{\mathscr{I}}, \widehat{\mathscr{R}})$ to prove:

 ($\leftarrow$) If $x \in \mathscr{I}$ and $y = h(x)$, then $F(x, y)$ (i.e., $y = F(x)$ in the alternative notation (*)), since $\widehat{\mathscr{I}} \subseteq F$. Let next $y = g_Q(x, o_1, \ldots, o_r)$ and $Q(a_1, \ldots, a_r, x)$, where also $F(a_i, o_i)$, for $i = 1, \ldots, r$. By (2), $\widehat{Q}(\langle a_1, o_1 \rangle, \ldots, \langle a_r, o_r \rangle, \langle x, g_Q(x, o_1, \ldots, o_r) \rangle)$; thus – F being closed under all the rules in $\widehat{\mathscr{R}} - F(x, g_Q(b, o_1, \ldots, o_r))$ holds; in short, $F(x, y)$ or $y = F(x)$.

($\rightarrow$) Now we assume that $F(x, y)$ holds and we want to infer the right hand side (of *iff*) in (1). We employ Metatheorem I.2.6.

Case 1. Let $\langle x, y \rangle$ be F-derivable[†] in $n = 1$ step. Then $\langle x, y \rangle \in \widehat{\mathscr{T}}$. Thus $y = h(x)$.

Case 2. Suppose next that $\langle x, y \rangle$ is F-derivable within $n + 1$ steps, namely, we have a derivation

$$\langle x_1, y_1 \rangle, \langle x_2, y_2 \rangle, \ldots, \langle x_n, y_n \rangle, \langle x, y \rangle \tag{4}$$

where $\widehat{Q}(\langle a_1, o_1 \rangle, \ldots, \langle a_r, o_r \rangle, \langle x, y \rangle)$ and $Q(a_1, \ldots, a_r, x)$ (see (2)), and each of $\langle a_1, o_1 \rangle, \ldots, \langle a_r, o_r \rangle$ appears in the above derivation, to the left of $\langle x, y \rangle$. This entails (by (2)) that $y = g_Q(x, o_1, \ldots, o_r)$. Since the $\langle a_i, o_i \rangle$ appear in (4), $F(a_i, o_i)$ holds, for $i = 1, \ldots, r$. Thus, $\langle x, y \rangle$ satisfies the right hand side of *iff* in (1), once more.

Uniqueness part. Let the function K also satisfy (1). We show, by induction on $\mathrm{Cl}(\mathscr{T}, \mathscr{R})$, that

For all $x \in \mathrm{Cl}(\mathscr{T}, \mathscr{R})$ and all $y \in Y$,　　$y = F(x)$ **iff** $y = K(x)$　(5)

($\rightarrow$) Let $x \in \mathscr{T}$, and $y = F(x)$. By lack of ambiguity, the case conditions of (1) are mutually exclusive. Thus, it must be that $y = h(x)$. But then, $y = K(x)$ as well, since K satisfies (1) too.

Let now $Q(a_1, \ldots, a_r, x)$ and $y = F(x)$. By (1), there are (unique, as we now know) $o_1, \ldots, o_r$ such that $o_i = F(a_i)$ for $i = 1, \ldots, r$, and $y = g_Q(x, o_1, \ldots, o_r)$. By the I.H., $o_i = K(a_i)$. But then (1) yields $y = K(x)$ as well (since K satisfies (1)).

($\leftarrow$) Just interchange the letters F and K in the above argument.　□

The above clearly is valid for functions h and g_Q that may fail to be defined everywhere in their "natural" input sets. To be able to have this degree of generality without having to state additional definitions (such as *left fields*, *right fields*, *partial functions*, *total functions*, *nontotal functions*, Kleene "*weak equality*"), we have stated the recurrence (1) the way we did (to keep an eye on both the input and output side of things) rather than the "usual"

$$f(x) = \begin{cases} h(x) & \textbf{if } x \in \mathscr{T} \\ g_Q(x, f(a_1), \ldots, f(a_r)) & \textbf{if } Q(a_1, \ldots, a_r, x) \text{ holds} \end{cases}$$

[†] $\mathrm{Cl}(\widehat{\mathscr{T}}, \widehat{\mathscr{R}})$-derivable.

Of course, if all the g_Q and h are defined everywhere on their input sets (i.e., they are "total"), then f is defined everywhere on $Cl(\mathscr{T}, \mathscr{R})$ (see Exercise I.4).

I.3. Axioms and Rules of Inference

Now that we have our language, L, we will embark on using it to formally effect *deductions*. These deductions start at the *axioms*. Deductions employ "acceptable" purely syntactic – i.e., based on *form*, not on *meaning* – rules that allow us to write a formula down (to *deduce* it) solely because certain other formulas *that are syntactically related to it* were already deduced (i.e., already written down). These string-manipulation rules are called *rules of inference*. We describe in this section the axioms and the rules of inference that we will accept into our logical calculus and that are common to all theories.

We start with a precise definition of *tautologies* in our first order language L.

I.3.1 Definition (Prime Formulas in Wff. Propositional Variables). A formula $\mathscr{A} \in$ **Wff** is a *prime formula* or a *propositional variable* iff it is either of

Pri1. atomic,
Pri2. a formula of the form $((\exists x)\mathscr{A})$.

We use the lowercase letters p, q, r (with or without subscripts or primes) to denote arbitrary prime formulas (propositional variables) of our language.   □

That is, a prime formula has either no propositional connectives, or if it does, it hides them inside the scope of $(\exists x)$.

We may think of a propositional variable as a "blob" that a myopic being makes out of a formula described in I.3.1. The same being will see an arbitrary well-formed formula as a bunch of blobs, brackets, and Boolean connectives ($\neg, \vee$), "correctly connected" as stipulated below.[†]

I.3.2 Definition (Propositional Formulas). The set of propositional formulas over $\mathscr{V}$, denoted here by **Prop**, is the smallest set such that:

(1) Every propositional variable (over $\mathscr{V}$) is in **Prop**.
(2) If $\mathscr{A}$ and $\mathscr{B}$ are in **Prop**, then so are $(\neg \mathscr{A})$ and $(\mathscr{A} \vee \mathscr{B})$.

We use the lowercase letters p, q, r (with or without subscripts or primes) to denote arbitrary prime formulas (propositional variables) of our language.   □

[†] Interestingly, our myope *can* see the brackets and the Boolean connectives.

I.3.3 Metatheorem. Prop = Wff.

Proof. (⊆) We do induction on **Prop**. Every item in I.3.2(1) is in **Wff**. **Wff** satisfies I.3.2(2) (see I.1.8(b)). Done.

(⊇) We do induction on **Wff**. Every item in I.1.8(a) is a propositional variable (over $\mathscr{V}$), and hence is in **Prop**.

Prop trivially satisfies I.1.8(b). It also satisfies I.1.8(c), for if $\mathscr{A}$ is in **Prop**, then it is in **Wff** by the ⊆-direction, above. Then, by I.3.1, $((\exists x)\mathscr{A})$ is a propositional variable and hence in **Prop**. We are done once more. □

I.3.4 Definition (Propositional Valuations). We can arbitrarily assign a value of 0 or 1 to every $\mathscr{A}$ in **Wff** (or **Prop**) as follows:

(1) We *fix* an assignment of 0 or 1 to *every prime formula*. We can think of this as an arbitrary but fixed function $v : \{\text{all prime formulas over } L\} \to \{0, 1\}$ in the metatheory.

(2) We define by recursion *an extension of* v, denoted by $\bar{v}$:

$$\bar{v}((\neg\mathscr{A})) = 1 - \bar{v}(\mathscr{A})$$
$$\bar{v}((\mathscr{A} \vee \mathscr{B})) = \bar{v}(\mathscr{A}) \cdot \bar{v}(\mathscr{B})$$

where "·" above denotes number multiplication.

We call, traditionally, the values 0 and 1 by the names "true" and "false" respectively, and write **t** and **f** respectively.

We also call a valuation v a *truth (value) assignment*.

We use the jargon "$\mathscr{A}$ takes the truth value **t** (respectively, **f**) under a valuation v" to mean "$\bar{v}(\mathscr{A}) = 0$ (respectively, $\bar{v}(\mathscr{A}) = 1$)". □

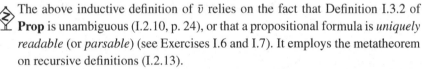 The above inductive definition of $\bar{v}$ relies on the fact that Definition I.3.2 of **Prop** is unambiguous (I.2.10, p. 24), or that a propositional formula is *uniquely readable* (or *parsable*) (see Exercises I.6 and I.7). It employs the metatheorem on recursive definitions (I.2.13).

The reader may think that all this about unique readability is just an annoying quibble. Actually it can be a matter of life and death. The ancient Oracle of Delphi had the nasty habit of issuing ambiguous – not uniquely readable, that is – pronouncements. One famous such pronouncement, rendered in English, went like this: "*You will go you will return not dying in the war*".[†] Given that ancient Greeks did not use punctuation, the above has two diametrically opposite meanings depending on whether you put a comma *before* or *after* "not".

[†] The original was "*Ιξεις αφιξεις ου θνηξεις εν πολεμ ω*".
_ι

The situation with formulas in **Prop** would have been as disastrous in the absence of brackets – which serve as punctuation – because unique readability would not be guaranteed: For example, for three distinct prime formulas p, q, r we could find a v such that $\bar{v}(p \to q \to r)$ is *different* depending on whether we meant to insert brackets around "$p \to q$" or around "$q \to r$" (can you find such a v?).

I.3.5 Remark (Truth Tables). Definition I.3.4 is often given in terms of *truth-functions*. For example, we could have defined (in the metatheory, of course) the function $F_\neg : \{\mathbf{t}, \mathbf{f}\} \to \{\mathbf{t}, \mathbf{f}\}$ by

$$F_\neg(x) = \begin{cases} \mathbf{t} & \text{if } x = \mathbf{f} \\ \mathbf{f} & \text{if } x = \mathbf{t} \end{cases}$$

We could then say that $\bar{v}((\neg\mathscr{A})) = F_\neg(\bar{v}(\mathscr{A}))$. One can similarly take care of all the connectives ($\vee$ and all the abbreviations) with the help of truth functions $F_\vee, F_\wedge, F_\to, F_\leftrightarrow$. These functions are conveniently given via so-called truth-tables as indicated below:

x	y	$F_\neg(x)$	$F_\vee(x, y)$	$F_\wedge(x, y)$	$F_\to(x, y)$	$F_\leftrightarrow(x, y)$
f	f	t	f	f	t	t
f	t	t	t	f	t	f
t	f	f	t	f	f	f
t	t	f	t	t	t	t

I.3.6 Definition (Tautologies, Satisfiable Formulas, Unsatisfiable Formulas in Wff). A formula $\mathscr{A} \in$ **Wff** (equivalently, in **Prop**) is a *tautology* iff for all valuations v one has $\bar{v}(\mathscr{A}) = \mathbf{t}$.

We call the set of all tautologies, as defined here, **Taut**. The symbol $\models_{\mathbf{Taut}} \mathscr{A}$ says "$\mathscr{A}$ is in **Taut**".

A formula $\mathscr{A} \in$ **Wff** (equivalently, in **Prop**) is *satisfiable* iff for some valuation v one has $\bar{v}(\mathscr{A}) = \mathbf{t}$. We say that v *satisfies* $\mathscr{A}$.

A *set* of formulas Γ is satisfiable iff for some valuation v, one has $\bar{v}(\mathscr{A}) = \mathbf{t}$ for every $\mathscr{A}$ in Γ. We say that v *satisfies* Γ.

A formula $\mathscr{A} \in$ **Wff** (equivalently, in **Prop**) is *unsatisfiable* iff for all valuations v one has $\bar{v}(\mathscr{A}) = \mathbf{f}$. A *set* of formulas Γ is unsatisfiable iff for all valuations v one has $\bar{v}(\mathscr{A}) = \mathbf{f}$ for some $\mathscr{A}$ in Γ.

I.3.7 Definition (Tautologically Implies, for Formulas in Wff). Let $\mathscr{A}$ and Γ be respectively any formula and any set of formulas (over L).

The symbol $\Gamma \models_{\text{Taut}} \mathscr{A}$, pronounced "$\Gamma$ tautologically implies $\mathscr{A}$", means that *every truth assignment v that satisfies Γ also satisfies $\mathscr{A}$.* □

"Satisfiable" and "unsatisfiable" are terms introduced here in the *propositional* or *Boolean* sense. These terms have a more complicated meaning when we decide to "see" the object variables and quantifiers that occur in formulas.

We have at once

I.3.8 Lemma.[†] $\Gamma \models_{\text{Taut}} \mathscr{A}$ *iff* $\Gamma \cup \{\neg \mathscr{A}\}$ *is unsatisfiable (in the propositional sense).*

If $\Gamma = \emptyset$ then $\Gamma \models_{\text{Taut}} \mathscr{A}$ says just $\models_{\text{Taut}} \mathscr{A}$, since the hypothesis "every truth assignment v that satisfies Γ", in the definition above, is vacuously satisfied. For that reason we almost never write $\emptyset \models_{\text{Taut}} \mathscr{A}$ and write instead $\models_{\text{Taut}} \mathscr{A}$.

I.3.9 Exercise. For any formula $\mathscr{A}$ and any two valuations v and v', $\bar{v}(\mathscr{A}) = \bar{v}'(\mathscr{A})$ if v and v' agree on all the propositional variables that occur in $\mathscr{A}$.

In the same manner, $\Gamma \models_{\text{Taut}} \mathscr{A}$ is oblivious to v-variations that do not affect the variables that occur in Γ and $\mathscr{A}$ (see Exercise I.8). □

Before presenting the axioms, we need to introduce the concept of *substitution.*

I.3.10 Tentative Definition (Substitutions of Terms). Let $\mathscr{A}$ be a formula, x an (object) variable, and t a term. $\mathscr{A}[x \leftarrow t]$ denotes the result of "replacing" *all free occurrences* of x in $\mathscr{A}$ by the term t, *provided no variable of t was "captured" (by a quantifier) during substitution.*

[†] The word "lemma" has Greek origin, "$\lambda \acute{\eta} \mu \mu \alpha$", plural "lemmata" (some people say "lemmas") from "$\lambda \acute{\eta} \mu \mu \alpha \tau \alpha$". It derives from the verb "$\lambda \alpha \mu \beta \acute{\alpha} \nu \omega$" (to take) and thus means "taken thing". In mathematical reasoning a lemma is a provable auxiliary statement that is *taken* and used as a stepping stone in lengthy mathematical arguments – invoked therein by name, as in "...by Lemma such and such..." – much as "subroutines" (or "procedures") are taken and used as auxiliary stepping stones to elucidate lengthy computer programs. Thus our purpose in having lemmata is to shorten proofs by breaking them up into modules.

If the proviso is valid, then we say that "*t* is *substitutable for x* (in $\mathscr{A}$)", or that "*t* is *free for x* (in $\mathscr{A}$)". If the proviso is not valid, then the substitution is *undefined.*   □

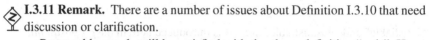 **I.3.11 Remark.** There are a number of issues about Definition I.3.10 that need discussion or clarification.

Reasonable people will be satisfied with the above definition "as is". However, there are some obscure points (enclosd in quotation marks above).

(1) What is this about "capture"? Well, suppose that $\mathscr{A} \equiv (\exists x)\neg x = y$. Let $t \equiv x$.[†] Then $\mathscr{A}[y \leftarrow t] \equiv (\exists x)\neg x = x$, which says something altogether different than the original. Intuitively, this is unexpected (and undesirable): $\mathscr{A}$ codes a statement about the free variable y, i.e., a statement about all objects which could be "values" (or meanings) of y. One would have expected that, in particular, $\mathscr{A}[y \leftarrow x]$ – *if the substitution were allowed* – would make this very same statement about the values of x. It does not.[‡] What happened is that x was *captured by the quantifier* upon substitution, thus distorting $\mathscr{A}$'s original meaning.

(2) Are we sure that the term "replace" is mathematically precise?

(3) Is $\mathscr{A}[x \leftarrow t]$ always a formula, if $\mathscr{A}$ is?

A re-visitation of I.3.10 via an inductive definition (by induction on terms and formulas) settles (1)–(3) at once (in particular, the informal terms "replace" and "capture" do not appear in the inductive definition). We define (again) the symbol $\mathscr{A}[x \leftarrow t]$, for any formula $\mathscr{A}$, variable x, and term t, this time by induction on terms and formulas:

First off, let us define $s[x \leftarrow t]$, where s is also a term, by cases:

$$s[x \leftarrow t] \equiv \begin{cases} t & \text{if } s \equiv x \\ a & \text{if } s \equiv a, \text{a constant} \\ & \qquad\qquad \text{(symbol)} \\ y & \text{if } s \equiv y, \text{a variable } \not\equiv x \\ fr_1[x \leftarrow t]r_2[x \leftarrow t]\ldots r_n[x \leftarrow t] & \text{if } s \equiv fr_1 \ldots r_n \end{cases}$$

Pause. Is $s[x \leftarrow t]$ always a term? That this is so follows directly by induction on terms, using the definition by cases above and the I.H. that each of $r_i[x \leftarrow t]$, $i = 1, \ldots, n$, is a term.

[†] Recall that in I.1.4 (p. 13) we defined the symbol "$\equiv$" to be equality on strings.

[‡] The original says that for any object y there is an object that is different from it; $\mathscr{A}[y \leftarrow x]$ says that there is an object that is different from itself.

We turn now to formulas. The symbols P, r, s (with or without subscripts) below denote a predicate of arity n, a term, and a term (respectively):

$$\mathscr{A}[x \leftarrow t] \equiv \begin{cases} s[x \leftarrow t] = r[x \leftarrow t] & \text{if } \mathscr{A} \equiv s = r \\ Pr_1[x \leftarrow t]r_2[x \leftarrow t]\dots & \text{if } \mathscr{A} \equiv Pr_1 \dots r_n \\ \quad r_n[x \leftarrow t] \\ (\mathscr{B}[x \leftarrow t] \vee \mathscr{C}[x \leftarrow t]) & \text{if } \mathscr{A} \equiv (\mathscr{B} \vee \mathscr{C}) \\ (\neg(\mathscr{B}[x \leftarrow t])) & \text{if } \mathscr{A} \equiv (\neg \mathscr{B}) \\ \mathscr{A} & \text{if } \mathscr{A} \equiv ((\exists y).\mathscr{B}) \text{ and } y \equiv x \\ ((\exists y)(\mathscr{B}[x \leftarrow t])) & \text{if } \mathscr{A} \equiv ((\exists y).\mathscr{B}) \text{ and } y \not\equiv x \\ & \text{and } y \text{ does not occur in } t \end{cases}$$

In all cases above, the left hand side is defined iff the right hand side is.

Pause. We have eliminated "replaces" and "captured". But is $\mathscr{A}[x \leftarrow t]$ a formula (whenever it is defined)? (See Exercise I.9) □

I.3.12 Definition (Simultaneous Substitution). The symbol

$$\mathscr{A}[y_1, \dots, y_r \leftarrow t_1, \dots, t_r]$$

or, equivalently, $\mathscr{A}[\vec{y}_r \leftarrow \vec{t}_r]$ – where $\vec{y}_r$ is an abbreviation of $y_1, \dots, y_r$ – denotes *simultaneous substitution* of the terms $t_1, \dots, t_r$ into the variables $y_1, \dots, y_r$ in the following sense: Let $\vec{z}_r$ be variables that do not occur at all (either as free or bound) in any of $\mathscr{A}, \vec{t}_r$. Then $\mathscr{A}[\vec{y}_r \leftarrow \vec{t}_r]$ is short for

$$\mathscr{A}[y_1 \leftarrow z_1] \dots [y_r \leftarrow z_r][z_1 \leftarrow t_1] \dots [z_r \leftarrow t_r] \tag{1}$$

□

Exercise I.10 shows that we obtain the same string in (1) above, regardless of our choice of *new variables* $\vec{z}_r$.

More Conventions. The symbol $[x \leftarrow t]$ lies in the metalanguage. This metasymbol has the highest priority, so that, e.g., $\mathscr{A} \vee \mathscr{B}[x \leftarrow t]$ means $\mathscr{A} \vee (\mathscr{B}[x \leftarrow t])$, $(\exists x)\mathscr{B}[x \leftarrow t]$ means $(\exists x)(\mathscr{B}[x \leftarrow t])$, etc.

The reader is reminded about the conventions regarding the metanotations $\mathscr{A}[\vec{x}_r]$ and $\mathscr{A}(\vec{x}_r)$ (see I.1.11). *In the context* of those notations, if $t_1, \dots, t_r$ are terms, the symbol $\mathscr{A}[t_1, \dots, t_r]$ abbreviates $\mathscr{A}[\vec{y}_r \leftarrow \vec{t}_r]$.

We are ready to introduce the (logical) axioms and rules of inference.

Schemata.[†] Some of the axioms below will actually be *schemata*. A *formula schema*, or *formula form*, is a string $\mathscr{G}$ of the metalanguage that contains *syntactic variables*, such as $\mathscr{A}, P, f, a, t, x$.

Whenever we replace all these syntactic variables that occur in $\mathscr{G}$ by specific formulas, predicates, functions, constants, terms, or variables respectively, we obtain a specific well-formed formula, a so-called *instance of the schema*. For example, an instance of $(\exists x)x = a$ is $(\exists v_{12})v_{12} = 0$ (in the language of Peano arithmetic). An instance of $\mathscr{A} \to \mathscr{A}$ is $v_{101} = v_{114} \to v_{101} = v_{114}$.

I.3.13 Definition (Axioms and Axiom Schemata). The *logical axioms* are *all* the formulas in the group **Ax1** and *all* the possible instances of the schemata in the remaining groups:

Ax1. All formulas in **Taut**.
Ax2. (*Schema*)

$$\mathscr{A}[x \leftarrow t] \to (\exists x)\mathscr{A} \qquad \text{for any term } t$$

By I.3.10–I.3.11, *the notation already imposes a condition on t, that it is substitutable for x.*

N.B. We often see the above written as

$$\mathscr{A}[t] \to (\exists x)\mathscr{A}[x]$$

or even

$$\mathscr{A}[t] \to (\exists x)\mathscr{A}$$

Ax3. (*Schema*) For *each* object variable x, the formula $x = x$.
Ax4. (*Leibniz's characterization of equality – first order version. Schema*) For any formula $\mathscr{A}$, object variable x, and terms t and s, the formula

$$t = s \to (\mathscr{A}[x \leftarrow t] \leftrightarrow \mathscr{A}[x \leftarrow s])$$

N.B. The above is written usually as

$$t = s \to (\mathscr{A}[t] \leftrightarrow \mathscr{A}[s])$$

We must remember that the notation already *requires* that t and s be free for x.

We will denote the above set of logical axioms by Λ. □

[†] Plural of *schema*. This is of Greek origin, $\sigma\chi\tilde{\eta}\mu\alpha$, meaning – e.g., in geometry – figure or configuration or even formation.

The logical axioms for equality are not the strongest possible, but they are adequate for the job. What Leibniz *really* proposed was the schema $t = s \leftrightarrow (\forall P)(P[t] \leftrightarrow P[s])$, which says, intuitively, that "two objects t and s are equal iff, for every 'property P', both have P or neither has P".

Unfortunately, our system of notation (first-order language) does not allow quantification over predicate symbols (which can have as "values" arbitrary "properties"). But is not **Ax4** read "for all formulas $\mathscr{A}$" anyway? Yes, but with one qualification: "For all formulas $\mathscr{A}$ *that we can write down in our system of notation*", and, alas, we cannot write *all* possible formulas of *real mathematics* down, because they are too many.[†]

While the symbol "$=$" is suggestive of equality, it is *not* its shape that qualifies it as equality. It is the two axioms, **Ax3** and **Ax4**, that make the symbol *behave* as we expect equality to behave, and any other symbol of any other shape (e.g., Enderton (1972) uses "$\approx$") satisfying these two axioms *qualifies* as *formal equality* that is intended to codify the metamathematical standard "$=$".

I.3.14 Remark. In **Ax2** and **Ax4** we imposed the condition that t (and s) must be substitutable in x. Here is why:

Take $\mathscr{A}$ to stand for $(\forall y)x = y$ and $\mathscr{B}$ to stand for $(\exists y)\neg x = y$. Then, *temporarily suspending the restriction on substitutability*, $\mathscr{A}[x \leftarrow y] \rightarrow (\exists x)\mathscr{A}$ is

$$(\forall y)y = y \rightarrow (\exists x)(\forall y)x = y$$

and $x = y \rightarrow (\mathscr{B} \leftrightarrow \mathscr{B}[x \leftarrow y])$ is

$$x = y \rightarrow ((\exists y)\neg x = y \leftrightarrow (\exists y)\neg y = y)$$

neither of which, obviously, is "valid".[‡]

There is a remedy in the metamathematics: Move the quantified variable(s) out of harm's way, by renaming them so that no quantified variable in $\mathscr{A}$ has the same name as any (free, of course) variable in t (or s).

This renaming is formally correct (i.e., it does not change the meaning of the formula), as we will see in the *variant* (meta)theorem (I.4.13). Of course,

---

[†] "Uncountably many", in a precise technical sense developed in the chapter on cardinality in volume 2 (see p. 62, of this volume for a brief informal "course" in cardinality). This is due to Cantor's theorem, which implies that there are uncountably many subsets of $\mathbb{N}$. Each such subset A gives rise to the formula $x \in A$ *in the metalanguage.*

On the other hand, set theory's formal system of notation, using just $\in$ and U as start-up (nonlogical) symbols, is only rich enough to write down a countably infinite set of formulas (cf. p. 62). Thus, our notation will fail to denote uncountably many "real formulas" $x \in A$.

[‡] Speaking intuitively is enough for now. Validity will be defined carefully pretty soon.

it is always possible to effect this renaming, since we have countably many variables, and only finitely many appear free in t (and s) and $\mathscr{A}$. This trivial remedy allows us to render the conditions in **Ax2** and **Ax4** harmless. Essentially, a t (or s) is always substitutable *after renaming*.   □

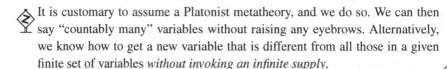

It is customary to assume a Platonist metatheory, and we do so. We can then say "countably many" variables without raising any eyebrows. Alternatively, we know how to get a new variable that is different from all those in a given finite set of variables *without invoking an infinite supply*.

I.3.15 Definition (Rules of Inference). The following are the two *rules of inference*. These rules are relations in the sense of Section I.2, with inputs from the set **Wff** and outputs also in **Wff**. They are written traditionally as "fractions". We call the "numerator" the *premise(s)* and the "denominator" the *conclusion*.

We say that a rule of inference is *applied* to the formula(s) in the numerator, and that it *yields* (or *results in*) the formula in the denominator.

Inf1. *Modus ponens*, or *MP*. For any formulas $\mathscr{A}$ and $\mathscr{B}$,

$$\frac{\mathscr{A},\ \mathscr{A} \to \mathscr{B}}{\mathscr{B}}$$

Inf2. ∃-*introduction* – pronounced *E-introduction*. For any formulas $\mathscr{A}$ and $\mathscr{B}$ such that x is not free in $\mathscr{B}$,

$$\frac{\mathscr{A} \to \mathscr{B}}{(\exists x)\mathscr{A} \to \mathscr{B}}$$

N.B. Recall the conventions on eliminating brackets!   □

It is immediately clear that the definition above meets our requirement that the rules of inference be "algorithmic", in the sense that *whether* they are applicable or *how* they are applicable can be decided and carried out in a finite number of steps by just looking at the *form* of (potential input) formulas (not at the "meaning" of such formulas).

We next define Γ-theorems, that is, formulas we can *prove from* the *set* of formulas Γ (this Γ may be empty).

I.3.16 Definition (Γ-Theorems). The set of Γ-*theorems*, **Thm**$_\Gamma$, is the least inclusive subset of **Wff** that satisfies:

Th1. Λ ⊆ **Thm**$_\Gamma$ (cf. I.3.13).
Th2. Γ ⊆ **Thm**$_\Gamma$. We call every member of Γ a *nonlogical axiom*.
Th3. **Thm**$_\Gamma$ is *closed under* each rule **Inf1–Inf2**.

The metalinguistic statement $\mathscr{A} \in \mathbf{Thm}_\Gamma$ is traditionally written as $\Gamma \vdash \mathscr{A}$, and we say that $\mathscr{A}$ *is proved from* Γ or that it is a Γ-*theorem*.

We also say that $\mathscr{A}$ is *deduced* by Γ, or that Γ *deduces* $\mathscr{A}$.

If $\Gamma = \emptyset$, then rather than $\emptyset \vdash \mathscr{A}$ we write $\vdash \mathscr{A}$. We often say in this case that $\mathscr{A}$ is *absolutely provable* (or *provable with no nonlogical axioms*).

We often write $\mathscr{A}, \mathscr{B}, \dots, \mathscr{D} \vdash \mathscr{E}$ for $\{\mathscr{A}, \mathscr{B}, \dots, \mathscr{D}\} \vdash \mathscr{E}$. □

I.3.17 Definition (Γ-Proofs). We just saw that $\mathbf{Thm}_\Gamma$ is $\mathrm{Cl}(\mathscr{I}, \mathscr{R})$, where $\mathscr{I}$ is the set of all logical and nonlogical axioms, and $\mathscr{R}$ contains just the two rules of inference. An $(\mathscr{I}, \mathscr{R})$-derivation is also called a Γ-proof (or just proof, if Γ is understood). □

I.3.18 Remark. (1) It is clear that if each of $\mathscr{A}_1, \dots, \mathscr{A}_n$ has a Γ-proof and $\mathscr{B}$ has an $\{\mathscr{A}_1, \dots, \mathscr{A}_n\}$-proof, then $\mathscr{B}$ has a Γ-proof. Indeed, simply concatenate all of the given Γ-proofs (in any sequence). Append to the right of that sequence the given $\{\mathscr{A}_1, \dots, \mathscr{A}_n\}$-proof (that ends with $\mathscr{B}$). Then the entire sequence is a Γ-proof, and ends with $\mathscr{B}$.

We refer to this phenomenon as the *transitivity of* $\vdash$.

N.B. Transitivity of $\vdash$ allows one to invoke previously proved (by him or others) *theorems* in the course of a proof. Thus, *practically*, a Γ-proof is a sequence of formulas in which each formula is an axiom, is a known Γ-theorem, or is obtained by applying a rule of inference on previous formulas of the sequence.

(2) If $\Gamma \subseteq \Delta$ and $\Gamma \vdash \mathscr{A}$, then also $\Delta \vdash \mathscr{A}$, as follows from I.3.16 or I.3.17. In particular, $\vdash \mathscr{A}$ implies $\Gamma \vdash \mathscr{A}$ for any Γ.

(3) It is immediate from the definitions that for any formulas $\mathscr{A}$ and $\mathscr{B}$,

$$\mathscr{A}, \mathscr{A} \to \mathscr{B} \vdash \mathscr{B} \qquad (i)$$

and if, moreover, x is not free in $\mathscr{B}$,

$$\mathscr{A} \to \mathscr{B} \vdash (\exists x)\mathscr{A} \to \mathscr{B} \qquad (ii)$$

Some texts (e.g., Schütte (1977)) give the rules in the format of (i)–(ii) above. □

The axioms and rules provide us with a *calculus*, that is, a means to "calculate" proofs and theorems. In the interest of making the calculus more user-friendly – and thus more easily applicable to mathematical theories of interest, such as Peano arithmetic or set theory – we are going to develop in the next section a number of *derived principles*. These principles are largely of the form

$\mathscr{A}_1, \ldots, \mathscr{A}_n \vdash \mathscr{B}$. We call such a (provable in the metatheory) principle a *derived rule of inference*, since, by transitivity of $\vdash$, it can be used as a proof-step in a Γ-proof. By contrast, the rules **Inf1–Inf2** are "basic" or "primary"; they are given outright.

 We can now fix our understanding of the concept of a formal or mathematical *theory*.

A *(first order) formal (mathematical) theory* over a language L, or just *theory over L*, or just *theory*, is a tuple (of "ingredients") $\mathfrak{T} = (L, \Lambda, \mathbf{I}, \mathscr{T})$, where L is a first order language, Λ is a set of *logical axioms*, $\mathbf{I}$ is a set of rules of inference, and $\mathscr{T}$ a *non-empty* subset of **Wff** that is required to contain Λ (i.e., $\Lambda \subseteq \mathscr{T}$) and be closed under the rules **I**.

Equivalently, one may simply require that $\mathscr{T}$ is *closed under* $\vdash$, that is, for *any* $\Gamma \subseteq \mathscr{T}$ and any formula $\mathscr{A}$, if $\Gamma \vdash \mathscr{A}$, then $\mathscr{A} \in \mathscr{T}$. This is, furthermore, equivalent to requiring that

$$\mathscr{A} \in \mathscr{T} \quad \text{iff} \quad \mathscr{T} \vdash \mathscr{A} \tag{1}$$

Indeed, the *if* direction follows from closure under $\vdash$, while the *only if* direction is a consequence of Definition I.3.16.

$\mathscr{T}$ is the set of the formulas *of the theory*,[†] and we often say "a theory $\mathscr{T}$", taking everything else for granted.

If $\mathscr{T} = \textbf{Wff}$, then the theory $\mathfrak{T}$ is called *inconsistent* or *contradictory*. Otherwise it is called *consistent*.

Throughout our exposition we fix Λ and **I** as in Definitions I.3.13 and I.3.15.

By (1), $\mathscr{T} = \textbf{Thm}_{\mathscr{T}}$. This observation suggests that we call theories such as the ones we have just defined *axiomatic theories*, in that a set Γ always exists so that $\mathscr{T} = \textbf{Thm}_\Gamma$ (if at a loss, we can just take $\Gamma = \mathscr{T}$).

We are mostly interested in theories $\mathfrak{T}$ for which there is a "small" set Γ ("small" by comparison with $\mathscr{T}$) such that $\mathscr{T} = \textbf{Thm}_\Gamma$. We say that $\mathfrak{T}$ is *axiomatized* by Γ. Naturally, we call $\mathscr{T}$ the *set of theorems*, and Γ the set of *nonlogical* axioms of $\mathfrak{T}$.

If, moreover, Γ is *recognizable* (i.e., we can tell "algorithmically" whether or not a formula $\mathscr{A}$ is in Γ), then we say that $\mathfrak{T}$ is *recursively axiomatized*.

Examples of recursively axiomatized theories are ZFC set theory and Peano arithmetic. On the other hand, if we take $\mathscr{T}$ to be *all* the formulas of arithmetic that are true when interpreted "in the intended way"[‡] over $\mathbb{N}$ – the so-called

---

[†] As opposed to "of the language", which is all of **Wff**.
[‡] That is, the symbol "0" of the language is interpreted as *the* $0 \in \mathbb{N}$, "*Sx*" as $x + 1$, "($\exists x$)" as "there is an $x \in \mathbb{N}$", etc.

complete arithmetic – then there is *no* recognizable Γ such that $\mathscr{T} = \mathbf{Thm}_\Gamma$. We say that complete arithmetic is not *recursively axiomatizable*.[†]

Pause. Why does complete arithmetic form a theory? Because work of Section I.5 – in particular, the soundness theorem – entails that it is closed under $\vdash$.

We tend to further abuse language and call axiomatic theories by the name of their (set of) nonlogical axioms Γ. Thus if $\mathfrak{T} = (L, \Lambda, \mathbf{I}, \mathscr{T})$ is a first order theory and $\mathscr{T} = \mathbf{Thm}_\Gamma$, then we may say interchangeably "theory $\mathfrak{T}$", "theory $\mathscr{T}$" or "theory Γ".

If $\Gamma = \emptyset$, then we have a *pure* or *absolute* theory (i.e., we are "just doing logic, not math"). If $\Gamma \neq \emptyset$, then we have an *applied* theory.

Argot. A *final note on language versus metalanguage, and theory versus metatheory.* When are we speaking the metalanguage and when are we speaking the formal language?

The answer is, respectively, "almost always" and "almost never". As it has been remarked before, *in principle*, we are speaking the *formal language* exactly when we are pronouncing or writing down a *string* from **Term** or **Wff**. Otherwise we are (speaking or writing) in the *metalanguage*. It appears that we (and everybody else who has written a book in logic or set theory) is speaking and writing within the metalanguage with a frequency approaching 100%.

The formalist is clever enough to simplify notation at all times. We will seldom be caught writing down a member of **Wff** in this book, and, on the rare occasions we may do so, it will only be to serve as *an illustration of why one should avoid writing down such formulas*: because they are too long and hard to read and understand.

We will be speaking the formal language with a heavy "accent" and using many "idioms" borrowed from "real" (meta)mathematics and English. We will call our dialect *argot*, following Manin (1977).

The important thing to remember is when we are *working* in the theory,[‡] and this is precisely when we generate theorems. That is, it does not matter if a theorem (and much of the what we write down during the proof) is written in *argot*.

Two examples:

(1) One is working *in* formal number theory (or formal arithmetic) if one states and proves (say, from the Peano axioms) that "every natural number $n > 1$

[†] The trivial solution – that is, taking $\Gamma = \mathscr{T}$ – will not do, for it turns out that $\mathscr{T}$ is not recognizable.

[‡] Important, because arguing in the theory restricts us to use *only* its axioms (and earlier proved theorems; cf. I.3.18) and its rules of inference – nothing extraneous to these syntactic tools is allowed.

has a prime factor". Note how this theorem is stated in *argot*. Below we give its translation into the formal language of arithmetic:[†]

$$(\forall n)(S0 < n \rightarrow (\exists x)(\exists y)(n = x \times y \wedge$$
$$S0 < x \wedge (\forall m)(\forall r)(x = m \times r \rightarrow m = S0 \vee m = x))) \tag{1}$$

(2) One is working in formal logic if one is writing a proof of $(\exists v_{13})v_{13} = v_{13}$.

Suppose though that our activity consists of effecting definitions, introducing axioms, or analyzing the behaviour or capability of $\mathfrak{T}$, e.g., proving some derived rule $\mathscr{A}_1, \ldots, \mathscr{A}_n \vdash \mathscr{B}$ – that is, a *theorem schema* – or investigating consistency[‡] or "relative consistency".[§] Then we are operating in the *metatheory*, that is, in "real" mathematics.

One of the most important problems posed in the metatheory is

"Given a theory $\mathfrak{T}$ and a formula $\mathscr{A}$. Is $\mathscr{A}$ a theorem of $\mathfrak{T}$?"

This is Hilbert's *Entscheidungsproblem*, or *decision problem*. Hilbert believed that every recursively axiomatized theory ought to admit a "general" solution, by more or less mechanical means, to its decision problem. The techniques of Gödel and the insight of Church showed that this problem is, in general, algorithmically unsolvable.

As we have already stated (p. 36), metamathematics exists outside and independently of our effort to build this or that formal system. All its methods are – in principle – available to us for use in the analysis of the behaviour of a formal system.

Pause. But how much of real mathematics are we allowed to use, *reliably*, to study or speak about the "simulator" that the formal system is?[¶] For example, have we not overstepped our license by using induction (and, implicitly, the entire *infinite* set $\mathbb{N}$) in our Platonist metatheory, specifically in the recursive or inductive definitions of terms, well-formed formulas, theorems, etc.?

The quibble here is largely "political". Some people argue (a major proponent of this was Hilbert) as follows: Formal mathematics was meant to crank out "true" statements of mathematics, but no "false" ones, and this freedom

---

[†] Well, almost. In the interest of brevity, all the variable names used in the displayed formula (1) are metasymbols.

[‡] That is, whether or not $\mathscr{T} = \mathbf{Wff}$.

[§] That is, "if Γ is consistent" – where we are naming the theory by its nonlogical axioms – "does it stay so after we have added some formula $\mathscr{A}$ as a nonlogical axiom?"

[¶] The methods or scope of the metamathematics that a logician uses – in the investigation of some formal system – are often *restricted* for technical or philosophical reasons.

from contradiction ought to be *verifiable*. Now, as we are so verifying in the metatheory (i.e., outside the formal system) shouldn't the metatheory itself be "above suspicion" (of contradiction, that is)? Naturally.

Hilbert's suggestion for achieving this "above suspicion" status was, essentially, to utilize in the metatheory only a small fragment of "reality" that is so simple and close to intuition that it does not need itself a "certificate" (via formalization) for its freedom from contradiction. In other words, restrict the metamathematics.[†] Such a fragment of the metatheory, he said, should have nothing to do with the infinite, in particular with the *entire set* $\mathbb{N}$ and all that it entails (e.g., inductive definitions and proofs).[‡]

If it were not for Gödel's incompleteness results, this position – that metamathematical techniques must be *finitary* – might have prevailed. However, Gödel proved it to be futile, and most mathematicians have learnt to feel comfortable with infinitary metamathematical techniques, or at least with $\mathbb{N}$ and induction.[§] Of course, it would be reckless to use as metamathematical tools "mathematics" of suspect consistency (e.g., the *full* naïve theory of sets).

It is worth pointing out that one could fit (with some effort) our inductive definitions within Hilbert's style. But we will not do so. First, one would have to abandon the elegant (and now widely used) approach with *closures*, and use instead the concept of *derivations* of Section I.2. Then one would somehow have to effect and study derivations without the benefit of the *entire* set $\mathbb{N}$. Bourbaki (1966b, p. 15) does so with his *constructions formatives*. Hermes (1973) is another author who does so, with his "term-" and "formula-calculi" (such calculi being, essentially, *finite* descriptions of derivations).

Bourbaki (but not Hermes) avoids induction over *all of* $\mathbb{N}$. In his metamathematical discussions of terms and formulas[¶] that are derived by a derivation

---

[†] Otherwise we would need to formalize the metamathematics – in order to "certify" it – and next the metametamathematics, and so on. For if "metaM" is to authoritatively check "M" for consistency, then it too must be consistent; so let us formalize "metaM" and let "metametaM" check it; . . . a never ending story.

[‡] See Hilbert and Bernays (1968, pp. 21–29) for an elaborate scheme that constructs "concrete number objects" – *Ziffern* or "numerals" – "|", "||", "|||", etc., that stand for "1", "2", "3", etc., complete with a "concrete mathematical induction" proof technique on these objects, and even the beginnings of their recursion theory. Of course, at any point, only finite sets of such objects were considered.

[§] Some proponents of infinitary techniques in metamathematics have used very strong words in describing the failure of "Hilbert's program". Rasiowa and Sikorski (1963) write in their introduction: "However Gödel's results exposed the fiasco of Hilbert's finitistic methods as far as consistency is concerned."

[¶] For example, in *loc. cit.*, p. 18, where he proves that, in our notation, $\mathscr{A}[x \leftarrow y]$ and $t[x \leftarrow y]$ are a formula and term respectively.

$d_1, \ldots, d_n$, he restricts his induction arguments on the segment $\{0, 1, \ldots, n\}$, that is, he takes an I.H. on $k < n$ and proceeds to $k + 1$.

I.4. Basic Metatheorems

We are dealing with an arbitrary theory $\mathfrak{T} = (L, \Lambda, \mathbf{I}, \mathscr{T})$, such that Λ is the set of logical axioms (I.3.13) and $\mathbf{I}$ are the inference rules (I.3.15). We also let Γ be an appropriate set of nonlogical axioms, i.e., $\mathscr{T} = \mathbf{Thm}_\Gamma$.

I.4.1 Metatheorem (Post's "Extended" Tautology Theorem). *If* $\mathscr{A}_1, \ldots,$ $\mathscr{A}_n \models_{\mathbf{Taut}} \mathscr{B}$ *then* $\mathscr{A}_1, \ldots, \mathscr{A}_n \vdash \mathscr{B}$.   □

Proof. The assumption yields that

$$\models_{\mathbf{Taut}} \mathscr{A}_1 \to \cdots \to \mathscr{A}_n \to \mathscr{B} \tag{1}$$

Thus, since the formula in (1) is in Λ, using Definition I.3.16, we have

$$\mathscr{A}_1, \cdots, \mathscr{A}_n \vdash \mathscr{A}_1 \to \cdots \to \mathscr{A}_n \to \mathscr{B} \tag{2}$$

Applying modus ponens to (2) n times, we deduce $\mathscr{B}$.   □

 I.4.1 is an omnipresent *derived rule.*

I.4.2 Definition. $\mathscr{A}$ and $\mathscr{B}$ *provably equivalent in* $\mathfrak{T}$ means that $\Gamma \vdash \mathscr{A} \leftrightarrow \mathscr{B}$.

I.4.3 Metatheorem. *Any two theorems* $\mathscr{A}$ *and* $\mathscr{B}$ *of* $\mathfrak{T}$ *are provably equivalent in* $\mathfrak{T}$.

Proof. By I.4.1, $\Gamma \vdash \mathscr{A}$ yields $\Gamma \vdash \mathscr{B} \to \mathscr{A}$. Similarly, $\Gamma \vdash \mathscr{B}$ yields $\Gamma \vdash \mathscr{A} \to \mathscr{B}$. One more application of I.4.1 yields $\Gamma \vdash \mathscr{A} \leftrightarrow \mathscr{B}$.   □

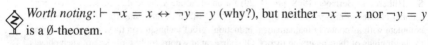

Worth noting: $\vdash \neg x = x \leftrightarrow \neg y = y$ (why?), but neither $\neg x = x$ nor $\neg y = y$ is a $\emptyset$-theorem.

 I.4.4 Remark (Hilbert Style Proofs). In practice we write proofs "vertically", that is, as numbered vertical sequences (or lists) of formulas. The numbering helps the annotational comments that we insert to the right of each formula that we list, as the following proof demonstrates.

A metatheorem admits a *meta*proof, strictly speaking. The following is a derived rule (or theorem schema) and thus belongs to the metatheory (and so does its proof).

Another point of view is possible, however: The syntactic symbols x, $\mathscr{A}$, and $\mathscr{B}$ below stand for a *specific* variable and *specific* formulas that we just forgot to write down explicitly. Then one can think of the proof as a (formal) Hilbert style proof. □

I.4.5 Metatheorem ($\forall$-Introduction – Pronounced "A-Introduction"). *If x does not occur free in $\mathscr{A}$, then $\mathscr{A} \to \mathscr{B} \vdash \mathscr{A} \to (\forall x)\mathscr{B}$.*

Proof.

(1)	$\mathscr{A} \to \mathscr{B}$	given
(2)	$\neg\mathscr{B} \to \neg\mathscr{A}$	(1) and I.4.1
(3)	$(\exists x)\neg\mathscr{B} \to \neg\mathscr{A}$	(2) and $\exists$-introduction
(4)	$\mathscr{A} \to \neg(\exists x)\neg\mathscr{B}$	(3) and I.4.1
(5)	$\mathscr{A} \to (\forall x)\mathscr{B}$	(4), introducing the $\forall$-abbreviation □

I.4.6 Metatheorem (Specialization). *For any formula $\mathscr{A}$ and term t, $\vdash (\forall x)\mathscr{A} \to \mathscr{A}[t]$.*

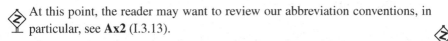 At this point, the reader may want to review our abbreviation conventions, in particular, see **Ax2** (I.3.13).

Proof.

(1)	$\neg\mathscr{A}[t] \to (\exists x)\neg\mathscr{A}$	in Λ
(2)	$\neg(\exists x)\neg\mathscr{A} \to \mathscr{A}[t]$	(1) and I.4.1
(3)	$(\forall x)\mathscr{A} \to \mathscr{A}[t]$	(2), introducing the $\forall$-abbreviation □

I.4.7 Corollary. *For any formula $\mathscr{A}$, $\vdash (\forall x)\mathscr{A} \to \mathscr{A}$.*

Proof. $\mathscr{A}[x \leftarrow x] \equiv \mathscr{A}$. □

Pause. Why is $\mathscr{A}[x \leftarrow x]$ the same string as $\mathscr{A}$?

I.4.8 Metatheorem (Generalization). *For any Γ and any $\mathscr{A}$, if $\Gamma \vdash \mathscr{A}$, then $\Gamma \vdash (\forall x)\mathscr{A}$.*

Proof. Choose $y \not\equiv x$. Then we continue any given proof of $\mathscr{A}$ (from Γ) as follows:

(1)	$\mathscr{A}$	proved from Γ
(2)	$y = y \rightarrow \mathscr{A}$	(1) and I.4.1
(3)	$y = y \rightarrow (\forall x)\mathscr{A}$	(2) and $\forall$-introduction
(4)	$y = y$	in Λ
(5)	$(\forall x)\mathscr{A}$	(3), (4), and MP $\square$

I.4.9 Corollary. *For any Γ and any $\mathscr{A}$, $\Gamma \vdash \mathscr{A}$ iff $\Gamma \vdash (\forall x)\mathscr{A}$.*

Proof. By I.4.7, I.4.8, and modus ponens. $\hspace{2cm}\square$

I.4.10 Corollary. *For any $\mathscr{A}$, $\mathscr{A} \vdash (\forall x)\mathscr{A}$ and $(\forall x)\mathscr{A} \vdash \mathscr{A}$.* $\hspace{1cm}\square$

The above corollary motivates the following definition. It also justifies the common mathematical practice of the "implied universal quantifier". That is, we often state "$\ldots x \ldots$" when we mean "$(\forall x)\ldots x \ldots$".

I.4.11 Definition (Universal Closure). Let $y_1, \ldots, y_n$ be the list of all free variables of $\mathscr{A}$. The *universal closure* of $\mathscr{A}$ is the formula $(\forall y_1)(\forall y_2)\cdots(\forall y_n)\mathscr{A}$ – often written more simply as $(\forall y_1 y_2 \ldots y_n)\mathscr{A}$ or even $(\forall \vec{y}_n)\mathscr{A}$. $\hspace{1cm}\square$

 By I.4.10, a formula deduces and is deduced by its universal closure.

Pause. We said *the* universal closure. Hopefully, the remark immediately above is undisturbed by permutation of $(\forall y_1)(\forall y_2)\cdots(\forall y_n)$. Is it? (Exercise I.11).

I.4.12 Corollary (Substitution of Terms). $\mathscr{A}[x_1, \ldots, x_n] \vdash \mathscr{A}[t_1, \ldots, t_n]$ *for any terms $t_1, \ldots, t_n$.*

The reader may wish to review I.3.12 and the remark following it.

Proof. We illustrate the proof for $n = 2$. What makes it interesting is the requirement to have "simultaneous substitution". To that end we first substitute into x_1 and x_2 *new* variables z, w – i.e., not occurring in either $\mathscr{A}$ or in the t_i. The proof is the following sequence. Comments justify, in each case, the presence of the formula immediately to the left by virtue of the presence of the immediately preceding formula.

$\mathcal{A}[x_1, x_2]$	starting point
$(\forall x_1).\mathcal{A}[x_1, x_2]$	generalization
$\mathcal{A}[z, x_2]$	specialization; $x_1 \leftarrow z$
$(\forall x_2).\mathcal{A}[z, x_2]$	generalization
$\mathcal{A}[z, w]$	specialization; $x_2 \leftarrow w$

Now $z \leftarrow t_1$, $w \leftarrow t_2$, *in any order*, is the same as "simultaneous substitution I.3.12":

$(\forall z).\mathcal{A}[z, w]$	generalization
$\mathcal{A}[t_1, w]$	specialization; $z \leftarrow t_1$
$(\forall w).\mathcal{A}[t_1, w]$	generalization
$\mathcal{A}[t_1, t_2]$	specialization; $w \leftarrow t_2$ □

I.4.13 Metatheorem (The Variant, or Dummy-Renaming, Metatheorem).
For any formula $(\exists x).\mathcal{A}$, *if z does* not *occur in it (i.e., is neither free nor bound),*
then $\vdash (\exists x).\mathcal{A} \leftrightarrow (\exists z).\mathcal{A}[x \leftarrow z]$.

We often write this (under the stated conditions) as $\vdash (\exists x).\mathcal{A}[x] \leftrightarrow (\exists z).\mathcal{A}[z]$.
By the way, another way to state the conditions is "if z does *not* occur in $\mathcal{A}$
(i.e., is neither free nor bound in $\mathcal{A}$), and is different from x". Of course, if
$z \equiv x$, then there is nothing to prove.

Proof. Since z is substitutable in x under the stated conditions, $\mathcal{A}[x \leftarrow z]$ is
defined. Thus, by **Ax2**,

$$\vdash \mathcal{A}[x \leftarrow z] \rightarrow (\exists x).\mathcal{A}$$

By ∃-introduction – since z is not free in $(\exists x).\mathcal{A}$ – we also have

$$\vdash (\exists z).\mathcal{A}[x \leftarrow z] \rightarrow (\exists x).\mathcal{A} \tag{1}$$

We note that x is not free in $(\exists z).\mathcal{A}[x \leftarrow z]$ and is free for z in $\mathcal{A}[x \leftarrow z]$. Indeed,
$\mathcal{A}[x \leftarrow z][z \leftarrow x] \equiv \mathcal{A}$. Thus, by **Ax2**,

$$\vdash \mathcal{A} \rightarrow (\exists z).\mathcal{A}[x \leftarrow z]$$

Hence, by ∃-introduction,

$$\vdash (\exists x).\mathcal{A} \rightarrow (\exists z).\mathcal{A}[x \leftarrow z] \tag{2}$$

Tautological implication from (1) and (2) concludes the argument. □

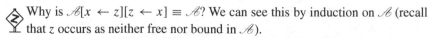 Why is $\mathscr{A}[x \leftarrow z][z \leftarrow x] \equiv \mathscr{A}$? We can see this by induction on $\mathscr{A}$ (recall that z occurs as neither free nor bound in $\mathscr{A}$).

If $\mathscr{A}$ is atomic, then the claim is trivial. The claim also clearly "propagates" with the propositional formation rules, that is, I.1.8(b).

Consider then the case that $\mathscr{A} \equiv (\exists w)\mathscr{B}$. Note that $w \equiv x$ is possible under our assumptions, but $w \equiv z$ is not. If $w \equiv x$, then $\mathscr{A}[x \leftarrow z] \equiv \mathscr{A}$; in particular, z is not free in $\mathscr{A}$; hence $\mathscr{A}[x \leftarrow z][z \leftarrow x] \equiv \mathscr{A}$ as well.

So let us work with $w \not\equiv x$. By I.H., $\mathscr{B}[x \leftarrow z][z \leftarrow x] \equiv \mathscr{B}$. Now

$$
\begin{aligned}
\mathscr{A}[x \leftarrow z][z \leftarrow x] &\equiv ((\exists w)\mathscr{B})[x \leftarrow z][z \leftarrow x] \\
&\equiv ((\exists w)\mathscr{B}[x \leftarrow z])[z \leftarrow x] && \text{see I.3.11; } w \not\equiv z \\
&\equiv ((\exists w)\mathscr{B}[x \leftarrow z][z \leftarrow x]) && \text{see I.3.11; } w \not\equiv x \\
&\equiv ((\exists w)\mathscr{B}) && \text{I.H.} \\
&\equiv \mathscr{A}
\end{aligned}
$$

 By I.4.13, the issue of substitutability becomes moot. Since we have an infinite supply of variables (to use, for example, as bound variables), we can always change the names of *all* the bound variables in $\mathscr{A}$ so that the new names are different from all the free variables in $\mathscr{A}$ or t. In so doing we obtain a formula $\mathscr{B}$ that is (absolutely) provably equivalent to the original.

Then $\mathscr{B}[x \leftarrow t]$ *will* be defined (t will be substitutable in x). Thus, the moral is: *any term t is free for x in $\mathscr{A}$ after an appropriate 'dummy' renaming.*

By the way, this is one of the reasons we want an infinite supply (or an extendible finite set, for the finitist) of formal variables.

I.4.14 Definition. In the following we will often discuss two (or more) theories at once. Let $\mathfrak{T} = (L, \Lambda, \mathbf{I}, \mathscr{T})$ and $\mathfrak{T}' = (L', \Lambda, \mathbf{I}, \mathscr{T}')$ be two theories, such that $\mathscr{V} \subseteq \mathscr{V}'$. This enables $\mathfrak{T}'$ to be "aware" of all the formulas of $\mathfrak{T}$ (but *not* vice versa, since L' contains additional nonlogical symbols).

We say that $\mathfrak{T}'$ is an *extension* of $\mathfrak{T}$ (in symbols, $\mathfrak{T} \leq \mathfrak{T}'$) iff $\mathscr{T} \subseteq \mathscr{T}'$.

Let $\mathscr{A}$ be a formula over L (so that both theories are aware of it). The symbols $\vdash_\mathfrak{T} \mathscr{A}$ and $\vdash_{\mathfrak{T}'} \mathscr{A}$ are synonymous with $\mathscr{A} \in \mathscr{T}$ and $\mathscr{A} \in \mathscr{T}'$ respectively.

Note that we did not explicitly mention the nonlogical axioms Γ or Γ' to the left of $\vdash$, since the subscript of $\vdash$ takes care of that information.

We say that the extension is *conservative* iff for any $\mathscr{A}$ over L, whenever $\vdash_{\mathfrak{T}'} \mathscr{A}$ it is also the case that $\vdash_\mathfrak{T} \mathscr{A}$. That is, when it comes to formulas over the language (L) that *both* theories understand, then the new theory does not do any better than the old in producing theorems. $\square$

I.4.15 Metatheorem (Metatheorem on Constants). *Let us extend a language L of a theory $\mathfrak{T}$ by adding new constant symbols $e_1, \ldots, e_n$ to the alphabet $\mathscr{V}$, resulting in the alphabet $\mathscr{V}'$, language L', and theory $\mathfrak{T}'$.*

Furthermore, assume that $\Gamma' = \Gamma$*, that is, we did not add any new nonlogical axioms.*

Then $\vdash_{\mathfrak{T}'} \mathscr{A}[e_1, \ldots, e_n]$ *implies* $\vdash_{\mathfrak{T}} \mathscr{A}[x_1, \ldots, x_n]$ *for any variables* $x_1, \ldots, x_n$ *that occur* nowhere in $\mathscr{A}[e_1, \ldots, e_n]$, as either free or bound variables.

Proof. Fix a set of variables $x_1, \ldots, x_n$ as described above. We do induction on $\mathfrak{T}'$-theorems.

Basis. $\mathscr{A}[e_1, \ldots, e_n]$ is a logical axiom (over L'); hence so is $\mathscr{A}[x_1, \ldots, x_n]$, *over* L – because of the restriction on the x_i – thus $\vdash_{\mathfrak{T}} \mathscr{A}[x_1, \ldots, x_n]$. Note that $\mathscr{A}[e_1, \ldots, e_n]$ cannot be nonlogical under our assumptions.

Pause. What does the restriction on the x_i have to do with the claim above?

Modus ponens. Here $\vdash_{\mathfrak{T}'} \mathscr{B}[e_1, \ldots, e_n] \rightarrow \mathscr{A}[e_1, \ldots, e_n]$ and $\vdash_{\mathfrak{T}'}$ $\mathscr{B}[e_1, \ldots, e_n]$. By I.H., $\vdash_{\mathfrak{T}} \mathscr{B}[y_1, \ldots, y_n] \rightarrow \mathscr{A}[y_1, \ldots, y_n]$ and $\vdash_{\mathfrak{T}}$ $\mathscr{B}[y_1, \ldots, y_n]$, where $y_1, \ldots, y_n$ occur nowhere in $\mathscr{B}[e_1, \ldots, e_n] \rightarrow \mathscr{A}[e_1, \ldots, e_n]$ as either free or bound variables. By modus ponens, $\vdash_{\mathfrak{T}}$ $\mathscr{A}[y_1, \ldots, y_n]$; hence $\vdash_{\mathfrak{T}} \mathscr{A}[x_1, \ldots, x_n]$ by I.4.12 (and I.4.13).

$\exists$-*introduction.* We have $\vdash_{\mathfrak{T}'} \mathscr{B}[e_1, \ldots, e_n] \rightarrow \mathscr{C}[e_1, \ldots, e_n]$, z is not free in $\mathscr{C}[e_1, \ldots, e_n]$, and $\mathscr{A}[e_1, \ldots, e_n] \equiv (\exists z)\mathscr{B}[e_1, \ldots, e_n] \rightarrow \mathscr{C}[e_1, \ldots, e_n]$. By I.H., if $w_1, \ldots, w_n$ – *distinct from* z – occur nowhere in $\mathscr{B}[e_1, \ldots, e_n] \rightarrow$ $\mathscr{C}[e_1, \ldots, e_n]$ as either free or bound, then we get $\vdash_{\mathfrak{T}} \mathscr{B}[w_1, \ldots, w_n] \rightarrow$ $\mathscr{C}[w_1, \ldots, w_n]$. By $\exists$-introduction we get $\vdash_{\mathfrak{T}} (\exists z)\mathscr{B}[w_1, \ldots, w_n] \rightarrow$ $\mathscr{C}[w_1, \ldots, w_n]$. By I.4.12 and I.4.13 we get $\vdash_{\mathfrak{T}} (\exists z)\mathscr{B}[x_1, \ldots, x_n] \rightarrow$ $\mathscr{C}[x_1, \ldots, x_n]$, i.e., $\vdash_{\mathfrak{T}} \mathscr{A}[x_1, \ldots, x_n]$. $\square$

I.4.16 Corollary. *Let us extend a language* L *of a theory* $\mathfrak{T}$ *by adding* new constant symbols $e_1, \ldots, e_n$ *to the alphabet* $\mathscr{V}$*, resulting to the alphabet* $\mathscr{V}'$*, language* L'*, and theory* $\mathfrak{T}'$*.*

Furthermore, assume that $\Gamma' = \Gamma$*, that is, we did not add any new nonlogical axioms.*

Then $\vdash_{\mathfrak{T}'} \mathscr{A}[e_1, \ldots, e_n]$ *iff* $\vdash_{\mathfrak{T}} \mathscr{A}[x_1, \ldots, x_n]$*, for any choice of variables* $x_1, \ldots, x_n$. $\square$

Proof. If part: Trivially, $\vdash_{\mathfrak{T}} \mathscr{A}[x_1, \ldots, x_n]$ implies $\vdash_{\mathfrak{T}'} \mathscr{A}[x_1, \ldots, x_n]$, hence $\vdash_{\mathfrak{T}'} \mathscr{A}[e_1, \ldots, e_n]$ by I.4.12.

Only-if part: Choose variables $y_1, \ldots, y_n$ that occur nowhere in $\mathscr{A}[e_1, \ldots, e_n]$ as either free or bound. By I.4.15, $\vdash_{\mathfrak{T}} \mathscr{A}[y_1, \ldots, y_n]$; hence, by I.4.12 and I.4.13, $\vdash_{\mathfrak{T}} \mathscr{A}[x_1, \ldots, x_n]$. $\square$

I.4.17 Remark. Thus, the extension $\mathfrak{T}'$ of $\mathfrak{T}$ is *conservative*, for, if $\mathscr{A}$ is over L, then $\mathscr{A}[e_1, \ldots, e_n] \equiv \mathscr{A}$. Therefore, if $\vdash_{\mathfrak{T}'} \mathscr{A}$, then $\vdash_{\mathfrak{T}'} \mathscr{A}[e_1, \ldots, e_n]$; hence $\vdash_{\mathfrak{T}} \mathscr{A}[x_1, \ldots, x_n]$, that is, $\vdash_{\mathfrak{T}} \mathscr{A}$.

A more emphatic way to put the above is this: $\mathfrak{T}'$ is not aware of any new *nonlogical* facts that $\mathfrak{T}$ did not already "know" although by a different name. If $\mathfrak{T}'$ can prove $\mathscr{A}[e_1, \ldots, e_n]$, then $\mathfrak{T}$ can prove the same "statement", however, using (any) names (other than the e_i) that are meaningful in its own language; namely, it can prove $\mathscr{A}[x_1, \ldots, x_n]$. □

The following corollary stems from the proof (rather than the statement) of I.4.15 and I.4.16, and is important.

I.4.18 Corollary. *Let $e_1, \ldots, e_n$ be constants that do not appear in the nonlogical axioms Γ. Then, if $x_1, \ldots, x_n$ are any variables, and if $\Gamma \vdash \mathscr{A}[e_1, \ldots, e_n]$, it is also the case that $\Gamma \vdash \mathscr{A}[x_1, \ldots, x_n]$.*

I.4.19 Metatheorem (The Deduction Theorem). *For any closed formula $\mathscr{A}$, arbitrary formula $\mathscr{B}$, and set of formulas Γ, if $\Gamma + \mathscr{A} \vdash \mathscr{B}$, then $\Gamma \vdash \mathscr{A} \rightarrow \mathscr{B}$.*

N.B. $\Gamma + \mathscr{A}$ denotes the augmentation of Γ by adding the formula $\mathscr{A}$. In the present metatheorem $\mathscr{A}$ is a single (but unspecified) formula. However, the notation extends to the case where $\mathscr{A}$ is a schema, in which case it means the augmentation of Γ by adding all the instances of the schema.

A converse of the metatheorem is also true trivially: That is, $\Gamma \vdash \mathscr{A} \rightarrow \mathscr{B}$ implies $\Gamma + \mathscr{A} \vdash \mathscr{B}$. This direction immediately follows by modus ponens and does not require the restriction on $\mathscr{A}$.

Proof. The proof is by induction on $\Gamma + \mathscr{A}$ theorems.

Basis. Let $\mathscr{B}$ be logical or nonlogical (but, in the latter case, assume $\mathscr{B} \not\equiv \mathscr{A}$). Then $\Gamma \vdash \mathscr{B}$.

Since $\mathscr{B} \models_{\text{Taut}} \mathscr{A} \rightarrow \mathscr{B}$, it follows by I.4.1 that $\Gamma \vdash \mathscr{A} \rightarrow \mathscr{B}$.

Now, if $\mathscr{B} \equiv \mathscr{A}$, then $\mathscr{A} \rightarrow \mathscr{B}$ is a logical axiom (group **Ax1**); hence $\Gamma \vdash \mathscr{A} \rightarrow \mathscr{B}$ once more.

Modus ponens. Let $\Gamma + \mathscr{A} \vdash \mathscr{C}$, and $\Gamma + \mathscr{A} \vdash \mathscr{C} \rightarrow \mathscr{B}$.

By I.H., $\Gamma \vdash \mathscr{A} \rightarrow \mathscr{C}$ and $\Gamma \vdash \mathscr{A} \rightarrow \mathscr{C} \rightarrow \mathscr{B}$.

Since $\mathscr{A} \rightarrow \mathscr{C}, \mathscr{A} \rightarrow \mathscr{C} \rightarrow \mathscr{B} \models_{\text{Taut}} \mathscr{A} \rightarrow \mathscr{B}$, we have $\Gamma \vdash \mathscr{A} \rightarrow \mathscr{B}$.

∃-introduction. Let $\Gamma + \mathscr{A} \vdash \mathscr{C} \rightarrow \mathscr{D}$, and $\mathscr{B} \equiv (\exists x)\mathscr{C} \rightarrow \mathscr{D}$, where x is not free in $\mathscr{D}$. By I.H., $\Gamma \vdash \mathscr{A} \rightarrow \mathscr{C} \rightarrow \mathscr{D}$. By I.4.1, $\Gamma \vdash \mathscr{C} \rightarrow \mathscr{A} \rightarrow \mathscr{D}$; hence $\Gamma \vdash (\exists x)\mathscr{C} \rightarrow \mathscr{A} \rightarrow \mathscr{D}$ by ∃-introduction ($\mathscr{A}$ is closed). One more application of I.4.1 yields $\Gamma \vdash \mathscr{A} \rightarrow (\exists x)\mathscr{C} \rightarrow \mathscr{D}$. □

I.4.20 Remark. (1) Is the restriction that, $\mathscr{A}$ must be closed important? Yes. Let $\mathscr{A} \equiv x = a$, where "a" is some constant. Then, even though $\mathscr{A} \vdash (\forall x)\mathscr{A}$

by generalization, it is not always true that $\vdash \mathscr{A} \to (\forall x)\mathscr{A}$. This follows from soundness considerations (next section). Intuitively, assuming that our logic "doesn't lie" (that is, it proves no "invalid" formulas), we immediately infer that $x = a \to (\forall x)x = a$ *cannot* be absolutely provable, for it is a "lie". It fails at least over $\mathbb{N}$, if a is interpreted to be "0".

(2) I.4.16 adds flexibility to applications of the deduction theorem:

$$\vdash_{\mathfrak{T}} (\mathscr{A} \to \mathscr{B})[x_1, \ldots, x_n] \tag{$*$}$$

where $[x_1, \ldots, x_n]$ is the list of *all free variables just in* $\mathscr{A}$, is equivalent (by I.4.16) to

$$\vdash_{\mathfrak{T}'} (\mathscr{A} \to \mathscr{B})[e_1, \ldots, e_n] \tag{$**$}$$

where $e_1, \ldots, e_n$ are new constants added to $\mathscr{V}$ (with no effect on nonlogical axioms: $\Gamma = \Gamma'$).

Now, since $\mathscr{A}[e_1, \ldots, e_n]$ is closed, proving

$$\Gamma' + \mathscr{A}[e_1, \ldots, e_n] \vdash \mathscr{B}[e_1, \ldots, e_n]$$

establishes ($**$), and hence also ($*$).

In practice, one does not perform this step explicitly, but ensures that, throughout the $\Gamma + \mathscr{A}$ proof, *whatever free variables were present in* $\mathscr{A}$ "behaved like constants", or, as we also say, were *frozen*.

(3) In some expositions the deduction theorem is *not* constrained by requiring that $\mathscr{A}$ be closed (e.g., Bourbaki (1966b) and more recently Enderton (1972)).

Which version is right? Both are *in their respective contexts*. If all the rules of inference are "propositional" (e.g., as in Bourbaki (1966b) and Enderton (1972), who only employ modus ponens) – that is, they do not meddle with quantifiers – then the deduction theorem is unconstrained. If, on the other hand, the rules of inference manipulate object variables via quantification, then one cannot avoid constraining the application of the deduction theorem, lest one want to derive (the invalid) $\vdash \mathscr{A} \to (\forall x)\mathscr{A}$ from the valid $\mathscr{A} \vdash (\forall x)\mathscr{A}$.

This also entails that approaches such as in Bourbaki (1966b) and Enderton (1972) do *not* allow "full" generalization "$\mathscr{A} \vdash (\forall x)\mathscr{A}$". They only allow a "weaker" rule, "if $\vdash \mathscr{A}$, then $\vdash (\forall x)\mathscr{A}$".[†]

(4) This divergence of approach in choosing rules of inference has some additional repercussions: One has to be careful in defining the semantic counterpart

---

[†] Indeed, they allow a bit more generally, namely, the rule "if $\Gamma \vdash \mathscr{A}$ *with a side condition*, then $\Gamma \vdash (\forall x)\mathscr{A}$. The side condition is that the formulas of Γ do not have free occurrences of x." Of course, Γ can be always taken to be finite (why?), so that this condition is not unrealistic.

of ⊢, namely, ⊨ (see next section). One wants the two symbols to "track each other" faithfully (Gödel's completeness theorem).[†]

I.4.21 Corollary (Proof by Contradiction). *Let $\mathscr{A}$ be closed. Then $\Gamma \vdash \mathscr{A}$ iff $\Gamma + \neg \mathscr{A}$ is inconsistent.*

Proof. If part: Given that $\mathscr{T} = \mathbf{Wff}$, where $\mathscr{T}$ is the theory $\Gamma + \neg \mathscr{A}$. In particular, $\Gamma + \neg \mathscr{A} \vdash \mathscr{A}$. By the deduction theorem, $\Gamma \vdash \neg \mathscr{A} \to \mathscr{A}$. But $\neg \mathscr{A} \to \mathscr{A} \models_{\mathbf{Taut}} \mathscr{A}$.

Only-if part: Given that $\Gamma \vdash \mathscr{A}$. Hence $\Gamma + \neg \mathscr{A} \vdash \mathscr{A}$ as well (recall I.3.18(2)). Of course, $\Gamma + \neg \mathscr{A} \vdash \neg \mathscr{A}$. Since $\mathscr{A}, \neg \mathscr{A} \models_{\mathbf{Taut}} \mathscr{B}$ for an arbitrary $\mathscr{B}$, we are done.   □

Pause. Is it necessary to assume that $\mathscr{A}$ is closed in I.4.21? Why?

The following is important enough to merit stating. It follows from the type of argument we employed in the only-if part above.

I.4.22 Metatheorem. $\mathfrak{T}$ *is inconsistent iff for some $\mathscr{A}$, both $\vdash_{\mathfrak{T}} \mathscr{A}$ and $\vdash_{\mathfrak{T}} \neg \mathscr{A}$ hold.*

We also list below a number of "quotable" proof techniques. These techniques are routinely used by mathematicians, and will be routinely used by us. The proofs of all the following metatheorems are delegated to the reader.

I.4.23 Metatheorem (Distributivity or Monotonicity of ∃). *For any x, $\mathscr{A}$, $\mathscr{B}$,*

$$\mathscr{A} \to \mathscr{B} \vdash (\exists x)\mathscr{A} \to (\exists x)\mathscr{B}$$

Proof. See Exercise I.12.   □

I.4.24 Metatheorem (Distributivity or Monotonicity of ∀). *For any x, $\mathscr{A}$, $\mathscr{B}$,*

$$\mathscr{A} \to \mathscr{B} \vdash (\forall x)\mathscr{A} \to (\forall x)\mathscr{B}$$

Proof. See Exercise I.13.   □

The term "monotonicity" is inspired by thinking of "→" as "≤". How? Well, we have the tautology

$$(\mathscr{A} \to \mathscr{B}) \leftrightarrow (A \vee B \leftrightarrow B) \tag{i}$$

[†] In Mendelson (1987), ⊨ is defined inconsistently with ⊢.

If we think of "$\mathscr{A} \vee \mathscr{B}$" as "max($\mathscr{A}, \mathscr{B}$)", then the right hand side in (*i*) above says that $\mathscr{B}$ is the maximum of $\mathscr{A}$ and $\mathscr{B}$, or that $\mathscr{A}$ is "less than or equal to" $\mathscr{B}$. The above metatheorems say that both ∃ and ∀ preserve this "inequality".

I.4.25 Metatheorem (Equivalence Theorem, or Leibniz Rule). *Let* $\Gamma \vdash$ $\mathscr{A} \leftrightarrow \mathscr{B}$, *and let* $\mathscr{C}'$ *be obtained from* $\mathscr{C}$ *by replacing* some *– possibly, but not necessarily, all – occurrences of a subformula* $\mathscr{A}$ *of* $\mathscr{C}$ *by* $\mathscr{B}$.
Then $\Gamma \vdash \mathscr{C} \leftrightarrow \mathscr{C}'$, *i.e.,*

$$\frac{\mathscr{A} \leftrightarrow \mathscr{B}}{\mathscr{C} \leftrightarrow \mathscr{C}'}$$

is a derived rule.

Proof. The proof is by induction on formulas $\mathscr{C}$. See Exercise I.15.   □

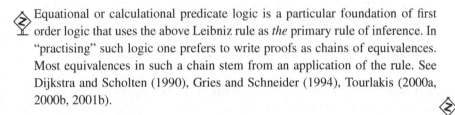 Equational or calculational predicate logic is a particular foundation of first order logic that uses the above Leibniz rule as *the* primary rule of inference. In "practising" such logic one prefers to write proofs as chains of equivalences. Most equivalences in such a chain stem from an application of the rule. See Dijkstra and Scholten (1990), Gries and Schneider (1994), Tourlakis (2000a, 2000b, 2001b).

I.4.26 Metatheorem (Proof by Cases). *Suppose that* $\Gamma \vdash \mathscr{A}_1 \vee \cdots \vee \mathscr{A}_n$, *and* $\Gamma \vdash \mathscr{A}_i \rightarrow \mathscr{B}$ *for* $i = 1, \ldots$. *Then* $\Gamma \vdash \mathscr{B}$.

Proof. Immediate, by I.4.1.   □

Proof by cases usually benefits from the application of the deduction theorem. That is, having established $\Gamma \vdash \mathscr{A}_1 \vee \cdots \vee \mathscr{A}_n$, one then proceeds to adopt, in turn, each $\mathscr{A}_i$ ($i = 1, \ldots, n$) as a new nonlogical axiom (with its variables "frozen"). In each "case" ($\mathscr{A}_i$) one proceeds to prove $\mathscr{B}$.
At the end of all this one has established $\Gamma \vdash \mathscr{B}$.

In practice we normally use the following *argot:*
"We will consider cases $\mathscr{A}_i$, for $i = 1, \ldots, n$.[†]

Case $\mathscr{A}_1$.   ... therefore, $\mathscr{B}$.[‡]

...

Case $\mathscr{A}_n$.   ... therefore, $\mathscr{B}$."

[†] To legitimize this splitting into cases, we must, of course, show $\Gamma \vdash \mathscr{A}_1 \vee \cdots \vee \mathscr{A}_n$.
[‡] That is, we add the axiom $\mathscr{A}_1$ to Γ, freezing its variables, and we then prove $\mathscr{B}$.

I.4.27 Metatheorem (Proof by Auxiliary Constant). *Suppose that for arbitrary $\mathscr{A}$ and $\mathscr{B}$ over the language L we know*

(1) $\Gamma \vdash (\exists x)\mathscr{A}[x]$

(2) $\Gamma + \mathscr{A}[a] \vdash \mathscr{B}$, *where a is a new constant not in the language L of Γ. Furthermore assume that in the proof of $\mathscr{B}$ all the free variables of $\mathscr{A}[a]$ were frozen. Then $\Gamma \vdash \mathscr{B}$.*

Proof. Exercise I.21.   □

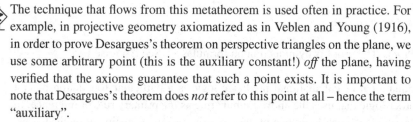

The technique that flows from this metatheorem is used often in practice. For example, in projective geometry axiomatized as in Veblen and Young (1916), in order to prove Desargues's theorem on perspective triangles on the plane, we use some arbitrary point (this is the auxiliary constant!) *off* the plane, having verified that the axioms guarantee that such a point exists. It is important to note that Desargues's theorem does *not* refer to this point at all – hence the term "auxiliary".

In this example, from projective geometry, "$\mathscr{B}$" is Desargues's theorem, "$(\exists x)\mathscr{A}[x]$" asserts that there are points outside the plane, a is an arbitrary such point, and the proof (2) starts with words like "Let a be a point off the plane" – which is *argot* for "*add the axiom $\mathscr{A}[a]$*".

I.5. Semantics; Soundness, Completeness, Compactness

So what do all these symbols mean? We show in this section how to "decode" the formal statements (formulas) into informal statements of "real" mathematics. Conversely, this will entail an understanding of how to code statements of real mathematics in our formal language.

The rigorous[†] definition of semantics for first order languages is due to Tarski and is often referred to as "Tarski semantics". The flavour of the particular definition given below is that of Shoenfield (1967), and it accurately reflects our syntactic choices – most importantly, the choice to allow full generalization $\mathscr{A} \vdash (\forall x)\mathscr{A}$. In particular, we will define the semantic counterpart of $\vdash$, namely, $\models$, pronounced "logically implies", to ensure that $\Gamma \vdash \mathscr{A}$ iff $\Gamma \models \mathscr{A}$. This is the content of Gödel's completeness theorem, which we prove in this section.

This section will place some additional demands on the reader's recollection of notation and facts from informal set theory. We will, among other things,

---

[†] One often says "The formal definition of semantics ...", but the word "formal" is misleading here, for we are actually defining semantics in the metatheory, not in the formal theory.

make use of notation from naïve set theory, such as

$$A^n \qquad (\text{or } \underbrace{A \times \cdots \times A}_{n \text{ times}})$$

for the set of ordered n-tuples of members of A.

We will also use the symbols $\subseteq$, $\cup$, $\bigcup_{a \in I}$.[†]

In some passages – delimited by ☝☝ warning signs – these demands will border on the unreasonable.

For example, in the proof of the Gödel-Mal'cev completeness-compactness result we will need some elementary understanding of ordinals – used as indexing tools – and cardinality. Some readers may not have such background. This prerequisite material can be attained by consulting a set theory book (e.g., the second volume of these lectures).

I.5.1 Definition. Given a language $L = (\mathscr{V}, \textbf{Term}, \textbf{Wff})$, a *structure* $\mathfrak{M} = (M, \mathscr{I})$ *appropriate for* L is such that $M \neq \emptyset$ is a set (the *domain* or *underlying set* or *universe*[‡]) and $\mathscr{I}$ ("$\mathscr{I}$" for *interpretation*) is a mapping that assigns

(1) to each constant a of $\mathscr{V}$ a unique member $a^{\mathscr{I}} \in M$,
(2) to each function f of $\mathscr{V}$ – of arity n – a unique (total)[§] function $f^{\mathscr{I}}$: $M^n \to M$,
(3) to each predicate P of $\mathscr{V}$ – of arity n – a unique set $P^{\mathscr{I}} \subseteq M^n$.[¶] $\qquad \square$

I.5.2 Remark. The structure $\mathfrak{M}$ is often given more verbosely, in conformity with practice in algebra. Namely, one "unpacks" the $\mathscr{I}$ into a list $a^{\mathscr{I}}, b^{\mathscr{I}}, \ldots$; $f^{\mathscr{I}}, g^{\mathscr{I}}, \ldots; P^{\mathscr{I}}, Q^{\mathscr{I}}, \ldots$ and writes instead $\mathfrak{M} = (M; a^{\mathscr{I}}, b^{\mathscr{I}}, \ldots; f^{\mathscr{I}}, g^{\mathscr{I}}, \ldots; P^{\mathscr{I}}, Q^{\mathscr{I}}, \ldots)$. Under this understanding, a structure is an underlying set (universe), M, along with a *list* of "concrete" constants, functions, and relations that "interpret" corresponding "abstract" items of the language.

Under the latter notational circumstances we often use the symbols $a^{\mathfrak{M}}$, $f^{\mathfrak{M}}$, $P^{\mathfrak{M}}$ – rather than $a^{\mathscr{I}}$, $f^{\mathscr{I}}$, $P^{\mathscr{I}}$ – to indicate the interpretations in $\mathfrak{M}$ of the constant a, function f, and predicate P respectively.

[†] If we have a set of sets $\{S_a, S_b, S_c, \ldots\}$, where the indices $a, b, c, \ldots$ all come out of an "index set" I, then the symbol $\bigcup_{i \in I} S_i$ stands for the collection of *all* those objects x that are found in *at least one* of the sets S_i. It is a common habit to write $\bigcup_{i=0}^{\infty} S_i$ instead of $\bigcup_{i \in \mathbb{N}} S_i$. $A \cup B$ is the same as $\bigcup_{i \in \{1,2\}} S_i$, where we have let $S_1 = A$ and $S_2 = B$.

[‡] Often the qualification "of discourse" is added to the terms "domain" and "universe".

[§] Requiring $f^{\mathscr{I}}$ to be total is a traditional convention. By the way, *total* means that $f^{\mathscr{I}}$ is defined everywhere on M^n.

[¶] Thus $P^{\mathscr{I}}$ is an n-ary relation with inputs and outputs in M.

We have said above "structure appropriate for L", thus emphasizing the generality of the language and therefore our ability to interpret what we say in it in many different ways. Often though, e.g., as in formal arithmetic or set theory, we have a structure in mind to begin with, and then build a formal language to formally codify statements about the objects in the structure. Under these circumstances, in effect, we define a language appropriate *for the structure*. We use the symbol $L_{\mathfrak{M}}$ to indicate that the language was built to fit the structure $\mathfrak{M}$. □

I.5.3 Definition. We routinely add new nonlogical symbols to a language L to obtain a language L'. We say that L' is an *extension* of L and that L is a *restriction* of L'. Suppose that $\mathfrak{M} = (M, \mathscr{I})$ is a structure for L, and let $\mathfrak{M}' = (M, \mathscr{I}')$ be a structure with the *same* underlying set M, but with $\mathscr{I}$ extended to $\mathscr{I}'$ so that the latter gives meaning to all *new* symbols while it gives the *same* meaning, as $\mathscr{I}$ does, to the symbols of L.

We call $\mathfrak{M}'$ an *expansion* (rather than "extension") of $\mathfrak{M}$, and $\mathfrak{M}$ a *reduct* (rather than "restriction") of $\mathfrak{M}'$. We may (often) write $\mathscr{I} = \mathscr{I}' \restriction L$ to indicate that the "mapping" $\mathscr{I}'$ – restricted to L (symbol " $\restriction$ ") – equals $\mathscr{I}$. We may also write $\mathfrak{M} = \mathfrak{M}' \restriction L$ instead. □

I.5.4 Definition. Given L and a structure $\mathfrak{M} = (M, \mathscr{I})$ appropriate for L. $L(\mathfrak{M})$ denotes the language obtained from L by adding to $\mathscr{V}$ a unique new name $\bar{i}$ for each object $i \in M$.

This amends both sets **Term**, **Wff** into **Term**$(\mathfrak{M})$, **Wff**$(\mathfrak{M})$. Members of the latter sets are called $\mathfrak{M}$-terms and $\mathfrak{M}$-formulas respectively.

We extend the mapping $\mathscr{I}$ to the new constants by: $\bar{i}^{\mathscr{I}} = i$ for all $i \in M$ (where the "$=$" here is metamathematical: equality on M). □

 All we have done here is to allow ourselves to do substitutions like $[x \leftarrow i]$ formally. We do, instead, $[x \leftarrow \bar{i}]$. One next gives "meaning" to *all closed* terms in $L(\mathfrak{M})$. The following uses definition by recursion (I.2.13) and relies on the fact that the rules that define terms are unambiguous.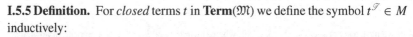

I.5.5 Definition. For *closed* terms t in **Term**$(\mathfrak{M})$ we define the symbol $t^{\mathscr{I}} \in M$ inductively:

(1) If t is any of a (original constant) or $\bar{i}$ (imported constant), then $t^{\mathscr{I}}$ has already been defined.
(2) If t is the string $ft_1 \ldots t_n$, where f is n-ary, and $t_1, \ldots, t_n$ are *closed* $\mathfrak{M}$-terms, we define $t^{\mathscr{I}}$ to be the object (of M) $f^{\mathscr{I}}(t_1^{\mathscr{I}}, \ldots, t_n^{\mathscr{I}})$. □

Finally, we give meaning to *all closed $\mathfrak{M}$-formulas*, again by recursion (over **Wff**).

I.5.6 Definition. For *any closed* formula $\mathscr{A}$ in **Wff**($\mathfrak{M}$) we define the symbol $\mathscr{A}^{\mathscr{I}}$ inductively. In all cases, $\mathscr{A}^{\mathscr{I}} \in \{\mathbf{t}, \mathbf{f}\}$.

(1) If $\mathscr{A} \equiv t = s$, where t and s are *closed* $\mathfrak{M}$-terms, then $\mathscr{A}^{\mathscr{I}} = \mathbf{t}$ iff $t^{\mathscr{I}} = s^{\mathscr{I}}$. (The last two occurrences of "=" are metamathematical.)

(2) If $\mathscr{A} \equiv Pt_1 \ldots t_n$, where P is an n-ary predicate and the t_i are *closed* $\mathfrak{M}$-terms, then $\mathscr{A}^{\mathscr{I}} = \mathbf{t}$ iff $\langle t_1^{\mathscr{I}}, \ldots, t_n^{\mathscr{I}} \rangle \in P^{\mathscr{I}}$ or $P^{\mathscr{I}}(t_1^{\mathscr{I}}, \ldots, t_n^{\mathscr{I}})$ "holds". (Or "is true"; see p. 19. Of course, the last occurrence of "=" is metamathematical.)

(3) If $\mathscr{A}$ is any of the *sentences* $\neg\mathscr{B}, \mathscr{B} \vee \mathscr{C}$, then $\mathscr{A}^{\mathscr{I}}$ is determined by the usual truth tables (see p. 30) using the values $\mathscr{B}^{\mathscr{I}}$ and $\mathscr{C}^{\mathscr{I}}$. That is, $(\neg\mathscr{B})^{\mathscr{I}} = F_{\neg}(\mathscr{B}^{\mathscr{I}})$ and $(\mathscr{B} \vee \mathscr{C})^{\mathscr{I}} = F_{\vee}(\mathscr{B}^{\mathscr{I}}, \mathscr{C}^{\mathscr{I}})$. (The last two occurrences of "=" are metamathematical.)

(4) If $\mathscr{A} \equiv (\exists x)\mathscr{B}$, then $\mathscr{A}^{\mathscr{I}} = \mathbf{t}$ iff $(\mathscr{B}[x \leftarrow \bar{i}])^{\mathscr{I}} = \mathbf{t}$ *for some* $i \in M$. (The last two occurrences of "=" are metamathematical.) $\square$

We have "imported" constants from M into L in order to be able to state the semantics of $(\exists x)\mathscr{B}$ above in the simple manner we just did (following Shoenfield (1967)).

We often state the semantics of $(\exists x)\mathscr{B}$ by writing

$$((\exists x)\mathscr{B}[x])^{\mathscr{I}} \text{ is true} \quad \text{iff} \quad (\exists i \in M)(\mathscr{B}[\bar{i}])^{\mathscr{I}} \text{ is true}$$

I.5.7 Definition. Let $\mathscr{A} \in$ **Wff**, and $\mathfrak{M}$ be a structure as above.

An $\mathfrak{M}$-*instance of* $\mathscr{A}$ is an $\mathfrak{M}$-sentence $\mathscr{A}(\bar{i}_1, \ldots, \bar{i}_k)$ (that is, all the free variables of $\mathscr{A}$ have been replaced by imported constants).

We say that $\mathscr{A}$ is *valid in* $\mathfrak{M}$, or that $\mathfrak{M}$ is a *model of* $\mathscr{A}$, iff *for all* $\mathfrak{M}$-instances $\mathscr{A}'$ of $\mathscr{A}$ it is the case that $\mathscr{A}'^{\mathscr{I}} = \mathbf{t}$.[†] Under these circumstances we write $\models_{\mathfrak{M}} \mathscr{A}$.

For any set of formulas Γ from **Wff**, the expression $\models_{\mathfrak{M}} \Gamma$, pronounced "$\mathfrak{M}$ *is a model of* Γ", means that *for all* $\mathscr{A} \in \Gamma$, $\models_{\mathfrak{M}} \mathscr{A}$.

A formula $\mathscr{A}$ is *universally valid* or *logically valid* (we often say just *valid*) iff *every structure appropriate for the language is a model of* $\mathscr{A}$.

Under these circumstances we simply write $\models \mathscr{A}$.

If Γ is a set of formulas, then we say it is *satisfiable* iff it has a model. It is *finitely satisfiable* iff *every finite subset of* Γ has a model.[‡] $\square$

The definition of validity of $\mathscr{A}$ in a structure $\mathfrak{M}$ corresponds with the normal mathematical practice. It says that a formula is true (in a given "context" $\mathfrak{M}$) just in case it is so for all possible values of the free variables.

† We henceforth discontinue our pedantic "(The last occurrence of "=" is metamathematical.)".
‡ These two concepts are often defined just for sentences.

I.5.8 Definition. We say that Γ *logically implies* $\mathcal{A}$, in symbols $\Gamma \models \mathcal{A}$, meaning that *every model of* Γ *is also a model of* $\mathcal{A}$.   □

I.5.9 Definition (Soundness). A theory (identified by its nonlogical axioms) Γ is *sound* iff, for all $\mathcal{A} \in \mathbf{Wff}$, $\Gamma \vdash \mathcal{A}$ implies $\Gamma \models \mathcal{A}$, that is, iff all the theorems of the theory are logically implied by the nonlogical axioms.   □

Clearly then, a *pure* theory $\mathfrak{T}$ is sound iff $\vdash_{\mathfrak{T}} \mathcal{A}$ implies $\models \mathcal{A}$ for all $\mathcal{A} \in \mathbf{Wff}$. That is, all its theorems are universally valid.

Towards the soundness result[†] below we look at two tedious (but easy) lemmata.

I.5.10 Lemma. *Given a term t, variables $x \not\equiv y$, where y does not occur in t, and a constant a. Then, for any term s and formula $\mathcal{A}$, $s[x \leftarrow t][y \leftarrow a] \equiv s[y \leftarrow a][x \leftarrow t]$ and $\mathcal{A}[x \leftarrow t][y \leftarrow a] \equiv \mathcal{A}[y \leftarrow a][x \leftarrow t]$.*

Proof. Induction on s:
  Basis:

$$s[x \leftarrow t][y \leftarrow a] \equiv \begin{cases} \text{if } s \equiv x & \text{then } t \\ \text{if } s \equiv y & \text{then } a \\ \text{if } s \equiv z & \text{where } x \not\equiv z \not\equiv y, \text{ then } z \\ \text{if } s \equiv b & \text{then } b \end{cases}$$

$$\equiv s[y \leftarrow a][x \leftarrow t]$$

For the induction step let $s \equiv f r_1 \ldots r_n$, where f has arity n. Then

$$\begin{aligned} s[x \leftarrow t][y \leftarrow a] &\equiv f r_1[x \leftarrow t][y \leftarrow a] \ldots r_n[x \leftarrow t][y \leftarrow a] \\ &\equiv f r_1[y \leftarrow a][x \leftarrow t] \ldots r_n[y \leftarrow a][x \leftarrow t] \quad \text{by I.H.} \\ &\equiv s[y \leftarrow a][x \leftarrow t] \end{aligned}$$

Induction on $\mathcal{A}$:
  Basis:

$$\mathcal{A}[x \leftarrow t][y \leftarrow a]$$

$$\equiv \begin{cases} \text{if } \mathcal{A} \equiv P r_1 \ldots r_n \text{ then} \\ \qquad P r_1[x \leftarrow t][y \leftarrow a] \ldots r_n[x \leftarrow t][y \leftarrow a] \\ \qquad \equiv P r_1[y \leftarrow a][x \leftarrow t] \ldots r_n[y \leftarrow a][x \leftarrow t] \\ \text{if } \mathcal{A} \equiv r = s \text{ then} \\ \qquad r[x \leftarrow t][y \leftarrow a] = s[x \leftarrow t][y \leftarrow a] \\ \qquad \equiv r[y \leftarrow a][x \leftarrow t] = s[y \leftarrow a][x \leftarrow t] \end{cases}$$

$$\equiv \mathcal{A}[y \leftarrow a][x \leftarrow t]$$

[†] Also nicknamed "the easy half of Gödel's completeness theorem".

The property we are proving, trivially, propagates with Boolean connectives. Let us do the induction step just in the case where $\mathscr{A} \equiv (\exists w).\mathscr{B}$. If $w \equiv x$ or $w \equiv y$, then the result is trivial. Otherwise,

$$\mathscr{A}[x \leftarrow t][y \leftarrow a] \equiv ((\exists w).\mathscr{B})[x \leftarrow t][y \leftarrow a]$$
$$\equiv ((\exists w).\mathscr{B}[x \leftarrow t][y \leftarrow a])$$
$$\equiv ((\exists w).\mathscr{B}[y \leftarrow a][x \leftarrow t]) \qquad \text{by I.H.}$$
$$\equiv ((\exists w).\mathscr{B})[y \leftarrow a][x \leftarrow t]$$
$$\equiv \mathscr{A}[y \leftarrow a][x \leftarrow t] \qquad \qquad \square$$

I.5.11 Lemma. *Given a structure* $\mathfrak{M} = (M, \mathscr{I})$, *a term* s, *and a formula* $\mathscr{A}$, *both over* $L(\mathfrak{M})$. *Suppose each of* s *and* $\mathscr{A}$ *have* at most one *free variable,* x.

Let t *be a* closed *term over* $L(\mathfrak{M})$ *such that* $t^{\mathscr{I}} = i \in M$. *Then* $(s[x \leftarrow t])^{\mathscr{I}} = (s[x \leftarrow \bar{i}])^{\mathscr{I}}$ *and* $(\mathscr{A}[x \leftarrow t])^{\mathscr{I}} = (\mathscr{A}[x \leftarrow \bar{i}])^{\mathscr{I}}$. *Of course, since* t *is closed,* $\mathscr{A}[x \leftarrow t]$ *is defined.*

Proof. Induction on s:

Basis. $s[x \leftarrow t] \equiv s$ if $s \in \{y, a, \bar{j}\}$ ($y \not\equiv x$). Hence $(s[x \leftarrow t])^{\mathscr{I}} = s^{\mathscr{I}} = (s[x \leftarrow \bar{i}])^{\mathscr{I}}$ in this case. If $s \equiv x$, then $s[x \leftarrow t] \equiv t$ and $s[x \leftarrow \bar{i}] \equiv \bar{i}$, and the claim follows once more.

For the induction step let $s \equiv fr_1 \ldots r_n$, where f has arity n. Then

$$(s[x \leftarrow t])^{\mathscr{I}} = f^{\mathscr{I}}((r_1[x \leftarrow t])^{\mathscr{I}}, \ldots, (r_n[x \leftarrow t])^{\mathscr{I}})$$
$$= f^{\mathscr{I}}((r_1[x \leftarrow \bar{i}])^{\mathscr{I}}, \ldots, (r_n[x \leftarrow \bar{i}])^{\mathscr{I}}) \qquad \text{by I.H.}$$
$$= (s[x \leftarrow \bar{i}])^{\mathscr{I}}$$

Induction on $\mathscr{A}$:

Basis. If $\mathscr{A} \equiv Pr_1 \ldots r_n$, then[†]

$$(\mathscr{A}[x \leftarrow t])^{\mathscr{I}} = P^{\mathscr{I}}((r_1[x \leftarrow t])^{\mathscr{I}}, \ldots, (r_n[x \leftarrow t])^{\mathscr{I}})$$
$$= P^{\mathscr{I}}((r_1[x \leftarrow \bar{i}])^{\mathscr{I}}, \ldots, (r_n[x \leftarrow \bar{i}])^{\mathscr{I}})$$
$$= (\mathscr{A}[x \leftarrow \bar{i}])^{\mathscr{I}}$$

Similarly if $\mathscr{A} \equiv r = s$.

The property we are proving, clearly, propagates with Boolean connectives. Let us do the induction step just in the case where $\mathscr{A} = (\exists w).\mathscr{B}$. If $w \equiv x$, the result is trivial. Otherwise, we note that – since t is closed – w does not occur

[†] For a metamathematical relation Q, as is usual (p. 19), $Q(a, b, \ldots) = \mathbf{t}$, or just $Q(a, b, \ldots)$, stands for $\langle a, b, \ldots \rangle \in Q$.

in t, and proceed as follows:

$(\mathscr{A}[x \leftarrow t])^{\mathscr{I}} = \mathbf{t}$ iff $(((\exists w).\mathscr{B})[x \leftarrow t])^{\mathscr{I}} = \mathbf{t}$

iff $(((\exists w).\mathscr{B}[x \leftarrow t]))^{\mathscr{I}} = \mathbf{t}$

iff $(\mathscr{B}[x \leftarrow t][w \leftarrow \overline{j}])^{\mathscr{I}} = \mathbf{t}$ for some $j \in M$, by I.5.6(4)

iff $(\mathscr{B}[w \leftarrow \overline{j}][x \leftarrow t])^{\mathscr{I}} = \mathbf{t}$ for some $j \in M$, by I.5.10

iff $((\mathscr{B}[w \leftarrow \overline{j}])[x \leftarrow t])^{\mathscr{I}} = \mathbf{t}$ for some $j \in M$

iff $((\mathscr{B}[w \leftarrow \overline{j}])[x \leftarrow \overline{i}])^{\mathscr{I}} = \mathbf{t}$ for some $j \in M$, by I.H.

iff $(\mathscr{B}[w \leftarrow \overline{j}][x \leftarrow \overline{i}])^{\mathscr{I}} = \mathbf{t}$ for some $j \in M$

iff $(\mathscr{B}[x \leftarrow \overline{i}][w \leftarrow \overline{j}])^{\mathscr{I}} = \mathbf{t}$ for some $j \in M$, by I.5.10

iff $(((\exists w).\mathscr{B}[x \leftarrow \overline{i}]))^{\mathscr{I}} = \mathbf{t}$ by I.5.6(4)

iff $(((\exists w).\mathscr{B})[x \leftarrow \overline{i}])^{\mathscr{I}} = \mathbf{t}$

iff $(\mathscr{A}[x \leftarrow \overline{i}])^{\mathscr{I}} = \mathbf{t}$ □

I.5.12 Metatheorem (Soundness). *Any first order theory (identified by its non-logical axioms) Γ, over some language L, is sound.*

Proof. By induction on Γ-theorems, $\mathscr{A}$, we prove that $\Gamma \models \mathscr{A}$. That is, we fix a structure for L, say $\mathfrak{M}$, and assume that $\models_{\mathfrak{M}} \Gamma$. We then proceed to show that $\models_{\mathfrak{M}} \mathscr{A}$.

Basis. $\mathscr{A}$ is a nonlogical axiom. Then our conclusion is part of the assumption, by I.5.7.

Or $\mathscr{A}$ is a logical axiom. There are a number of cases:

Case 1. $\models_{\text{Taut}} \mathscr{A}$. We fix an $\mathfrak{M}$-instance of $\mathscr{A}$, say $\mathscr{A}'$, and show that $\mathscr{A}'^{\mathscr{I}} = \mathbf{t}$. Let $p_1, \ldots, p_n$ be *all the propositional variables* (alias *prime formulas*) occurring in $\mathscr{A}'$. Define a valuation v by setting $v(p_i) = p_i^{\mathscr{I}}$ for $i = 1, \ldots, n$. Clearly, $\mathbf{t} = \bar{v}(\mathscr{A}') = \mathscr{A}'^{\mathscr{I}}$ (the first "=" because $\models_{\text{Taut}} \mathscr{A}'$, the second because after prime formulas were taken care of, all that remains to be done for the evaluation of $\mathscr{A}'^{\mathscr{I}}$ is to apply Boolean connectives – see I.5.6(3)).

Pause. Why is $\models_{\text{Taut}} \mathscr{A}'$?

Case 2. $\mathscr{A} \equiv \mathscr{B}[t] \rightarrow (\exists x).\mathscr{B}$. Again, we look at an $\mathfrak{M}$-instance $\mathscr{B}'[t'] \rightarrow (\exists x).\mathscr{B}'$. We want $(\mathscr{B}'[t'] \rightarrow (\exists x).\mathscr{B}')^{\mathscr{I}} = \mathbf{t}$, but suppose instead that

$$(\mathscr{B}'[t'])^{\mathscr{I}} = \mathbf{t} \tag{1}$$

and

$$((\exists x).\mathscr{B}')^{\mathscr{I}} = \mathbf{f} \tag{2}$$

Let $t'^{\mathscr{I}} = i$ ($i \in M$). By I.5.11 and (1), $(\mathscr{B}'[\overline{i}])^{\mathscr{I}} = \mathbf{t}$. By I.5.6(4), $((\exists x).\mathscr{B}')^{\mathscr{I}} = \mathbf{t}$, contradicting (2).

Case 3. $\mathcal{A} \equiv x = x$. Then an arbitrary $\mathfrak{M}$-instance is $\bar{i} = \bar{i}$ for some $i \in M$. By I.5.6(1), $(\bar{i} = \bar{i})^{\mathscr{I}} = \mathbf{t}$.

Case 4. $\mathcal{A} \equiv t = s \rightarrow (\mathscr{B}[t] \leftrightarrow \mathscr{B}[s])$. Once more, we take an arbitrary $\mathfrak{M}$-instance, $t' = s' \rightarrow (\mathscr{B}'[t'] \leftrightarrow \mathscr{B}'[s'])$. Suppose that $(t' = s')^{\mathscr{I}} = \mathbf{t}$. That is, $t'^{\mathscr{I}} = s'^{\mathscr{I}} = $ (let us say) i (in M). But then

$$\begin{aligned}(\mathscr{B}'[t'])^{\mathscr{I}} &= (\mathscr{B}'[\bar{i}])^{\mathscr{I}} && \text{by I.5.11}\\ &= (\mathscr{B}'[s'])^{\mathscr{I}} && \text{by I.5.11}\end{aligned}$$

Hence $(\mathscr{B}[t] \leftrightarrow \mathscr{B}[s])^{\mathscr{I}} = \mathbf{t}$.

For the induction step we have two cases:

Modus ponens. Let $\mathscr{B}$ and $\mathscr{B} \rightarrow \mathcal{A}$ be Γ-theorems. Fix an $\mathfrak{M}$-instance $\mathscr{B}' \rightarrow \mathcal{A}'$. Since $\mathscr{B}', \mathscr{B}' \rightarrow \mathcal{A}' \models_{\text{Taut}} \mathcal{A}'$, the argument here is entirely analogous to the case $\mathcal{A} \in \Lambda$ (hence we omit it).

$\exists$-*introduction.* Let $\mathcal{A} \equiv (\exists x).\mathscr{B} \rightarrow \mathscr{C}$ and $\Gamma \vdash \mathscr{B} \rightarrow \mathscr{C}$, where x is not free in $\mathscr{C}$. By the I.H.

$$\models_{\mathfrak{M}} \mathscr{B} \rightarrow \mathscr{C} \tag{3}$$

Let $(\exists x).\mathscr{B}' \rightarrow \mathscr{C}'$ be an $\mathfrak{M}$-instance such that (despite expectations) $((\exists x).\mathscr{B}')^{\mathscr{I}} = \mathbf{t}$ but

$$\mathscr{C}'^{\mathscr{I}} = \mathbf{f} \tag{4}$$

Thus

$$\mathscr{B}'[\bar{i}]^{\mathscr{I}} = \mathbf{t} \tag{5}$$

for some $i \in M$. Since x is not free in $\mathscr{C}$, $\mathscr{B}'[\bar{i}] \rightarrow \mathscr{C}'$ is a false (by (4) and (5)) $\mathfrak{M}$-instance of $\mathscr{B} \rightarrow \mathscr{C}$, contradicting (3). $\quad\square$

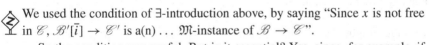 We used the condition of $\exists$-introduction above, by saying "Since x is not free in $\mathscr{C}$, $\mathscr{B}'[\bar{i}] \rightarrow \mathscr{C}'$ is a(n) ... $\mathfrak{M}$-instance of $\mathscr{B} \rightarrow \mathscr{C}$".

So the condition was useful. But is it essential? Yes, since, for example, if $x \not\equiv y$, then $x = y \rightarrow x = y \not\models (\exists x)x = y \rightarrow x = y$.

As a corollary of soundness we have the consistency of pure theories:

I.5.13 Corollary. *Any first order pure theory is consistent.*

Proof. Let $\mathfrak{T}$ be a pure theory over some language L. Since $\not\models \neg x = x$, it follows that $\not\vdash_{\mathfrak{T}} \neg x = x$, thus $\mathscr{T} \neq \mathbf{Wff}$. $\quad\square$

I.5.14 Corollary. *Any first order theory that has a model is consistent.*

Proof. Let $\mathfrak{T}$ be a first theory over some language L, and $\mathfrak{M}$ a model of $\mathfrak{T}$. Since $\not\models_{\mathfrak{M}} \neg x = x$, it follows that $\not\vdash_{\mathfrak{T}} \neg x = x$, thus $\mathscr{T} \neq \mathbf{Wff}$. $\quad\square$

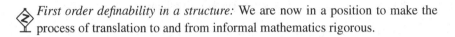 *First order definability in a structure:* We are now in a position to make the process of translation to and from informal mathematics rigorous.

I.5.15 Definition. Let L be a first order language, and $\mathfrak{M}$ a structure for L. A *set* (synonymously, *relation*) $S \subseteq M^n$ is *(first order) definable in $\mathfrak{M}$ over L* iff for some formula $\mathscr{S}(y_1, \ldots, y_n)$ (see p. 18 for a reminder on round-bracket notation) and for all i_j, $j = 1, \ldots, n$, in M,

$$\langle i_1, \ldots, i_n \rangle \in S \quad \text{iff} \quad \models_{\mathfrak{M}} \mathscr{S}(\bar{i}_1, \ldots, \bar{i}_n)$$

We often just say "definable in $\mathfrak{M}$".

A function $f : M^n \to M$ is definable in $\mathfrak{M}$ over L iff the relation $y = f(x_1, \ldots, x_n)$ is so definable. $\quad\square$

N.B. Some authors say "(first order) *expressible*" (Smullyan (1992)) rather than "(first order) definable" in a structure.

In the context of $(\mathfrak{M})$, the above definition gives precision to statements such as "we code (or translate) an informal statement into the formal language" or "the (formal language) formula $\mathscr{A}$ informally 'says' . . .", since any (informal) "statement" (or relation) that depends on the informal variables $x_1, \ldots, x_n$ has the form "$\langle x_1, \ldots, x_n \rangle \in S$" for some (informal) set S. It also captures the essence of the statement.

"The (informal) statement $\langle x_1, \ldots, x_n \rangle \in S$ can be *written* (or *made*) in the *formal language*."

What "makes" the statement, *in the formal language*, is the formula $\mathscr{S}$ that first order defines it.

I.5.16 Example. The informal statement "z is a prime" has a formal translation

$$S0 < z \wedge (\forall x)(\forall y)(z = x \times y \to x = z \vee x = S0)$$

over the language of elementary number theory, where the nonlogical symbols are $0, S, +, \times, <$ and the *definition* (translation) is effected in the *standard structure* $\mathfrak{N} = (\mathbb{N}; 0; S, +, \times; <)$, where "$S$" satisfies, for all $n \in \mathbb{N}$, $S(n) = n + 1$ and interprets "S" (see I.5.2, p. 53, for the "unpacked" notation we have just used to denote the structure $\mathfrak{N}$). We have used the variable name "z" both formally and informally. $\quad\square$

It must be said that translation is not just an art or skill. There are theoretical limitations to translation. The trivial limitation is that if M is an infinite set and, say, L has a finite set of nonlogical symbols (as is the case in number theory and set theory), then we cannot define *all* $S \subseteq M$, simply because we do not have enough first order formulas to do so.

There are *non-trivial* limitations too. Some sets are not first order definable because they are "far too complex". See Section I.9.

This is a good place to introduce a common notational *argot* that allows us to write "mixed-mode" formulas that have a formal part (over some language L) but may contain "informal" constants (names of, to be sure, but names that have *not* formally been imported into L) from some structure $\mathfrak{M}$ appropriate for L.

I.5.17 Informal Definition. Let L be a first order language and $\mathfrak{M} = (M, \mathscr{I})$ a structure for L. Let $\mathscr{A}$ be a formula with at most $x_1, \ldots, x_n$ free, and $i_1, \ldots, i_n$ members of M. The notation $\mathscr{A}[\![i_1, \ldots, i_n]\!]$ is an abbreviation of $(\mathscr{A}[\overline{i_1}, \ldots, \overline{i_n}])^{\mathscr{I}}$. $\square$

This *argot* allows one to substitute informal objects into variables outright, by-passing the procedure of importing formal names for such objects into the language. It is noteworthy that mixed mode formulas can be defined *directly* by induction on formulas – that is, without forming $L(\mathfrak{M})$ first – as follows:

Let L and $\mathfrak{M}$ be as above. Let $x_1, \ldots, x_n$ contain all the free variables that appear in a term t or formula $\mathscr{A}$ over L (not over $L(\mathfrak{M})$!). Let $i_1, \ldots, i_n$ be arbitrary in M.

For terms we define

$$t[\![i_1, \ldots i_n]\!] = \begin{cases} i_j & \text{if } t \equiv x_j \ (1 \le j \le n) \\ a^{\mathscr{I}} & \text{if } t \equiv a \\ f^{\mathscr{I}}(t_1[\![i_1, \ldots, i_n]\!], \ldots, \\ \qquad t_r[\![i_1, \ldots, i_n]\!]) & \text{if } t \equiv ft_1 \ldots t_r \end{cases}$$

For formulas we let

$$\mathscr{A}[\![i_1, \ldots i_n]\!] = \begin{cases} t[\![i_1, \ldots i_n]\!] = s[\![i_1, \ldots i_n]\!] & \text{if } \mathscr{A} \equiv t = s \\ P^{\mathscr{I}}(t_1[\![i_1, \ldots, i_n]\!], \ldots, \\ \qquad t_r[\![i_1, \ldots, i_n]\!]) & \text{if } \mathscr{A} \equiv Pt_1 \ldots t_r \\ \neg(\mathscr{B}[\![i_1, \ldots i_n]\!]) & \text{if } \mathscr{A} \equiv \neg\mathscr{B} \\ (\mathscr{B}[\![i_1, \ldots i_n]\!] \lor \mathscr{C}[\![i_1, \ldots i_n]\!]) & \text{if } \mathscr{A} \equiv \mathscr{B} \lor \mathscr{C} \\ (\exists a \in M).\mathscr{B}[\![a, i_1, \ldots i_n]\!] & \text{if } \mathscr{A} \equiv (\exists z).\mathscr{B}[z, \vec{x}_n] \end{cases}$$

where '$(\exists a \in M) \ldots$" is short for "$(\exists a)(a \in M \wedge \ldots)$". The right hand side has no free (informal) variables; thus it evaluates to one of **t** or **f**.

We now turn to the "hard half" of Gödel's completeness theorem, that our syntactic proof apparatus can faithfully mimic "proofs by logical implication". That is, the syntactic apparatus is "complete".

I.5.18 Definition. A theory over L (designated by its nonlogical axioms) Γ is *semantically complete* iff $\Gamma \models \mathscr{A}$ implies $\Gamma \vdash \mathscr{A}$ for any formula $\mathscr{A}$.   □

The term "semantically complete" is not being used much. There is a competing *syntactic* notion of completeness, that of *simple completeness*, also called just *completeness*. The latter is the notion one has normally in mind when saying "...a complete theory...". More on this shortly.

We show the semantic completeness of *every* first order theory by proving, using the technique of Henkin (1952), the consistency theorem below. The completeness theorem will then be derived as a corollary.

I.5.19 Metatheorem (Consistency Theorem). *If a (first order) theory $\mathfrak{T}$ is consistent, then it has a model.*

We will first give a proof (via a sequence of lemmata) for the case of "countable languages" L, that is, languages that have a *countable alphabet*. We will then amend the proof to include the uncountable case.

A crash course on countable sets: A set A is *countable*[†] if it is empty or (in the opposite case) if there is a way to arrange all its members in an infinite sequence, in a "row of locations", utilizing one location for each member of $\mathbb{N}$. It *is* allowed to repeatedly list any element of A, so that finite sets are countable. Technically, this enumeration is a (total) function $f : \mathscr{N} \to A$ whose range (set of outputs) equals A (that is, f is *onto*). We say that $f(n)$ is the nth element of A in the enumeration f. We often write f_n instead of $f(n)$ and then call n a "subscript" or "index".

We can convert a multi-row enumeration

$$(f_{i,j})_{i,j \text{ in } \mathbb{N}}$$

[†] Naïvely speaking. The definition is similar to the formal one that is given in volume 2. Here we are just offering a quick-review service – in the metamathematical domain, just in case the reader needs it – in preparation for the proof of the consistency theorem.

into a single row enumeration quite easily. Technically, we say that the set $\mathbb{N} \times \mathbb{N}$ – the set of "double subscripts" (i, j) – is countable. This is shown diagrammatically below. The "linearization" or "unfolding" of the infinite matrix of rows is effected by walking along the arrows:

$$
\begin{array}{llll}
(0,0) & (0,1) & (0,2) & (0,3) \;\; \ldots \\
& \nearrow & \nearrow & \nearrow \\
(1,0) & (1,1) & (1,2) \\
& \nearrow & \nearrow \\
(2,0) & (2,1) \\
& \nearrow \\
(3,0) \\
\vdots
\end{array}
$$

This observation yields a very useful fact regarding strings over countable sets (alphabets): If $\mathscr{V}$ is countable, then the set of all strings *of length* 2 over $\mathscr{V}$ is also countable. Why? Because the arbitrary string of length 2 is of the form $d_i d_j$ where d_i and d_j represent the ith and j elements of the enumeration of $\mathscr{V}$ respectively. Unfolding the infinite matrix exactly as above, we get a single-row enumeration of these strings.

By induction on the length $n \geq 2$ of strings we see that the set of strings of any length $n \geq 2$ is also countable. Indeed, a string of length $n+1$ is a string ab, where a has length n and $b \in \mathscr{V}$. By the I.H. the set of all a's can be arranged in a single row (countable), and we are done exactly as in the case of the $d_i d_j$ above.

Finally, let us collect *all* the strings over $\mathscr{V}$ into a set S. Is S countable? Yes. We can arrange S, at first, into an infinite matrix of strings $m_{i,j}$, that is, the *jth string of length i*. Then we employ our matrix-unfolding trick above.

Suppose now that we start with a countable set A. Is every subset of A countable? Yes. If $B \subseteq A$, then the elements of B form a subsequence of the elements of A (in any given enumeration). Therefore, just drop the members of A that are not in B, and compact the subscripts.[†]

To prove the consistency theorem let us fix a countable language L and a first order theory $\mathfrak{T}$ over L with nonlogical axioms Γ. In the search for a model,

---

[†] By "compact the subscripts" we mean this: After dropping members of A that do not belong to B, the enumeration has gaps in general. For example, B might be the subsequence $a_{13}, a_{95}, a_{96}, a_{97}, a_{1001}, \ldots$. We just shift the members of B to the left, eliminating gaps, so that the above few listed members take the locations (subscripts), 0, 1, 2, 3, 4, ..., respectively. Admittedly, this explanation is not precise, but it will have to do for now.

I. Basic Logic

we start with a simple countable set, here we will take $\mathbb{N}$ itself, and endow it with enough structure (the $\mathscr{S}$-part) to obtain a model, $\mathfrak{M} = (M, \mathscr{S})$ of $\mathfrak{T}$.

Since, in particular, this will entail that a subset of $\mathbb{N}$ (called M in what follows) will be the domain of the structure, we start by importing all the constants $n \in \mathbb{N}$ into L. That is, we add to $\mathscr{V}$ a *new* constant symbol $\bar{n}$ *for each* $n \in \mathbb{N}$. The new alphabet is denoted by $\mathscr{V}(\mathbb{N})$, and the resulting language $L(\mathbb{N})$.

I.5.20 Definition. In general, let $L = (\mathscr{V}, \mathbf{Term}, \mathbf{Wff})$ be a first order language and M some set. We add to $\mathscr{V}$ a *new* constant symbol $\bar{i}$ *for each* $i \in M$. The new alphabet is denoted by $\mathscr{V}(M)$, and the new language by $L(M) = (\mathscr{V}(M), \mathbf{Term}(M), \mathbf{Wff}(M))$.

This concept originated with Henkin and Abraham Robinson. The augmented language $L(M)$ is called *the diagram language of M*. ☐

The above definition generalizes Definition I.5.4 and is useful when (as happens in our current context) we have a language L and a set (here $\mathbb{N}$), but not, as yet, a structure for L with domain $\mathbb{N}$ (or some subset thereof).

Of course, if $\mathfrak{M} = (M, \mathscr{S})$, then $L(M) = L(\mathfrak{M})$.

Two observations are immediate: One, Γ has not been affected by the addition of the new constants, and, two, $L(\mathbb{N})$ is still countable.[†] Thus, there are enumerations

$$\mathscr{F}_0, \mathscr{F}_1, \mathscr{F}_2, \ldots \text{ of all sentences in } \mathbf{Wff}(\mathbb{N}) \tag{1}$$

and

$$\mathscr{G}_1, \mathscr{G}_2, \mathscr{G}_3, \ldots \text{ of all sentences in } \mathbf{Wff}(\mathbb{N}) \text{ of the form } (\exists x)\mathscr{A} \tag{2}$$

where, in (2), every sentence $(\exists x)\mathscr{A}$ of $\mathbf{Wff}(\mathbb{N})$ is listed infinitely often.

Pause. How can we do this? Form an infinite matrix, each row of which is the same fixed enumeration of the $(\exists x)\mathscr{A}$ sentences. Then unfold the matrix into a single row.

With the preliminaries out of the way, we next define by induction (or recursion) over $\mathbb{N}$ an infinite sequence of theories, successively adding *sentences* over $L(\mathbb{N})$ as new nonlogical axioms: We set $\Gamma_0 = \Gamma$. For any $n \geq 0$, we define

---

[†] If A and B are countable, then so is $A \cup B$, since we can arrange the union as an infinite matrix, the 0th row being occupied by A-members while all the other rows being identical to some fixed enumeration of B.

Γ_{n+1} in two stages: We first let

$$\Delta_n = \begin{cases} \Gamma_n \cup \{\mathscr{F}_n\} & \text{if } \Gamma_n \nvdash \neg\mathscr{F}_n \\ \Gamma_n & \text{otherwise} \end{cases} \tag{3}$$

Then, we set

$$\Gamma_{n+1} = \begin{cases} \Delta_n \cup \{\mathscr{A}[x \leftarrow \bar{i}]\} & \text{if } \Delta_n \vdash \mathscr{G}_{n+1}, \text{ where } \mathscr{G}_{n+1} \equiv (\exists x)\mathscr{A} \\ \Delta_n & \text{otherwise} \end{cases}$$

The choice of i is important: It is the *smallest* i such that the constant $\bar{i}$ does not occur (as a substring) in any of the sentences

$$\mathscr{F}_0, \ldots, \mathscr{F}_n, \mathscr{G}_1, \ldots, \mathscr{G}_{n+1} \tag{4}$$

The sentence $\mathscr{A}[x \leftarrow \bar{i}]$ added to Γ_{n+1} is called a *special Henkin axiom*.[†] The constant $\bar{i}$ is the associated Henkin (also called *witnessing*) constant.

We now set

$$\Gamma_\infty = \bigcup_{n \in \mathbb{N}} \Gamma_n \tag{5}$$

This is a set of formulas over **Wff**($\mathbb{N}$) that defines a theory $\mathfrak{T}_\infty$ over **Wff**($\mathbb{N}$) (as the set of nonlogical axioms of $\mathfrak{T}_\infty$).

I.5.21 Lemma. *The theory $\mathfrak{T}_\infty$ is consistent.*

Proof. It suffices to show that each of the theories Γ_n is consistent.
    (Indeed, if $\vdash_{\mathfrak{T}_\infty} \neg x = x$,[‡] then $\mathscr{A}_1, \ldots, \mathscr{A}_m \vdash \neg x = x$, for some $\mathscr{A}_i (i = 1, \ldots, m)$ in Γ_∞, since proofs have finite length. Let Γ_n include all of $\mathscr{A}_1, \ldots, \mathscr{A}_m$.

Pause. Is there such a Γ_n?
    Then Γ_n will be inconsistent.)
    On to the main task. We know (assumption) that Γ_0 is consistent. We take the I.H. that Γ_n is consistent, and consider Γ_{n+1} next.
    First, we argue that Δ_n is consistent. If $\Delta_n = \Gamma_n$, then we are done by the I.H. If $\Delta_n = \Gamma_n \cup \{\mathscr{F}_n\}$, then inconsistency would entail (by I.4.21) $\Gamma \vdash \neg\mathscr{F}_n$, contradicting the hypothesis of the case we are at (top case of (3) above).

[†] Another possible choice for Henkin axiom is $(\exists x)\mathscr{A} \to \mathscr{A}[x \leftarrow \bar{i}]$.
[‡] $\neg x = x$ is our favourite "contradiction builder". See also I.4.22.

Next, we show that $\Delta_n \cup \{\mathscr{A}[x \leftarrow \bar{i}]\}$ is consistent, where $\mathscr{A}[x \leftarrow \bar{i}]$ is the added Henkin axiom – if indeed such an axiom *was* added.

Suppose instead that $\Delta_n \cup \{\mathscr{A}[x \leftarrow \bar{i}]\} \vdash \neg z = z$, for some variable z. Now $\bar{i}$ does not occur in any formulas in the set $\Delta_n \cup \{(\exists x)\mathscr{A}, \neg z = z\}$.

Since $(\exists x)\mathscr{A} \equiv \mathscr{G}_{n+1}$ and $\Delta_n \vdash \mathscr{G}_{n+1}$, we get $\Delta_n \vdash \neg z = z$ by I.4.27 (auxiliary constant metatheorem). This is no good, since Δ_n is supposed to be consistent.   □

I.5.22 Definition. A theory $\mathfrak{T}$ over L *decides* a sentence $\mathscr{A}$ iff one of $\vdash_{\mathfrak{T}} \mathscr{A}$ or $\vdash_{\mathfrak{T}} \neg \mathscr{A}$ holds. We say that $\mathscr{A}$ is *decidable by* $\mathfrak{T}$. In the case that $\vdash_{\mathfrak{T}} \neg \mathscr{A}$ holds, we say that $\mathfrak{T}$ refutes $\mathscr{A}$. $\mathfrak{T}$ is *simply complete*, or just *complete*, iff every sentence is decidable by $\mathfrak{T}$.   □

The definition is often offered in terms of consistent theories, since an inconsistent theory decides every formula anyway.

I.5.23 Lemma. $\mathfrak{T}_\infty$ *is simply complete.*

Proof. Let $\mathscr{A}$ be a sentence. Then $\mathscr{A} \equiv \mathscr{F}_n$ for some n. If $\Gamma_n \vdash \neg \mathscr{F}_n$, then we are done. If however Γ_n does *not* refute $\mathscr{F}_n$, then we are done by (3) (p. 65).   □

I.5.24 Lemma. $\mathfrak{T}_\infty$ *has the* witness property,[†] *namely, whenever* $\vdash_{\mathfrak{T}_\infty} (\exists x)\mathscr{A}$, *where* $(\exists x)\mathscr{A}$ *is a sentence over* $L(\mathbb{N})$, *then for some* $m \in \mathbb{N}$ *we have* $\vdash_{\mathfrak{T}_\infty} \mathscr{A}[x \leftarrow \overline{m}]$.

Proof. Since the proof $\vdash_{\mathfrak{T}_\infty} (\exists x)\mathscr{A}$ involves finitely many formulas from Γ_∞, there is an $n \geq 0$ such that $\Delta_n \vdash (\exists x)\mathscr{A}$. Now, $(\exists x)\mathscr{A} \equiv \mathscr{G}_{k+1}$ for some $k \geq n$, since $(\exists x)\mathscr{A}$ occurs in the sequence (2) (p. 64) infinitely often. But then $\Delta_k \vdash (\exists x)\mathscr{A}$ as well.

Pause. Why?

Hence, $\mathscr{A}[x \leftarrow \overline{m}]$ is added to Γ_{k+1} as a Henkin axiom, for an appropriate Henkin constant $\overline{m}$, and we are done.   □

I.5.25 Definition. We next define a relation, $\sim$, on $\mathbb{N}$ by

$$n \sim m \quad \text{iff} \quad \vdash_{\mathfrak{T}_\infty} \bar{n} = \overline{m} \tag{6}$$

□

[†] We also say that it is a *Henkin theory.*

$\sim$ has the following properties:

(a) *Reflexivity.* $n \sim n$ (all n): By $\vdash x = x$ and I.4.12.
(b) *Symmetry.* If $n \sim m$, then $m \sim n$ (all m, n): It follows from $\vdash x = y \rightarrow$ $y = x$ (Exercise I.16) and I.4.12.
(c) *Transitivity.* If $n \sim m$ and $m \sim k$, then $n \sim k$ (all m, n, k): It follows from $\vdash x = y \rightarrow y = z \rightarrow x = z$ (Exercise I.17) and I.4.12.

Let us define a function $f : \mathbb{N} \rightarrow \mathbb{N}$ by

$$f(n) = \text{smallest } m \text{ such that } m \sim n \tag{7}$$

By (a) above, f is totally defined. We also define a subset M of $\mathbb{N}$, by

$$M = \{f(n) : n \in \mathbb{N}\} \tag{8}$$

This M will be the domain of a structure that we will build, and show that it is a model of the original $\mathfrak{T}$.

First, we modify Γ_∞ "downwards".

I.5.26 Definition. The M-restriction of a formula $\mathscr{A}$ is the formula $\mathscr{A}^M$ obtained from $\mathscr{A}$ by replacing by $\overline{f(n)}$ *every occurrence of an* $\overline{n}$ *in* $\mathscr{A}$. □

We now let

$$\Gamma_\infty^M = \{\mathscr{A}^M : \mathscr{A} \in \Gamma_\infty\} \tag{9}$$

We have the following results regarding Γ_∞^M (or the associated theory $\mathfrak{T}_\infty^M$):

I.5.27 Remark. Before proceeding, we note that the language of $\mathfrak{T}_\infty^M$ is $L(M)$, and that $\mathscr{A} \equiv \mathscr{A}^M$ if $\mathscr{A}$ is over $L(M)$, since $f(m) = m$ for $m \in M$ (why?). □

I.5.28 Lemma. *Let* $\mathscr{A}$ *be over* $L(\mathbb{N})$. *If* $\vdash_{\mathfrak{T}_\infty} \mathscr{A}$, *then* $\vdash_{\mathfrak{T}_\infty^M} \mathscr{A}^M$.

Proof. Induction on $\mathfrak{T}_\infty$-theorems.

Basis. If $\mathscr{A} \in \Gamma_\infty$, then $\mathscr{A}^M \in \Gamma_\infty^M$ by (9). If $\mathscr{A} \in \Lambda$, then $\mathscr{A}^M \in \Lambda$ (why?)

Modus ponens. Let $\Gamma_\infty \vdash \mathscr{B}$ and $\Gamma_\infty \vdash \mathscr{B} \rightarrow \mathscr{A}$. By the I.H., $\Gamma_\infty^M \vdash \mathscr{B}^M$ and $\Gamma_\infty^M \vdash \mathscr{B}^M \rightarrow \mathscr{A}^M$, and we are done.

$\exists$-*introduction.* Let $\Gamma_\infty \vdash \mathscr{B} \rightarrow \mathscr{C}$, where x is not free in $\mathscr{C}$ and $\mathscr{A} \equiv (\exists x)\mathscr{B} \rightarrow \mathscr{C}$. By the I.H., $\Gamma_\infty^M \vdash \mathscr{B}^M \rightarrow \mathscr{C}^M$, and we are done by $\exists$-introduction. □

I.5.29 Lemma. $\mathfrak{T}_\infty^M \leq \mathfrak{T}_\infty$, *conservatively.*

Proof. Leaving the "conservatively" part aside for a moment, let us verify that for any $\mathscr{A}$ over $L(\mathbb{N})$

$$\Gamma_\infty \vdash \mathscr{A}^M \leftrightarrow \mathscr{A} \tag{$*$}$$

This follows from $\Gamma_\infty \vdash \overline{n} = \overline{f(n)}$ (recall that $n \sim f(n)$ by definition of f) for all $n \in \mathbb{N}$, and Exercise I.18.

Because of $(*)$, Γ_∞ can prove any $\mathscr{B} \in \Gamma_\infty^M$. Indeed, let $\mathscr{B} \equiv \mathscr{A}^M$ for some $\mathscr{A} \in \Gamma_\infty$ (by (9)). By $(*)$ and $\Gamma_\infty \vdash \mathscr{A}$, we obtain $\Gamma_\infty \vdash \mathscr{B}$.

Thus, $\mathfrak{T}_\infty^M \leq \mathfrak{T}_\infty$.

Pause. Do you believe this conclusion?

Turning to the "conservatively" part, this follows from Lemma I.5.28 and Remark I.5.27. □

I.5.30 Corollary. $\mathfrak{T}_\infty^M$ *is consistent.*

Proof. If it can prove $\neg x = x$, then so can $\mathfrak{T}_\infty$. □

I.5.31 Lemma. $\mathfrak{T}_\infty^M$ *is simply complete.*

Proof. Let the sentence $\mathscr{A}$ be over $L(M)$. It is decidable by $\mathfrak{T}_\infty$ (I.5.23). By I.5.28, $\mathfrak{T}_\infty^M$ decides $\mathscr{A}^M$. But that's $\mathscr{A}$. □

I.5.32 Lemma. $\mathfrak{T}_\infty^M$ *is a Henkin theory over* $L(M)$.

Proof. Let $\Gamma_\infty^M \vdash (\exists x)\mathscr{A}$, where $(\exists x)\mathscr{A}$ is a sentence. Then $\Gamma_\infty \vdash (\exists x)\mathscr{A}$ as well, hence $\Gamma_\infty \vdash \mathscr{A}[x \leftarrow \overline{n}]$, for some $n \in \mathbb{N}$ (by I.5.24).

It follows that $\Gamma_\infty^M \vdash \mathscr{A}^M[x \leftarrow \overline{f(n)}]$, and we are done, since $\mathscr{A} \equiv \mathscr{A}^M$ and $f(n) \in M$. □

I.5.33 Lemma. $\mathfrak{T}_\infty^M$ *distinguishes the constants of* M, *that is, if* $n \neq m$ (both in M), *then* $\vdash_{\mathfrak{T}_\infty^M} \neg \overline{n} = \overline{m}$.

Proof. By I.5.31, if $\vdash_{\mathfrak{T}_\infty^M} \neg \overline{n} = \overline{m}$ fails, then $\vdash_{\mathfrak{T}_\infty^M} \overline{n} = \overline{m}$; hence $n \sim m$. By the definition of $f(n)$ and M (p. 67), it follows that $n = m$ (since each set $\{m \in \mathbb{N} : m \sim n\}$ determined by n can have exactly one smallest element, and any two distinct such sets – determined by n and k – have no common elements (why?)). This contradicts the assumption that $n \neq m$. □

We have started with a consistent theory $\mathfrak{T}$ over L. We now have a *consistent, complete, Henkin* extension $\mathfrak{T}_\infty^M$ of $\mathfrak{T}$, over the language $L(M)$ ($M \subseteq \mathbb{N}$), which *distinguishes the constants of M*.

We are now ready to define our model $\mathfrak{M} = (M, \mathscr{I})$. We are pleased to note that the constants of M are already imported into the language, as required by Definitions I.5.5 and I.5.6.

For any predicate[†] P of L of arity k, we define for arbitrary $n_1, \dots, n_k$ in M

$$P^{\mathscr{I}}(n_1, \dots, n_k) = \mathbf{t} \quad \text{iff} \quad \Gamma_\infty^M \vdash P\overline{n}_1, \dots, \overline{n}_k \qquad (A)$$

that is, we are defining the set of k-tuples (relation) $P^{\mathscr{I}}$ by

$$P^{\mathscr{I}} = \left\{ \langle n_1, \dots, n_k \rangle : \Gamma_\infty^M \vdash P\overline{n}_1, \dots, \overline{n}_k \right\} \qquad (A')$$

Let next f be a function letter of arity k, and let $n_1, \dots, n_k$ be an input for $f^{\mathscr{I}}$. What is the appropriate output?[‡]

Well, first observe that $\Gamma_\infty^M \vdash (\exists x) f\overline{n}_1, \dots, \overline{n}_k = x$ (why?). By I.5.32, there is an $m \in M$ such that

$$\Gamma_\infty^M \vdash f\overline{n}_1, \dots, \overline{n}_k = \overline{m} \qquad (B)$$

We need this m to be unique. It is so, for if also $\Gamma_\infty^M \vdash f\overline{n}_1, \dots, \overline{n}_k = \overline{j}$, then (Exercise I.17) $\Gamma_\infty^M \vdash \overline{m} = \overline{j}$, and thus $m = j$ (if $m \neq j$, then also $\Gamma_\infty^M \vdash \neg \overline{m} = \overline{j}$, by I.5.33 – impossible, since Γ_∞^M is consistent).

For the input $n_1, \dots, n_k$ we set

$$f^{\mathscr{I}}(n_1, \dots, n_k) = m \qquad (B.1)$$

where m is uniquely determined by (B). This defines $f^{\mathscr{I}}$.

The case of constants is an interesting special case.[§] As above, we let $a^{\mathscr{I}}$ be the *unique* $m \in M$ such that

$$\Gamma_\infty^M \vdash a = \overline{m} \qquad (C)$$

The interesting, indeed crucial, observation (required by I.5.5) is that, for any imported constant $\overline{m}$, we have $\overline{m}^{\mathscr{I}} \equiv m$. Indeed, this follows from uniqueness and the trivial fact $\Gamma_\infty^M \vdash \overline{m} = \overline{m}$.

[†] Recall that we have added no new predicates and no new functions in going from L to $L(M)$. We have just added constants.

[‡] We have yet to determine $f^{\mathscr{I}}$. Here we are just "thinking out loud" towards suggesting a good $f^{\mathscr{I}}$.

[§] Actually, it is not a special case for us, since we did not allow 0-ary functions. But some authors do.

The following will be handy in the proof of the Main Lemma below:

$$\Gamma_\infty^M \vdash t = \overline{t^{\mathscr{I}}} \tag{D}$$

for any closed term t over $L(M)$. We prove (D) by induction on terms t.

Basis. If $t \equiv a$ (a original or imported), then (C) reads $\Gamma_\infty^M \vdash a = \overline{a^{\mathscr{I}}}$.

Let $t \equiv ft_1 \ldots t_k$. By I.5.5, $t^{\mathscr{I}} = f^{\mathscr{I}}(t_1^{\mathscr{I}}, \ldots, t_k^{\mathscr{I}})$. Set $n_i = t_i^{\mathscr{I}}$ for $i = 1, \ldots, k$. Then $t^{\mathscr{I}} = f^{\mathscr{I}}(n_1, \ldots, n_k)$. By (B) and $(B.1)$,

$$\Gamma_\infty^M \vdash f\overline{n}_1, \ldots, \overline{n}_k = \overline{f^{\mathscr{I}}(n_1, \ldots, n_k)}$$

By the I.H., $\Gamma_\infty^M \vdash t_i = \overline{t_i^{\mathscr{I}}}$, in other words $\Gamma_\infty^M \vdash t_i = \overline{n}_i$, for $i = 1, \ldots, k$. By **Ax4**, we obtain (via Exercise I.19)

$$\Gamma_\infty^M \vdash ft_1, \ldots, t_k = f\overline{n}_1, \ldots, \overline{n}_k$$

Thus (Exercise I.17),

$$\Gamma_\infty^M \vdash ft_1, \ldots, t_k = \overline{f^{\mathscr{I}}\left(t_1^{\mathscr{I}}, \ldots, t_k^{\mathscr{I}}\right)}$$

This concludes the proof of (D).

I.5.34 Main Lemma. *For every sentence $\mathscr{A}$ over $L(M)$, $\mathscr{A}^{\mathscr{I}} = \mathbf{t}$ iff $\Gamma_\infty^M \vdash \mathscr{A}$.*

Proof. This is proved by induction on formulas.

Basis. Case where $\mathscr{A} \equiv Pt_1 \ldots t_k$: Let $n_i = t_i^{\mathscr{I}}$ ($i = 1, \ldots, k$). By (A) above (p. 69), $P^{\mathscr{I}}(n_1, \ldots, n_k) = \mathbf{t}$ iff $\Gamma_\infty^M \vdash P\overline{n}_1 \ldots \overline{n}_k$ iff $\Gamma_\infty^M \vdash Pt_1 \ldots t_k$ – the second "iff" by **Ax4** (via Exercise I.19) and (D). We are done in this case, since $\mathscr{A}^{\mathscr{I}} = P^{\mathscr{I}}(t_1^{\mathscr{I}} \ldots t_k^{\mathscr{I}})$.

Case where $\mathscr{A} \equiv t = s$: Let also $n = t^{\mathscr{I}}$ and $m = s^{\mathscr{I}}$. Then $\mathscr{A}^{\mathscr{I}} = \mathbf{t}$ iff $t^{\mathscr{I}} = s^{\mathscr{I}}$ iff $n = m$ iff $\Gamma_\infty^M \vdash \overline{n} = \overline{m}$ (for the last "iff" the if part follows by consistency and I.5.33; the only-if part by **Ax3** and I.4.12).

The induction steps. Let $\mathscr{A} \equiv \neg\mathscr{B}$. Then $\mathscr{A}^{\mathscr{I}} = \mathbf{t}$ iff $\mathscr{B}^{\mathscr{I}} = \mathbf{f}$ iff (I.H.) $\Gamma_\infty^M \nvdash \mathscr{B}$ iff (completeness I.5.31 ($\rightarrow$) and consistency I.5.30 ($\leftarrow$)) $\Gamma_\infty^M \vdash \mathscr{A}$.

Let $\mathscr{A} \equiv \mathscr{B} \vee \mathscr{C}$. Consider first $\mathscr{B}^{\mathscr{I}} \vee \mathscr{C}^{\mathscr{I}} = \mathbf{t}$. Say $\mathscr{B}^{\mathscr{I}} = \mathbf{t}$. By I.H., $\Gamma_\infty^M \vdash \mathscr{B}$; hence $\Gamma_\infty^M \vdash \mathscr{B} \vee \mathscr{C}$ by tautological implication. Similarly if $\mathscr{C}^{\mathscr{I}} = \mathbf{t}$. Conversely, let $\Gamma_\infty^M \vdash \mathscr{B} \vee \mathscr{C}$. Then one of $\Gamma_\infty^M \vdash \mathscr{B}$ or $\Gamma_\infty^M \vdash \mathscr{C}$ must be the case: If neither holds, then $\Gamma_\infty^M \vdash \neg\mathscr{B}$ and $\Gamma_\infty^M \vdash \neg\mathscr{C}$ by completeness, hence Γ_∞^M is inconsistent (why?).

The final case: $\mathscr{A} \equiv (\exists x)\mathscr{B}$. Let $((\exists x)\mathscr{B})^{\mathscr{I}} = \mathbf{t}$. Then $(\mathscr{B}[x \leftarrow \overline{n}])^{\mathscr{I}} = \mathbf{t}$ for some $n \in M$. By I.H., $\Gamma_\infty^M \vdash \mathscr{B}[x \leftarrow \overline{n}]$; hence $\Gamma_\infty^M \vdash (\exists x)\mathscr{B}$ by **Ax2** and MP. Conversely, let $\Gamma_\infty^M \vdash (\exists x)\mathscr{B}$. By I.5.32, $\Gamma_\infty^M \vdash \mathscr{B}[x \leftarrow \overline{n}]$, where $\overline{n}$ is the appropriate Henkin constant. By I.H., $(\mathscr{B}[x \leftarrow \overline{n}])^{\mathscr{I}} = \mathbf{t}$; hence $\mathscr{A}^{\mathscr{I}} = \mathbf{t}$. □

Finally,

Proof (**of the Consistency Theorem**). Let $\mathscr{A}'$ be an $\mathfrak{M}$-instance of a formula $\mathscr{A}$ in Γ.

Note that $\mathscr{A}$ is over L.

From $\Gamma \subseteq \Gamma_\infty^M$ it follows that $\Gamma_\infty^M \vdash \mathscr{A}$, and hence $\Gamma_\infty^M \vdash \mathscr{A}'$, by I.4.12. By the Main Lemma, $\mathscr{A}'^{\mathscr{I}} = \mathbf{t}$.

Thus, the reduct $\mathfrak{M}' = (M, \mathscr{I} \restriction L)$ is a model of Γ. $\qquad\square$

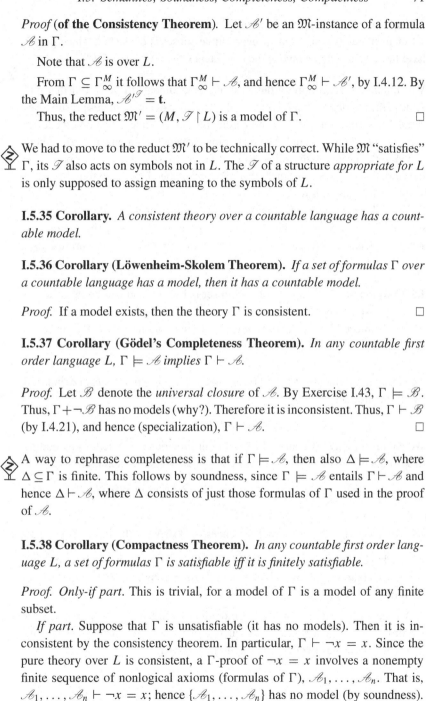

We had to move to the reduct $\mathfrak{M}'$ to be technically correct. While $\mathfrak{M}$ "satisfies" Γ, its $\mathscr{I}$ also acts on symbols not in L. The $\mathscr{I}$ of a structure *appropriate for L* is only supposed to assign meaning to the symbols of L.

I.5.35 Corollary. *A consistent theory over a countable language has a countable model.*

I.5.36 Corollary (Löwenheim-Skolem Theorem). *If a set of formulas Γ over a countable language has a model, then it has a countable model.*

Proof. If a model exists, then the theory Γ is consistent. $\qquad\square$

I.5.37 Corollary (Gödel's Completeness Theorem). *In any countable first order language L, $\Gamma \models \mathscr{A}$ implies $\Gamma \vdash \mathscr{A}$.*

Proof. Let $\mathscr{B}$ denote the *universal closure* of $\mathscr{A}$. By Exercise I.43, $\Gamma \models \mathscr{B}$. Thus, $\Gamma + \neg \mathscr{B}$ has no models (why?). Therefore it is inconsistent. Thus, $\Gamma \vdash \mathscr{B}$ (by I.4.21), and hence (specialization), $\Gamma \vdash \mathscr{A}$. $\qquad\square$

A way to rephrase completeness is that if $\Gamma \models \mathscr{A}$, then also $\Delta \models \mathscr{A}$, where $\Delta \subseteq \Gamma$ is finite. This follows by soundness, since $\Gamma \models \mathscr{A}$ entails $\Gamma \vdash \mathscr{A}$ and hence $\Delta \vdash \mathscr{A}$, where Δ consists of just those formulas of Γ used in the proof of $\mathscr{A}$.

I.5.38 Corollary (Compactness Theorem). *In any countable first order language L, a set of formulas Γ is satisfiable iff it is finitely satisfiable.*

Proof. Only-if part. This is trivial, for a model of Γ is a model of any finite subset.

If part. Suppose that Γ is unsatisfiable (it has no models). Then it is inconsistent by the consistency theorem. In particular, $\Gamma \vdash \neg x = x$. Since the pure theory over L is consistent, a Γ-proof of $\neg x = x$ involves a nonempty finite sequence of nonlogical axioms (formulas of Γ), $\mathscr{A}_1, \ldots, \mathscr{A}_n$. That is, $\mathscr{A}_1, \ldots, \mathscr{A}_n \vdash \neg x = x$; hence $\{\mathscr{A}_1, \ldots, \mathscr{A}_n\}$ has no model (by soundness). This contradicts the hypothesis. $\qquad\square$

Alternatively, we can prove the above by invoking "syntactic compactness": A set of formulas is consistent iff every finite subset is consistent, since proofs have finite length. Now invoke the consistency theorem and I.5.14.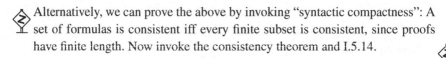

We conclude this section by outlining the amendments to our proof that will remove the restriction that L is countable. This plan of amendments presupposes some knowledge[†] of ordinals and cardinality (cf. volume 2) beyond what our "crash course" has covered. The reader may accept the statements proved here and skip the proofs with no loss of continuity. These statements, in particular the Gödel-Mal′cev compactness theorem, are applied later on to founding nonstandard analysis (following A. Robinson).

Let L be (possibly) uncountable, in the sense that the cardinality $\mathfrak{k}$ of $\mathscr{V}$ is $\geq \omega$. The cardinality of the set of all strings over $\mathscr{V}$ is also $\mathfrak{k}$ (for a proof see volume 2, Chapter VII). We now pick and fix for our discussion an arbitrary set N of cardinality $\mathfrak{k}$.

I.5.39 Remark. An example of such a set is $\mathfrak{k}$ itself, and can be taken as N. However, we can profit from greater generality: N can be *any* set of any type of (real) objects that we choose with some purpose in mind, as long as its cardinality is $\mathfrak{k}$. □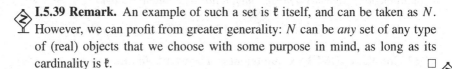

Therefore the elements of N can be arranged in a *transfinite sequence* (indexed by all the ordinals $\alpha < \mathfrak{k}$)

$$n_0, n_1, \ldots, n_\alpha, \ldots$$

We then form $L(N)$ (to parallel $L(\mathbb{N})$ of our previous construction) by adding to $\mathscr{V}$ a distinct name $\overline{n_\alpha}$ for each $n_\alpha \in N$. Thus, we have enumerations

$$\mathscr{F}_0, \mathscr{F}_1, \mathscr{F}_2, \ldots, \mathscr{F}_\alpha, \ldots \text{ of all sentences in } \mathbf{Wff}(N) \qquad (1')$$

and

$$\mathscr{G}_1, \mathscr{G}_2, \mathscr{G}_3, \ldots, \mathscr{G}_\alpha, \ldots \text{ of all sentences in } \mathbf{Wff}(N) \text{ of the form } (\exists x)\mathscr{A} \quad (2')$$

where, in (2′), every sentence $(\exists x)\mathscr{A}$ of $\mathbf{Wff}(N)$ is listed infinitely often, and the indices (subscripts) in both (1′) and (2′) run through all the ordinals $\alpha < \mathfrak{k}$ (omitting index 0 in the second listing).

We next proceed as expected: We define by induction (or recursion) over the ordinals $\alpha < \mathfrak{k}$ a transfinite sequence of theories (determined by Γ and additional sentences over $L(N)$ as nonlogical axioms[‡]):

[†] On an informal level, of course. All this is going on in the metatheory, just like the countable case.

[‡] Note that the formulas in Γ need not be closed.

We set $\Gamma_0 = \Gamma$. For any $\alpha < \mathfrak{k}$, we define Γ_α to be $\bigcup_{\beta<\alpha} \Gamma_\beta$ just in case α is a *limit ordinal*. If $\alpha = \beta + 1$ (a *successor*) then the definition is effected in two stages: We first let

$$\Delta_\beta = \begin{cases} \Gamma_\beta \cup \{\mathscr{F}_\beta\} & \text{if } \Gamma_\beta \not\vdash \neg\mathscr{F}_\beta \\ \Gamma_\beta & \text{otherwise} \end{cases} \tag{3'}$$

Then, we set $\Gamma_\alpha = \Delta_\beta \cup \{\mathscr{A}[x \leftarrow \bar{i}]\}$ *just in case* $\Delta_\beta \vdash \mathscr{G}_\alpha$, where $\mathscr{G}_\alpha \equiv (\exists x)\mathscr{A}$.

 The choice of i is important: $i = n_\alpha \in N$, where $\alpha < \mathfrak{k}$ is the *smallest* index such that the constant $\overline{n_\alpha}$ does *not* occur (as a substring) in any of the sentences

$$\mathscr{F}_0, \ldots, \mathscr{F}_\beta, \mathscr{G}_1, \ldots, \mathscr{G}_\alpha \tag{4'}$$

The sentence $\mathscr{A}[x \leftarrow \bar{i}]$ added to Γ_α is called a *special Henkin axiom*. The constant $\bar{i}$ is the associated Henkin or witnessing constant.

We now set

$$\Gamma_\mathfrak{k} = \bigcup_{\alpha<\mathfrak{k}} \Gamma_\alpha \tag{5'}$$

This is a set of formulas over $\mathbf{Wff}(N)$ that defines a theory $\mathfrak{T}_\mathfrak{k}$ over $\mathbf{Wff}(N)$ (as the set of nonlogical axioms of $\mathfrak{T}_\mathfrak{k}$). We next define $\sim$, on N this time, as before ((6) on p. 66):

$$n \sim m \quad \text{iff} \quad \vdash_{\mathfrak{T}_\mathfrak{k}} \bar{n} = \bar{m}$$

We note its properties, and proceed to define a subset M of N as in (7) and (8) (p. 67).

Since $M \subseteq N$, its cardinality is $\leq \mathfrak{k}$. After defining the M-restriction of a formula $\mathscr{A}$ as before, all the rest proceeds as in Lemmata I.5.28–I.5.33, replacing throughout $\mathfrak{T}_\infty$, $\mathfrak{T}_\infty^M$, and members $i \in \mathbb{N}$ by $\mathfrak{T}_\mathfrak{k}$, $\mathfrak{T}_\mathfrak{k}^M$, and members $i \in N$ respectively. We are then able to state:

I.5.40 Metatheorem (Consistency Theorem). *If a (first order) theory $\mathfrak{T}$ over a language L of cardinality $\mathfrak{k}$ is consistent, then it has a model of cardinality $\leq \mathfrak{k}$.*

 Terminology: The cardinality of a model is that of its domain.

I.5.41 Corollary (Completeness Theorem). *In any first order language L, $\Gamma \models \mathscr{A}$ implies $\Gamma \vdash \mathscr{A}$.*

I.5.42 Corollary (Gödel-Mal′cev Compactness Theorem). *In any first order language L, a set of formulas* Γ *is satisfiable iff it is finitely satisfiable.*

The Löwenheim-Skolem theorem takes the following form:

I.5.43 Corollary (Upward Löwenheim-Skolem Theorem). *If a set of formulas* Γ *over a language L of cardinality* $\mathfrak{k}$ *has an infinite model, then it has a model of any cardinality* $\mathfrak{n}$ *such that* $\mathfrak{k} \le \mathfrak{n}$.

Proof. Let $\mathfrak{K} = (K, \mathscr{I})$ be an infinite model of Γ. Pick a set N of cardinality $\mathfrak{n}$, and import its individuals c as new formal constants $\bar{c}$ into the language of Γ. The set $\overline{\Gamma} = \Gamma \cup \{\neg \bar{c} = \bar{d} : c \ne d \text{ on } N\}$ is finitely satisfiable. This is because every finite subset of $\overline{\Gamma}$ involves only finitely many of the sentences $\neg \bar{c} = \bar{d}$; thus there is capacity in K (as it is infinite) to extend $\mathscr{I}$ into $\mathscr{I}'$ (keeping K the same, but defining distinct $\bar{c}^{\mathscr{I}'}, \bar{d}^{\mathscr{I}'}$, etc.) to satisfy these sentences in an expanded structure $\mathfrak{K}' = (K, \mathscr{I}')$.

Hence $\overline{\Gamma}$ is consistent.

Following the construction given earlier, we take this N (and $\overline{\Gamma}$) as our starting point to build a simply complete, consistent extension $\mathfrak{T}_\mathfrak{n}$ of $\overline{\Gamma}$, and a model $\mathfrak{M}$ *for* $\overline{\Gamma}$, with domain some subset M of N. The choice of M follows the definition of "$\sim$" (see pp. 66 and 73). Under the present circumstances, "$\sim$" is "$=$" on N, for $\vdash_{\mathfrak{T}_\mathfrak{n}} \bar{c} = \bar{d}$ implies $c = d$ on N (otherwise $\neg \bar{c} = \bar{d}$ is an axiom – impossible, since $\mathfrak{T}_\mathfrak{n}$ is consistent). Thus $M = N$; hence the cardinality of M is $\mathfrak{n}$.

The reduct of $\mathfrak{M}$ on L is what we want. Of course, its cardinality is still right (M did not change). $\square$

Good as the above proof may be, it relies on the particular proof of the consistency theorem, and this is a drawback. Hence we offer another proof that does not have this defect.

Proof. (Once more) We develop a different argument, starting from the point where we concluded that $\overline{\Gamma}$ is consistent. Since the language of this theory has cardinality $\le \mathfrak{n}$,

Pause. Why?

we know by I.5.40 that

$$\text{we have a model } \mathfrak{M} = (M, \mathscr{I}) \text{ for } \overline{\Gamma} \text{ of cardinality } \le \mathfrak{n} \qquad (**)$$

Define now a function $f : N \to M$ by $f(n) = \bar{n}^{\mathscr{I}}$. Since all $n \in N$ have been imported into the language of $\bar{\Gamma}$, f is total. f is also 1-1: Indeed, if $c \neq d$ on N, then $\neg \bar{c} = \bar{d}$ is an axiom. Hence, $(\neg \bar{c} = \bar{d})^{\mathscr{I}} = \mathbf{t}$. That is, $f(c) \neq f(d)$ on M. But then (by results on cardinality in volume 2) the cardinality $\mathfrak{n}$ of N is $\leq$ the cardinality of M. By yet another result about cardinality,[†] (**) and what we have just concluded imply that N and M have the same cardinality.

At this point we take the reduct of $\mathfrak{M}$ on L, exactly as in the previous proof. $\qquad\square$

The above proof required more set theory. But it was independent of any knowledge of the proof of the consistency theorem.

I.5.44 Remark (about *Truth*). The completeness theorem shows that the syntactic apparatus of a first order (formal) logic totally captures the semantic notion of truth, *modulo* the acceptance as true of any given assumptions Γ. This justifies the habit of the mathematician (even the formalist: see Bourbaki (1966b, p. 21)) of saying – in the context of any given theory Γ – "it is true" (meaning "it is a Γ-theorem", or "it is Γ-proved"), "it is false" (meaning "the negation is a Γ-theorem"), "assume that $\mathscr{A}$ is true" (meaning "add the formula $\mathscr{A}$ – to Γ – as a nonlogical axiom"), and "assume that $\mathscr{A}$ is false" (meaning "add the formula $\neg \mathscr{A}$ – to Γ – as a nonlogical axiom").

Thus, "it is true" (in the context of a theory) means "it is true *in all its models*", hence provable: a theorem. We will not use this particular *argot*.

There is yet another *argot* use of "is true" (often said emphatically, "is *really* true"), meaning truth in some specific structure, the "intended model". Due to Gödel's first incompleteness theorem (Section I.9) this truth does *not* coincide with provability.

I.6. Substructures, Diagrams, and Applications

On p. 38 we saw one way of generating theories, as the sets of theorems, $\mathbf{Thm}_\Gamma$, proved from some set of (nonlogical) axioms, Γ. Another way of generating theories is by taking all the formulas that are valid in some *class* of structures.[‡]

[†] That $\mathfrak{k} \leq \mathfrak{l} \wedge \mathfrak{l} \leq \mathfrak{k} \to \mathfrak{k} = \mathfrak{l}$ for any cardinals $\mathfrak{k}$ and $\mathfrak{l}$.

[‡] In axiomatic set theory (e.g., as this is developed in volume 2 of these lectures) the term "class" means a "collection" that is not necessarily a set, by virtue of its enormous size. Examples of such collections are that of all ordinal numbers, that of all cardinal numbers, that of all the objects that set theory talks about, and many others. In the present case we allow for a huge collection of structures – for example, *all* structures – hence we have used the term "class", rather than "set", deliberately.

I.6.1 Definition. Let L be a first order language and $\mathscr{C}$ a class of structures appropriate for L. We define two symbols

$$\mathscr{T}(\mathscr{C}) \overset{\text{def}}{=} \{\mathscr{A} : \text{for all } \mathfrak{M} \in \mathscr{C}, \models_{\mathfrak{M}} \mathscr{A}\}$$

and

$$\text{Th}(\mathscr{C}) \overset{\text{def}}{=} \{\mathscr{A} : \mathscr{A} \text{ is closed and, for all } \mathfrak{M} \in \mathscr{C}, \models_{\mathfrak{M}} \mathscr{A}\}$$

If $\mathscr{C} = \{\mathfrak{M}\}$ then we write $\mathscr{T}(\mathfrak{M})$ and $\text{Th}(\mathfrak{M})$ rather than $\mathscr{T}(\{\mathfrak{M}\})$ and $\text{Th}(\{\mathfrak{M}\})$. □

For any class of structures, $\mathscr{C}$, $\mathscr{T}(\mathscr{C})$ is a theory in the sense of p. 38. This follows from an easy verification of

$$\mathscr{A} \in \mathscr{T}(\mathscr{C}) \quad \text{iff} \quad \mathscr{T}(\mathscr{C}) \vdash \mathscr{A} \tag{1}$$

We verify the *if* direction of (1), the opposite one being trivial. To prove $\mathscr{A} \in \mathscr{T}(\mathscr{C})$ we let $\mathfrak{M} \in \mathscr{C}$ and prove

$$\models_{\mathfrak{M}} \mathscr{A} \tag{2}$$

By soundness, it suffices to prove $\models_{\mathfrak{M}} \mathscr{T}(\mathscr{C})$, i.e.,

$$\models_{\mathfrak{M}} \mathscr{B} \quad \text{for all} \quad \mathscr{B} \in \mathscr{T}(\mathscr{C})$$

which holds by Definition I.6.1.

I.6.2 Example. For any structure $\mathfrak{M}$ for a language L, $\mathscr{T}(\mathfrak{M})$ is a complete theory: We want, for any sentence $\mathscr{A}$, $\mathscr{T}(\mathfrak{M}) \vdash \mathscr{A}$ or $\mathscr{T}(\mathfrak{M}) \vdash \neg\mathscr{A}$. By the previous observation, this translates to $\mathscr{A} \in \mathscr{T}(\mathfrak{M})$ or $(\neg\mathscr{A}) \in \mathscr{T}(\mathfrak{M})$. This holds, by I.6.1. □

I.6.3 Example. Let $\mathscr{K}$ be the (enormous) class of all structures for some language L. Then $\mathscr{T}(\mathscr{K})$ is the set of all universally valid formulas over L. □

I.6.4 Definition. Suppose that $\mathfrak{M}$ and $\mathfrak{K}$ are two structures for a language L. If it happens that $\mathscr{T}(\mathfrak{M}) = \mathscr{T}(\mathfrak{K})$, then we say that $\mathfrak{M}$ and $\mathfrak{K}$ are *elementarily equivalent* and write $\mathfrak{M} \equiv \mathfrak{K}$.

The context will fend off any mistaking of the symbol for *elementary equivalence*, $\equiv$, for that for string equality (cf. I.1.4). □

I.6.5 Proposition.[†] *Let $\mathfrak{M}$ and $\mathfrak{K}$ be two structures for L. Then, $\mathfrak{M} \equiv \mathfrak{K}$ iff* $\text{Th}(\mathfrak{M}) = \text{Th}(\mathfrak{K})$.

[†] Normally we use the word "Proposition" for a theorem that we do not want to make a big fuss about.

Proof. The only-if part is immediate from I.6.4. For the if part, let $\mathscr{A}$ be an arbitrary formula of L. Let $\mathscr{A}'$ be its universal closure (cf. I.4.11). Then $\models_{\mathfrak{M}} \mathscr{A}$ iff $\models_{\mathfrak{M}} \mathscr{A}'$ and $\models_{\mathfrak{K}} \mathscr{A}$ iff $\models_{\mathfrak{K}} \mathscr{A}'$; thus $\models_{\mathfrak{M}} \mathscr{A}$ iff $\models_{\mathfrak{K}} \mathscr{A}$ (since $\models_{\mathfrak{M}} \mathscr{A}'$ iff $\models_{\mathfrak{K}} \mathscr{A}'$). □

One way to obtain a structure $\mathfrak{K}$ that is elementarily equivalent to a structure $\mathfrak{M}$ is by a systematic renaming of all the members of the domain of $\mathfrak{M}$.

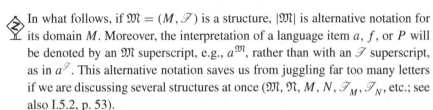

In what follows, if $\mathfrak{M} = (M, \mathscr{I})$ is a structure, $|\mathfrak{M}|$ is alternative notation for its domain M. Moreover, the interpretation of a language item a, f, or P will be denoted by an $\mathfrak{M}$ superscript, e.g., $a^{\mathfrak{M}}$, rather than with an $\mathscr{I}$ superscript, as in $a^{\mathscr{I}}$. This alternative notation saves us from juggling far too many letters if we are discussing several structures at once ($\mathfrak{M}$, $\mathfrak{N}$, M, N, $\mathscr{I}_M$, $\mathscr{I}_N$, etc.; see also I.5.2, p. 53).

I.6.6 Definition (Embeddings, Isomorphisms, and Substructures). Let $\mathfrak{M}$ and $\mathfrak{K}$ be structures for a language L, and $\phi : |\mathfrak{M}| \to |\mathfrak{K}|$ be a total (that is, everywhere defined on $|\mathfrak{M}|$) and 1-1[†] function. ϕ is a *structure embedding* – in which case we write $\phi : \mathfrak{M} \to \mathfrak{K}$ – just in case ϕ preserves all the nonlogical interpretations. This means the following:

(1) $a^{\mathfrak{K}} = \phi(a^{\mathfrak{M}})$ for all constants a of L

(2) $f^{\mathfrak{K}}(\phi(i_1), \dots, \phi(i_n)) = \phi(f^{\mathfrak{M}}(i_1, \dots, i_n))$ for all n-ary functions f of L and all $i_j \in |\mathfrak{M}|$

(3) $P^{\mathfrak{K}}(\phi(i_1), \dots, \phi(i_n)) = P^{\mathfrak{M}}(i_1, \dots, i_n)$ for all n-ary predicates P of L and all $i_j \in |\mathfrak{M}|$

In the last two cases "=" is metamathematical, comparing members of $|\mathfrak{K}|$ in the former and members of the truth set, $\{\mathbf{t}, \mathbf{f}\}$, in the latter.

An important special case occurs when $\phi : |\mathfrak{M}| \to |\mathfrak{K}|$ is the *inclusion map*, that is, $\phi(i) = i$ for all $i \in |\mathfrak{M}|$ – which entails $|\mathfrak{M}| \subseteq |\mathfrak{K}|$. We then say that $\mathfrak{M}$ is a *substructure* of $\mathfrak{K}$ and write $\mathfrak{M} \subseteq \mathfrak{K}$ (note the absence of $| \dots |$). Symmetrically, we say that $\mathfrak{K}$ is an *extension* of $\mathfrak{M}$.

If ϕ is onto as well – that is, $(\forall b \in |\mathfrak{K}|)(\exists a \in |\mathfrak{M}|)\phi(a) = b$ – then we say that it is a *structure isomorphism*, in symbols $\mathfrak{M} \underset{\phi}{\cong} \mathfrak{K}$, or just $\mathfrak{M} \cong \mathfrak{K}$.

It will be convenient to use the following abbreviation (Shoenfield (1967)): i^{ϕ} is short for $\phi(i)$ for all $i \in |\mathfrak{M}|$. □

[†] That is, $\phi(a) = \phi(b)$ implies $a = b$ for all a and b in $|\mathfrak{M}|$.

There is no special significance in using the letter ϕ for the embedding or isomorphism, other than its being a symbol other than what we normally use (generically) for formal function symbols (f, g, h).

One way to visualize what is going on in Definition I.6.6 is to employ a so-called *commutative diagram* (for simplicity, for a unary f)

$$|\mathfrak{M}| \ni i \overset{f^{\mathfrak{M}}}{\longrightarrow} f^{\mathfrak{M}}(i) \in |\mathfrak{M}|$$

$$\phi \downarrow \qquad\qquad \downarrow \phi$$

$$|\mathfrak{K}| \ni i^{\phi} \overset{f^{\mathfrak{K}}}{\longrightarrow} f^{\mathfrak{K}}(i^{\phi}) \in |\mathfrak{K}|$$

That the diagram commutes means that all possible compositions give the same result. Here $\phi(f^{\mathfrak{M}}(i)) = f^{\mathfrak{K}}(\phi(i))$.

When $\mathfrak{M} \subseteq \mathfrak{K}$ and ϕ is the inclusion map, then the diagram above simply says that $f^{\mathfrak{M}}$ is the *restriction* of $f^{\mathfrak{K}}$ on $|\mathfrak{M}|$, in symbols, $f^{\mathfrak{M}} = f^{\mathfrak{K}} \restriction |\mathfrak{M}|$. Condition (3) in Definition I.6.6 simply says that $P^{\mathfrak{M}}$ is the restriction of $P^{\mathfrak{K}}$ on $|\mathfrak{M}|^{n}$, i.e., $P^{\mathfrak{M}} = P^{\mathfrak{K}} \cap |\mathfrak{M}|^{n}$. One often indicates this last equality by $P^{\mathfrak{M}} = P^{\mathfrak{K}} \restriction |\mathfrak{M}|$.

I.6.7 Remark. Sometimes we have a structure $\mathfrak{M} = (M, \mathscr{I})$ for L and a 1-1 correspondence (total, 1-1, and onto) $\phi : M \to K$ for some set K. We can turn K into a structure for L, isomorphic to $\mathfrak{M}$, by defining its "$\mathscr{I}$"-part mimicking Definition I.6.6. We say that ϕ *induces an isomorphism* $\mathfrak{M} \cong_{\phi} \mathfrak{K}$.

We do this as follows:

(i) We set $a^{\mathfrak{K}} = \phi(a^{\mathfrak{M}})$ for all constants a of L

(ii) We set $f^{\mathfrak{K}}(\vec{i}_{n}) = \phi(f^{\mathfrak{M}}(\phi^{-1}(i_{1}), \ldots, \phi^{-1}(i_{n})))$ for all n-ary function symbols f of L and all i_{j} in K.[†]

(iii) We set $P^{\mathfrak{K}}(\vec{i}_{n}) = P^{\mathfrak{M}}(\phi^{-1}(i_{1}), \ldots, \phi^{-1}(i_{n}))$ for all n-ary predicate symbols P of L and all i_{j} in K.

It is now trivial to check via I.6.6 that for the $\mathfrak{K}$ so defined above, $\mathfrak{M} \cong_{\phi} \mathfrak{K}$ (Exercise I.47). □

The following shows that an embedding (and *a fortiori* an isomorphism) preserves meaning beyond that of the nonlogical symbols.

[†] By $\phi^{-1}(a)$, for $a \in K$, we mean the unique $b \in M$ such that $\phi(b) = a$.

I.6.8 Theorem. *Let* $\phi : \mathfrak{M} \to \mathfrak{K}$ *be an embedding of structures of a language* L. *Then:*

(1) *For every term* t *of* L *whose variables are among* $\vec{x}_n$, *and for all* $\vec{i}_n$ *in* $|\mathfrak{M}|$, *one has* $t[\![i_1^\phi, \dots, i_n^\phi]\!] = \phi(t[\![\vec{i}_n]\!])$.[†]

(2) *For every atomic formula* $\mathscr{A}$ *of* L *whose variables are among* $\vec{x}_n$, *and for all* $\vec{i}_n$ *in* $|\mathfrak{M}|$, *one has* $\models_\mathfrak{M} \mathscr{A}[\![\vec{i}_n]\!]$ *iff* $\models_\mathfrak{K} \mathscr{A}[\![i_1^\phi, \dots, i_n^\phi]\!]$.

(3) *If* ϕ *is an isomorphism, then we can remove the restriction* atomic *from (2) above.*

Strictly speaking, we are supposed to apply "$\models$" to a (formal) formula (cf. I.5.7). "$\mathscr{A}[\![\vec{i}_n]\!]$" is an abbreviated notation for a "concrete" (already interpreted) *sentence*. However, it is useful to extend our *argot* to allow the notation "$\models_\mathfrak{M} \mathscr{A}[\![\vec{i}_n]\!]$" – which says "the concrete sentence $\mathscr{A}[\![\vec{i}_n]\!]$ is true in $\mathfrak{M}$", in short, $(\mathscr{A}[\bar{i}_1, \dots, \bar{i}_n])^\mathfrak{M} = \mathbf{t}$ (cf. I.5.17) – as this notation readily discloses *where* $\mathscr{A}$ was interpreted and therefore where it is claimed to be true.

Proof. (1): We do induction on terms t (cf. I.5.17). If $t \equiv a$, a constant of L, then the left hand side is $a^\mathfrak{K}$ while the right hand side is $\phi(a^\mathfrak{M})$. These (both in $|\mathfrak{K}|$) are equal by I.6.6.

If $t \equiv x_j$ (for some x_j among the $\vec{x}_n$), then $t[\![i_1^\phi, \dots, i_n^\phi]\!] = i_j^\phi \in |\mathfrak{K}|$ while $t[\![i_1, \dots, i_n]\!] = i_j \in |\mathfrak{M}|$. Thus (1) of the theorem statement is proved in this case.

Finally, let $t \equiv ft_1 \dots t_r$. Then

$$\begin{aligned}
\phi((ft_1 \dots t_r)[\![\vec{i}_n]\!]) &= \phi(f^\mathfrak{M}(t_1[\![\vec{i}_n]\!], \dots, t_r[\![\vec{i}_n]\!])) \text{ (cf. I.5.17)} \\
&= f^\mathfrak{K}(\phi(t_1[\![\vec{i}_n]\!]), \dots, \phi(t_r[\![\vec{i}_n]\!])) \text{ (by I.6.6)} \\
&= f^\mathfrak{K}(t_1[\![i_1^\phi, \dots, i_n^\phi]\!], \dots, t_r[\![i_1^\phi, \dots, i_n^\phi]\!]) \text{ (by I.H.)} \\
&= (ft_1 \dots t_r)[\![i_1^\phi, \dots, i_n^\phi]\!] \text{ (by I.5.17)}
\end{aligned}$$

(2): We have two cases:

Case where $\mathscr{A} \equiv Pt_1 \dots t_r$. Then

$$\begin{aligned}
&\models_\mathfrak{M} (Pt_1 \dots t_r)[\![\vec{i}_n]\!] \\
&\text{iff } \models_\mathfrak{M} P^\mathfrak{M}(t_1[\![\vec{i}_n]\!], \dots, t_r[\![\vec{i}_n]\!]) \quad \text{(by I.5.17)} \\
&\text{iff } \models_\mathfrak{K} P^\mathfrak{K}(\phi(t_1[\![\vec{i}_n]\!]), \dots, \phi(t_r[\![\vec{i}_n]\!])) \quad \text{(by I.6.6)} \\
&\text{iff } \models_\mathfrak{K} P^\mathfrak{K}(t_1[\![i_1^\phi, \dots, i_n^\phi]\!], \dots, t_r[\![i_1^\phi, \dots, i_n^\phi]\!]) \quad \text{(by (1))} \\
&\text{iff } \models_\mathfrak{K} (Pt_1 \dots t_r)[\![i_1^\phi, \dots, i_n^\phi]\!] \quad \text{(by I.5.17)}
\end{aligned}$$

[†] Recall I.5.17. Of course, $t[\![\vec{i}_n]\!] \in |\mathfrak{M}|$, while $t[\![i_1^\phi, \dots, i_n^\phi]\!] \in |\mathfrak{K}|$.

Case where $\mathscr{A} \equiv t = s$. Then

$$\models_{\mathfrak{M}} (t = s)[\![\vec{i}_n]\!] \text{ iff } \models_{\mathfrak{M}} t[\![\vec{i}_n]\!] = s[\![\vec{i}_n]\!] \quad (\text{I.5.17})$$
$$\text{iff } \models_{\mathfrak{K}} \phi(t[\![\vec{i}_n]\!]) = \phi(s[\![\vec{i}_n]\!]) \quad (if \text{ by 1-1-ness})$$
$$\text{iff } \models_{\mathfrak{K}} t[\![i_1^\phi, \ldots, i_n^\phi]\!] = s[\![i_1^\phi, \ldots, i_n^\phi]\!] \quad (\text{by } (1))$$
$$\text{iff } \models_{\mathfrak{K}} (t = s)[\![i_1^\phi, \ldots, i_n^\phi]\!] \quad (\text{by I.5.17})$$

(3): It is an easy exercise to see that if $\mathscr{A}$ and $\mathscr{B}$ have the property claimed, then so do $\neg\mathscr{A}$ and $\mathscr{A} \vee \mathscr{B}$ without the additional assumption of ontoness. Ontoness helps $(\exists x)\mathscr{A}$:

$$\models_{\mathfrak{M}} ((\exists x)\mathscr{A})[\![\vec{i}_n]\!]$$
$$\text{iff } (\exists a \in |\mathfrak{M}|) \models_{\mathfrak{M}} \mathscr{A}[\![a, \vec{i}_n]\!] \quad (\text{by I.5.17})$$
$$\text{iff } (\exists a \in |\mathfrak{M}|) \models_{\mathfrak{K}} \mathscr{A}[\![a^\phi, i_1^\phi, \ldots, i_n^\phi]\!] \quad (\text{by I.H. on formulas})$$
$$\text{iff } (\exists b \in |\mathfrak{K}|) \models_{\mathfrak{K}} \mathscr{A}[\![b, i_1^\phi, \ldots, i_n^\phi]\!] \quad (if \text{ by ontoness})$$
$$\text{iff } \models_{\mathfrak{K}} ((\exists x)\mathscr{A})[\![i_1^\phi, \ldots, i_n^\phi]\!] \quad (\text{by I.5.17}) \qquad \square$$

I.6.9 Corollary. *If $\mathfrak{M}$ and $\mathfrak{K}$ are structures for L and $\mathfrak{M} \subseteq \mathfrak{K}$, then for all quantifier-free formulas $\mathscr{A}$ whose variables are among $\vec{x}_n$, and for all $\vec{i}_n$ in $|\mathfrak{M}|$, one has $\models_{\mathfrak{M}} \mathscr{A}[\![\vec{i}_n]\!]$ iff $\models_{\mathfrak{K}} \mathscr{A}[\![\vec{i}_n]\!]$.*

I.6.10 Corollary. *If $\mathfrak{M}$ and $\mathfrak{K}$ are structures for L and $\mathfrak{M} \underset{\phi}{\cong} \mathfrak{K}$, then $\mathfrak{M} \equiv \mathfrak{K}$.*
Proof. Let $\mathscr{A}(\vec{x}_n)$ be a formula. The sought

$$\models_{\mathfrak{M}} \mathscr{A}(\vec{x}_n) \quad \text{iff} \quad \models_{\mathfrak{K}} \mathscr{A}(\vec{x}_n)$$

translates to

$$(\forall i_1 \in |\mathfrak{M}|) \ldots (\forall i_n \in |\mathfrak{M}|) \models_{\mathfrak{M}} \mathscr{A}[\![\vec{i}_n]\!]$$
$$\text{iff} \quad (\forall j_1 \in |\mathfrak{K}|) \ldots (\forall j_n \in |\mathfrak{K}|) \models_{\mathfrak{K}} \mathscr{A}[\![\vec{j}_n]\!]$$

which is true,[†] by (3) of I.6.8, since every $j \in |\mathfrak{K}|$ is an i^ϕ (for some $i \in |\mathfrak{M}|$). $\qquad \square$

I.6.11 Corollary. *Let $\mathfrak{M}$ and $\mathfrak{K}$ be structures for L, $\mathfrak{M} \underset{\phi}{\cong} \mathfrak{K}$, and $\mathfrak{T}$ a theory over L. Then $\models_{\mathfrak{M}} \mathfrak{T}$ iff $\models_{\mathfrak{K}} \mathfrak{T}$.*

Condition (2) in I.6.8 is quite strong, as the following shows.

---

† Since we are arguing in the metatheory, we can say "true" rather than "provable".

I.6.12 Theorem. *Let* $\mathfrak{M}$ *and* $\mathfrak{K}$ *be structures for L, and* $\phi : |\mathfrak{M}| \to |\mathfrak{K}|$ *be a total function. If for every* atomic $\mathscr{A}$ *that contains at most the variables* $\vec{x}_n$ *and for all* $\vec{i}_n$ *in* $|\mathfrak{M}|$ *one has*

$$\models_{\mathfrak{M}} \mathscr{A}[\![\vec{i}_n]\!] \quad \textit{iff} \quad \models_{\mathfrak{K}} \mathscr{A}[\![i_1^\phi, \ldots, i_n^\phi]\!] \tag{1}$$

then $\phi : \mathfrak{M} \to \mathfrak{K}$ *is an embedding.*

Proof. First off, ϕ is 1-1. Indeed, since $x = y$ is atomic, then, by (1) above, for all i and j in $|\mathfrak{M}|$, $\models_{\mathfrak{M}} i = j$ iff $\models_{\mathfrak{K}} i^\phi = j^\phi$.[†] The if direction gives 1-1-ness.

We now check the three conditions of Definition I.6.6.

Let next $a^{\mathfrak{M}} = i$. Using the atomic formula $a = x$, we get (by (1))

$$\models_{\mathfrak{M}} (a = x)[\![i]\!] \quad \text{iff} \quad \models_{\mathfrak{K}} (a = x)[\![i^\phi]\!]$$

that is,

$$\models_{\mathfrak{M}} a^{\mathfrak{M}} = i \quad \text{iff} \quad \models_{\mathfrak{K}} a^{\mathfrak{K}} = i^\phi$$

Since $a^{\mathfrak{M}} = i$ is true, so is $a^{\mathfrak{K}} = i^\phi$. That is, $a^{\mathfrak{K}} = \phi(a^{\mathfrak{M}})$. (This part only needed the only-if direction of (1).)

Let now f be n-ary. Consider the atomic formula $f(\vec{x}_n) = x_{n+1}$. By (1),

$$\models_{\mathfrak{M}} (f(\vec{x}_n) = x_{n+1})[\![\vec{i}_{n+1}]\!] \quad \text{iff} \quad \models_{\mathfrak{K}} (f(\vec{x}_n) = x_{n+1})[\![i_1^\phi, \ldots, i_{n+1}^\phi]\!]$$

That is,

$$\models_{\mathfrak{M}} f^{\mathfrak{M}}(\vec{i}_n) = i_{n+1} \quad \text{iff} \quad \models_{\mathfrak{K}} f^{\mathfrak{K}}(i_1^\phi, \ldots, i_n^\phi) = i_{n+1}^\phi$$

Thus $f^{\mathfrak{K}}(i_1^\phi, \ldots, i_n^\phi) = \phi(f^{\mathfrak{M}}(\vec{i}_n))$. (This part too only needed the only-if direction of (1).)

Finally, let P be n-ary. By (1)

$$\models_{\mathfrak{M}} P(\vec{x}_n)[\![\vec{i}_n]\!] \quad \text{iff} \quad \models_{\mathfrak{K}} P(\vec{x}_n)[\![i_1^\phi, \ldots, i_n^\phi]\!]$$

That is,

$$\models_{\mathfrak{M}} P^{\mathfrak{M}}(\vec{i}_n) \quad \text{iff} \quad \models_{\mathfrak{K}} P^{\mathfrak{K}}(i_1^\phi, \ldots, i_n^\phi) \qquad \square$$

I.6.13 Corollary. *Let* $\mathfrak{M}$ *and* $\mathfrak{K}$ *be structures for L, and* $\phi : |\mathfrak{M}| \to |\mathfrak{K}|$ *be a total function. If, for every* atomic *and* negated *atomic* $\mathscr{A}$ *that contains at most the variables* $\vec{x}_n$ *and for all* $\vec{i}_n$ *in* $|\mathfrak{M}|$ *one has*

$$\models_{\mathfrak{M}} \mathscr{A}[\![\vec{i}_n]\!] \quad \textit{implies} \quad \models_{\mathfrak{K}} \mathscr{A}[\![i_1^\phi, \ldots, i_n^\phi]\!]$$

then $\phi : \mathfrak{M} \to \mathfrak{K}$ *is an embedding.*

[†] $(x = y)[\![i, j]\!] \equiv i = j$.

Proof. Let $\models_{\mathfrak{K}} \mathscr{A}[\![i_1^\phi, \dots, i_n^\phi]\!]$, $\mathscr{A}$ atomic. If $\not\models_{\mathfrak{M}} \mathscr{A}[\![\vec{i}_n]\!]$, then $\models_{\mathfrak{M}} \neg \mathscr{A}[\![\vec{i}_n]\!]$. $\neg\mathscr{A}$ is negated atomic. Thus, $\models_{\mathfrak{K}} \neg\mathscr{A}[\![i_1^\phi, \dots, i_n^\phi]\!]$, i.e., $\not\models_{\mathfrak{K}} \mathscr{A}[\![i_1^\phi, \dots, i_n^\phi]\!]$, contradicting the assumption. The "implies" has now been promoted to "iff".

$\square$

I.6.14 Corollary. *Let $\mathfrak{M}$ and $\mathfrak{K}$ be structures for L, and $|\mathfrak{M}| \subseteq |\mathfrak{K}|$. If, for every* atomic *and* negated atomic $\mathscr{A}$ *that contains at most the variables $\vec{x}_n$ and for all $\vec{i}_n$ in $|\mathfrak{M}|$ one has*

$$\models_{\mathfrak{M}} \mathscr{A}[\![\vec{i}_n]\!] \quad \text{implies} \quad \models_{\mathfrak{K}} \mathscr{A}[\![\vec{i}_n]\!]$$

then $\mathfrak{M} \subseteq \mathfrak{K}$.

Proof. $|\mathfrak{M}| \subseteq |\mathfrak{K}|$ means that we may take here $\phi : |\mathfrak{M}| \to |\mathfrak{K}|$ to be the inclusion map. $\square$

The converses of the above two corollaries hold (these, essentially, are I.6.8 and I.6.9).

Moreover, these corollaries lead to

I.6.15 Definition (Elementary Embeddings and Substructures). Let $\mathfrak{M}$ and $\mathfrak{K}$ be structures for L, and $\phi : |\mathfrak{M}| \to |\mathfrak{K}|$ a total function. If, for every formula $\mathscr{A}$ that contains at most the variables $\vec{x}_n$ and for all $\vec{i}_n$ in $|\mathfrak{M}|$, one has

$$\models_{\mathfrak{M}} \mathscr{A}[\![\vec{i}_n]\!] \quad \text{implies} \quad \models_{\mathfrak{K}} \mathscr{A}[\![i_1^\phi, \dots, i_n^\phi]\!] \tag{1}$$

then $\phi : \mathfrak{M} \to \mathfrak{K}$ is called an *elementary embedding*. We write $\phi : \mathfrak{M} \to_{\prec} \mathfrak{K}$ in this case.

If ϕ is the inclusion map $|\mathfrak{M}| \subseteq |\mathfrak{K}|$, then (1) becomes

$$\models_{\mathfrak{M}} \mathscr{A}[\![\vec{i}_n]\!] \quad \text{implies} \quad \models_{\mathfrak{K}} \mathscr{A}[\![\vec{i}_n]\!] \tag{2}$$

In this case we say that $\mathfrak{M}$ is an *elementary substructure* of $\mathfrak{K}$, or that the latter is an *elementary extension* of the former. In symbols, $\mathfrak{M} \prec \mathfrak{K}$. $\square$

I.6.16 Remark. The "implies" in (1) and (2) in the definition above is promoted to "iff" as in the proof of I.6.13.

By I.6.13 and I.6.14, an elementary embedding $\phi : \mathfrak{M} \to_{\prec} \mathfrak{K}$ is also an embedding $\phi : \mathfrak{M} \to \mathfrak{K}$, and an elementary substructure $\mathfrak{M} \prec \mathfrak{K}$ is also a substructure $\mathfrak{M} \subseteq \mathfrak{K}$.

If $\phi : \mathfrak{M} \to_{\prec} \mathfrak{K}$, then condition (1) in Definition I.6.15 (which, as we have already noted, is an equivalence) yields, for all sentences $\mathscr{A}$ over L,

$$\models_{\mathfrak{M}} \mathscr{A} \quad \text{iff} \quad \models_{\mathfrak{K}} \mathscr{A}$$

By I.6.5, $\mathfrak{M} \equiv \mathfrak{K}$. In particular, $\mathfrak{M} \prec \mathfrak{K}$ implies $\mathfrak{M} \equiv \mathfrak{K}$. □

I.6.17 Remark. In many arguments in model theory, starting with a structure $\mathfrak{M}$ for some language L, one constructs another structure $\mathfrak{K}$ for L and an embedding $\phi : \mathfrak{M} \to \mathfrak{K}$ or an elementary embedding $\phi : \mathfrak{M} \to_{\prec} \mathfrak{K}$. More often than not one would rather have a structure extension or an elementary extension of $\mathfrak{M}$ (i.e., preferring that ϕ be the inclusion map) respectively. This can be easily obtained as follows:

Let $\mathfrak{M} = (M, \dots)$ and $\mathfrak{K} = (K, \dots)$. Let N be a set disjoint from M,[†] of cardinality equal to that of $K - \phi[M]$.[‡] Thus there is a 1-1 correspondence $\psi : K - \phi[M] \to N$, from which we can build a 1-1 correspondence $\chi : K \to M \cup N$ by defining

$$\chi(x) = \begin{cases} \psi(x) & \text{if } x \in K - \phi[M] \\ \phi^{-1}(x) & \text{if } x \in \phi[M] \end{cases}$$

Using Remark I.6.7, we get a structure $\mathfrak{K}' = (M \cup N, \dots)$ such that

$$\mathfrak{K} \underset{\chi}{\cong} \mathfrak{K}' \tag{1}$$

We verify that if we had $\phi : \mathfrak{M} \to_{\prec} \mathfrak{K}$ initially, then we now have[§]

$$\chi \circ \phi : \mathfrak{M} \to_{\prec} \mathfrak{K}' \tag{2}$$

Well, let $\mathscr{A}$ be arbitrary over L with free variables among $\vec{x}_n$. Pick any $\vec{i}_n$ in M, and assume $\models_{\mathfrak{M}} \mathscr{A}[\![\vec{i}_n]\!]$. By hypothesis on ϕ and I.6.15, $\models_{\mathfrak{K}} \mathscr{A}[\![\phi(i_1), \dots, \phi(i_n)]\!]$. By I.6.8(3) and (1) above, $\models_{\mathfrak{K}'} \mathscr{A}[\![\chi(\phi(i_1)), \dots, \chi(\phi(i_n))]\!]$. This settles (2), by I.6.15. By definition of χ above, if $x \in M$, then, since $\phi(x) \in \phi[M]$, we have $\chi(\phi(x)) = \phi^{-1}(\phi(x)) = x$. That is, (2) is an elementary extension ($\chi \circ \phi$ is the identity on M).

The alternative initial result, an embedding $\phi : \mathfrak{M} \to \mathfrak{K}$, yields a structure extension $\chi \circ \phi : \mathfrak{M} \subseteq \mathfrak{K}'$, since the composition of embeddings is an embedding (Exercise I.48). □

[†] $N \cap M = \emptyset$.
[‡] $\phi[X] = \{\phi(x) : x \in X\}$. $X - Y = \{x \in X : x \notin Y\}$.
[§] $\chi \circ \phi$ denotes function composition. That is, $(\chi \circ \phi)(x) = \chi(\phi(x))$ for all $x \in M$.

The following criterion for elementary extensions evaluates truth in the bigger of the two structures. It is useful, among other places, in the proof of the downward Löwenheim-Skolem theorem (I.6.20 below).

I.6.18 Proposition. *Let $\mathfrak{M}$ and $\mathfrak{K}$ be structures for L. If $\mathfrak{M} \subseteq \mathfrak{K}$, and moreover, for every formula $\mathscr{A}$ whose free variables are among y and $\vec{x}_n$, and for all $\vec{b}_n$ in $|\mathfrak{M}|$,*

$$(\exists a \in |\mathfrak{K}|) \models_{\mathfrak{K}} \mathscr{A}[\![a, \vec{b}_n]\!] \quad \text{implies} \quad (\exists a \in |\mathfrak{M}|) \models_{\mathfrak{K}} \mathscr{A}[\![a, \vec{b}_n]\!] \qquad (3)$$

then $\mathfrak{M} \prec \mathfrak{K}$.

Proof. We need to prove (2) of I.6.15 using (3) above.

We do induction on formulas. We have already remarked in I.6.16 that the definition is an "iff", and it is more convenient to carry the induction through with an "iff" rather than an "implies".

First off, by $\mathfrak{M} \subseteq \mathfrak{K}$ and Corollary I.6.9, we are done for all quantifier-free formulas. This takes care of atomic formulas.

For the induction step, from $\mathscr{B}$ and $\mathscr{C}$ to $\neg\mathscr{B}$ and $\mathscr{B} \vee \mathscr{C}$, we note (with $\vec{i}$ arbitrary, in $|\mathfrak{M}|$)

$$\begin{aligned}
\models_{\mathfrak{M}} \neg\mathscr{B}[\![\vec{i}]\!] \quad &\text{iff} \quad \not\models_{\mathfrak{M}} \mathscr{B}[\![\vec{i}]\!] \\
&\text{iff} \quad \not\models_{\mathfrak{K}} \mathscr{B}[\![\vec{i}]\!] \quad \text{by I.H.} \\
&\text{iff} \quad \models_{\mathfrak{K}} \neg\mathscr{B}[\![\vec{i}]\!]
\end{aligned}$$

Similarly,

$$\begin{aligned}
\models_{\mathfrak{M}} (\mathscr{B} \vee \mathscr{C})[\![\vec{i}]\!] \quad &\text{iff} \quad \models_{\mathfrak{M}} \mathscr{B}[\![\vec{i}]\!] \text{ or } \models_{\mathfrak{M}} \mathscr{C}[\![\vec{i}]\!] \\
&\text{iff} \quad \models_{\mathfrak{K}} \mathscr{B}[\![\vec{i}]\!] \text{ or } \models_{\mathfrak{K}} \mathscr{C}[\![\vec{i}]\!] \quad \text{by I.H.} \\
&\text{iff} \quad \models_{\mathfrak{K}} (\mathscr{B} \vee \mathscr{C})[\![\vec{i}]\!]
\end{aligned}$$

It only remains to check existential formulas. The implication

$$\models_{\mathfrak{M}} ((\exists x)\mathscr{B})[\![\vec{i}]\!] \quad \text{implies} \quad \models_{\mathfrak{K}} ((\exists x)\mathscr{B})[\![\vec{i}]\!]$$

that is,

$$(\exists a \in |\mathfrak{M}|) \models_{\mathfrak{M}} \mathscr{B}[\![a, \vec{i}]\!] \quad \text{implies} \quad (\exists a \in |\mathfrak{K}|) \models_{\mathfrak{K}} \mathscr{B}[\![a, \vec{i}]\!]$$

being trivial by $|\mathfrak{M}| \subseteq |\mathfrak{K}|$, we check the opposite direction. Let $\vec{i}$ still be in $|\mathfrak{M}|$, and $(\exists a \in |\mathfrak{K}|) \models_{\mathfrak{K}} \mathscr{B}[\![a, \vec{i}]\!]$. By (3), $(\exists a \in |\mathfrak{M}|) \models_{\mathfrak{K}} \mathscr{B}[\![a, \vec{i}]\!]$. By I.H., $(\exists a \in |\mathfrak{M}|) \models_{\mathfrak{M}} \mathscr{B}[\![a, \vec{i}]\!]$. □

I.6.19 Example. The structure $\mathfrak{N}' = (\mathbb{N}, <, 0)$ is a reduct of the standard model of arithmetic. It is essentially a linearly ordered set, where we have also focused on a "distinguished element", 0, that also happens to be the minimum element of the order. Any nonempty $B \subseteq \mathbb{N}$ that contains 0 leads to a $\mathfrak{B} = (B, <, 0) \subseteq \mathfrak{N}'$, trivially.

If we ask for more, that $\mathfrak{B} \prec \mathfrak{N}'$, then we reduce our choices (of B that work) drastically. Indeed, let $\mathscr{A}(x, y)$ say that x and y are consecutive, $x < y$. That is, $\mathscr{A}(x, y)$ is

$$x < y \wedge \neg(\exists z)(x < z \wedge z < y)$$

Then

$$(\forall a \in \mathbb{N}) \models_{\mathfrak{N}'} ((\exists y)\mathscr{A}(x, y))[\![a]\!] \tag{1}$$

Thus, if $\mathfrak{B} \prec \mathfrak{N}'$, then we require $\mathfrak{B} \subseteq \mathfrak{N}'$ – in particular, $0 \in B$ – and (from (1), I.6.15 and I.6.16)

$$(\forall a \in B)(\exists b \in B) \models_{\mathfrak{B}} \mathscr{A}[\![a, b]\!] \tag{2}$$

$0 \in B$ and (2) yield (by metamathematical induction) that $B = \mathbb{N}$. □

The following is a useful tool in model theory.

I.6.20 Theorem (Downward Löwenheim-Skolem Theorem). *Assume that L has cardinality* at most $\mathfrak{m}$ *(that is, the set of nonlogical symbols does*[†]*), while the structure* $\mathfrak{M} = (M, \mathscr{I})$ *(for L) has cardinality* at least $\mathfrak{m}$*. Let $X \subseteq M$, and the cardinality of X be $\leq \mathfrak{m}$. Then there is a structure $\mathfrak{K}$ for L, of cardinality exactly $\mathfrak{m}$, that satisfies $X \subseteq |\mathfrak{K}|$ and $\mathfrak{K} \prec \mathfrak{M}$.*

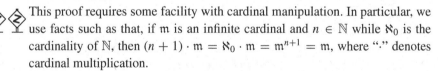 This proof requires some facility with cardinal manipulation. In particular, we use facts such as that, if $\mathfrak{m}$ is an infinite cardinal and $n \in \mathbb{N}$ while $\aleph_0$ is the cardinality of $\mathbb{N}$, then $(n + 1) \cdot \mathfrak{m} = \aleph_0 \cdot \mathfrak{m} = \mathfrak{m}^{n+1} = \mathfrak{m}$, where "·" denotes cardinal multiplication.

Proof. (Thinking out loud.) If $\mathfrak{K} = (K, \dots)$ – still to be defined – is to satisfy $\mathfrak{K} \prec \mathfrak{M}$, then we have to ensure (I.6.18), among other things, that for every $\mathscr{A}(y, \vec{x}_n)$ over L and for all $\vec{a}_n$ in K, if $\mathscr{A}[\![b, \vec{a}_n]\!]$ is true in $\mathfrak{M}$ (where $b \in M$), then some $b' \in K$ exists such that $\mathscr{A}[\![b', \vec{a}_n]\!]$ is *also* true in $\mathfrak{M}$.

[†] It is a set-theoretical fact that the union of a countable set and an infinite set has cardinality equal to that of the infinite set, or $\aleph_0 + \mathfrak{m} = \mathfrak{m}$. Recall that the set of logical symbols is countable.

In short, we need to obtain K as a *subset of M that includes X and is closed under all the relations*

$$Q_{\mathscr{A}} = \{\langle \vec{a}_n, b \rangle : \models_{\mathfrak{M}} (\mathscr{A}(y, \vec{x}_n))[\![b, \vec{a}_n]\!]\} \tag{1}$$

where b is the output part in $Q_{\mathscr{A}}$ above.

It is prudent to take two precautions:

First, in order to get the smallest cardinal possible for K, we build the $\subseteq$-smallest set described in italics above, that is, we take $K = Cl(X, \mathscr{R})$,[†] where $\mathscr{R}$ is the set of all the relations $Q_{\mathscr{A}}$ as $\mathscr{A}$ varies over all the formulas over L.

Second, in order to place a *lower bound* of m to the cardinality of K, we start, rather than with X, with a set Y of cardinality exactly m such that $X \subseteq Y \subseteq M$. Such a Y exists by the size estimates for X and M. In other words, we have just changed our mind, and now let $K = Cl(Y, \mathscr{R})$.

Actually, a bit more prudence is in order. Let us rename the $Cl(Y, \mathscr{R})$ above K'. We are still looking for a K that we will keep.

We will "cut down" each $Q_{\mathscr{A}}$ in (1) – with the help of the axiom of choice (AC) – making the relation *single-valued*[‡] in its output, b. (End of thinking out loud.)

We define

$$\tilde{Q}_{\mathscr{A}} = \{\langle \vec{a}_n, b \rangle : b \text{ picked by AC in } \{x : Q_{\mathscr{A}}(\vec{a}_n, x)\} \text{ if } (\exists x \in M)Q_{\mathscr{A}}(\vec{a}_n, x)\} \tag{2}$$

Let now $\tilde{\mathscr{R}}$ be the set of all $\tilde{Q}_{\mathscr{A}}$, and set $K = Cl(Y, \tilde{\mathscr{R}})$. This one we keep!

First off, trivially, $K \subseteq M$. To turn K into a substructure of $\mathfrak{M}$, we have to interpret every f, a, P – function, constant and predicate, respectively – of L as $f^{\mathfrak{M}} \upharpoonright K^n$ (n-ary case), leave as $a^{\mathfrak{M}}$, and $P^{\mathfrak{M}} | K$ (n-ary case), respectively.

For functions and constants we have more work: For the former we need to show that K is closed under all $f^{\mathfrak{M}} \upharpoonright K^n$. Let then f be n-ary in L and $\vec{a}_n$ be in K. Then $f^{\mathfrak{M}}(\vec{a}_n) = b$ and $b \in M$. We want $b \in K$.

Well, if we set $\mathscr{A} \equiv f(\vec{x}_n) = y$, then $\models_{\mathfrak{M}} (f(\vec{x}_n) = y)[\![\vec{a}_n, b]\!]$; hence $Q_{\mathscr{A}}(\vec{a}_n, b)$ is true (cf. (1)), and therefore $\tilde{Q}_{\mathscr{A}}(\vec{a}_n, b)$ is true, for b is unique, given the $\vec{a}_n$, so that it must be what AC picks in (2). Thus, since K is $\tilde{Q}_{\mathscr{A}}$-closed, $b \in K$. Similarly one proves the case for constants a of L, using the formula $a = y$.

Thus we do have a substructure, $\mathfrak{K} = (K, \dots)$, of $\mathfrak{M}$.

It remains to settle the property of Proposition I.6.18 and the cardinality claims.

[†] You may want to review Section I.2 at this point.
[‡] A fancy term for "function". This function need not be defined everywhere on M.

First off, let $\models_{\mathfrak{M}} \mathscr{A}[\![b, \vec{a}_n]\!]$, where $b \in M$, and the $\vec{a}_n$ are in K. For some b' picked by AC in (2), $\tilde{Q}_{\mathscr{A}}(\vec{a}_n, b')$ is true, that is, $\models_{\mathfrak{M}} \mathscr{A}[\![b', \vec{a}_n]\!]$. Since K is $\tilde{Q}_{\mathscr{A}}$-closed, $b' \in K$. Thus, the $\prec$-criterion for $\mathfrak{K} \prec \mathfrak{M}$ is met.

Finally, we compute the cardinality of K. This inductively defined set is being built by stages (see I.2.9),

$$K = \bigcup_{i \geq 0} K_i \qquad (3)$$

At stage 0 we have $K_0 = Y$. Having built K_i, $i \leq n$, we build K_{n+1} by appending to $\bigcup_{i \leq n} K_i$ all b such that $\tilde{Q}_{\mathscr{A}}(\vec{a}_r, b)$ holds. We do so for all $\tilde{Q}_{\mathscr{A}}(\vec{x}_r, y)$ and all possible $\vec{a}_r$ in $\bigcup_{i \leq n} K_i$.

We show that the cardinality of each K_n is $\leq \mathfrak{m}$. This is true for K_0. We take an I.H. for all $n \leq k$ and consider K_{k+1} next.

Now, the set of formulas over L has cardinality $\leq \mathfrak{m}$; hence

$$\text{the set } \widetilde{\mathscr{R}} \text{ as well has cardinality } \leq \mathfrak{m} \qquad (4)$$

We let $S = \bigcup_{i \leq k} K_i$ for convenience. By I.H.,

$$(\text{cardinality of } S) \leq (k+1) \cdot \mathfrak{m} = \mathfrak{m} \qquad (5)$$

For *each* $\tilde{Q}_{\mathscr{A}}(\vec{x}_{r+1}) \in \widetilde{\mathscr{R}}$, the cardinality of the set of *contributed outputs*, taking inputs in S, is at most equal to the cardinality of S^r, i.e., $\leq \mathfrak{m}^r = \mathfrak{m}$. The total contribution of *all* the $\tilde{Q}_{\mathscr{A}}$ is then, by (4), of cardinality $\leq \mathfrak{m} \cdot \mathfrak{m} = \mathfrak{m}$. Thus, using (5), the cardinality of K_{k+1} is $\leq \mathfrak{m} + \mathfrak{m} = \mathfrak{m}$. By (3), the cardinal of K is $\leq \aleph_0 \cdot \mathfrak{m} = \mathfrak{m}$. Since it is also $\geq \mathfrak{m}$ (by $Y \subseteq K$), we are done. $\qquad \square$

N.B. The set K' that we have discarded earlier also satisfies $\mathfrak{K}' = (K', \dots) \prec \mathfrak{M}$, and has cardinality at most that of M. We cannot expect though that it has cardinality $\leq \mathfrak{m}$.

I.6.21 Corollary. *Assume that L has cardinality at most $\mathfrak{m}$, while the structure $\mathfrak{M} = (M, \mathscr{I})$ (for L) has cardinality at least $\mathfrak{m}$. Let $\mathfrak{X} \prec \mathfrak{M}$, and the cardinality of $|\mathfrak{X}|$ be $\leq \mathfrak{m}$. Then there is a structure $\mathfrak{K}$ for L, of cardinality exactly $\mathfrak{m}$ that satisfies $\mathfrak{X} \prec \mathfrak{K} \prec \mathfrak{M}$.*

Proof. Working with $X = |\mathfrak{X}|$ exactly as in I.6.20, we get $\mathfrak{K}$. It is straightforward to check that $\mathfrak{X} \subseteq \mathfrak{K}$.[†] The $\prec$-criterion (I.6.18) for $\mathfrak{X} \prec \mathfrak{K}$ is: Given $\mathscr{A}$ with free

---

[†] For n-ary f in L, $f^{\mathfrak{X}} = f^{\mathfrak{M}} \restriction X^n$ by assumption. On the other hand, $f^{\mathfrak{K}} = f^{\mathfrak{M}} \restriction K^n$ by construction of $\mathfrak{K}$. But $X \subseteq K$.

variables among the y *and* $\vec{x}_n$, and $\vec{a}_n$ in X. Verify that if

$$\text{for some } b \in K, \qquad \models_{\mathfrak{K}} \mathscr{A}[\![b, \vec{a}_n]\!] \tag{1}$$

then

$$\models_{\mathfrak{K}} \mathscr{A}[\![b', \vec{a}_n]\!] \qquad \text{for some } b' \in X \tag{2}$$

Well, $\mathfrak{K} \subseteq \mathfrak{M}$ and (1) yield $\models_{\mathfrak{M}} \mathscr{A}[\![b, \vec{a}_n]\!]$. (2) follows from $\mathfrak{X} \prec \mathfrak{M}$. □

The following definition builds on I.5.4 and I.5.20.

I.6.22 Definition (Diagrams). Given a language L and a structure $\mathfrak{A}$ for L. Let $\emptyset \neq M \subseteq |\mathfrak{A}|$. We call $L(M)$, constructed as in I.5.20, the *diagram language* of M. We denote by $\mathfrak{A}_M$ the *expansion* of $\mathfrak{A}$ that has all else the same as $\mathfrak{A}$, except that it gives the "natural meaning" to the *new* constants, $\bar{i}$, of $L(M)$. Namely, $\bar{i}^{\mathfrak{A}_M} = i$ for all $i \in M$. $\mathfrak{A}_M$ is called the *diagram expansion* of $\mathfrak{A}$ (over $L(M)$).

We often write $\mathfrak{A}_M = \{|\mathfrak{A}|, \ldots, (i)_{i \in M}\}$ to indicate that constants $\bar{i}$ with interpretations i where added. The part "$\{|\mathfrak{A}|, \ldots\}$" is just $\mathfrak{A}$.

If $M = |\mathfrak{A}|$, then we write $L(\mathfrak{A})$ for $L(M)$, as in I.5.4.

The *basic diagram* of $\mathfrak{A}$, denoted by $D(\mathfrak{A})$, is the set of all *atomic* and *negated atomic* sentences of $L(\mathfrak{A})$ that are true in $\mathfrak{A}_{|\mathfrak{A}|}$.

The *elementary diagram* of $\mathfrak{A}$ is just $\mathrm{Th}(\mathfrak{A}_{|\mathfrak{A}|})$.

Suppose now that $\mathfrak{A}$ and M are as above, $\mathfrak{B}$ is a structure for L as well – possibly $\mathfrak{A} = \mathfrak{B}$ – and $\phi : M \to |\mathfrak{B}|$ is a *total* function. We may now wish to import the constants i of M into L, as $\bar{i}$, and expand $\mathfrak{B}$ so that, for each $i \in M$, $\bar{i}$ is interpreted as $\phi(i)$. This expansion of $\mathfrak{B}$ is denoted by $\mathfrak{B}_{\phi[M]}$.[†]

Thus, all else in $\mathfrak{B}_{\phi[M]}$ is the same as in $\mathfrak{B}$, except that $\bar{i}^{\mathfrak{B}_{\phi[M]}} = \phi(i)$.

We often write $\mathfrak{B}_{\phi[M]} = \{|\mathfrak{B}|, \ldots, (\phi(i))_{i \in M}\}$ to indicate that constants $\bar{i}$ with interpretations $\phi(i)$ were added.

Therefore, if $M \subseteq |\mathfrak{B}|$ and ϕ is the inclusion map, then $\mathfrak{B}_{\phi[M]}$ is the expansion of $\mathfrak{B}$, $\mathfrak{B}_M$. □

The particular way that we have approached the assignment of semantics is not the "usual" one.[‡] One often sees a manner of handling *substitution* of informal constants into formal variables that differs from that of I.5.4. Namely, other authors often do this by *assigning* (or "corresponding") values to variables, rather than *substituting* (formal names of) values into variables. This is achieved by a so-called *assignment function*, s, from the set of all variables $\{v_0, v_1, \ldots\}$

[†] In the choice of our notation we are following Keisler (1978).
[‡] It is the same as the one in Shoenfield (1967).

to the domain, $|\mathfrak{A}|$, of the structure into which we want to interpret a language L. Intuitively, $s(v_j) = a$ says that we have plugged the value a into v_j.

While we have not favoured this approach, we will grant it one thing: It does not compel us to extend the language L into the diagram language $L(\mathfrak{A})$ to effect the interpretation. Such extensions may be slightly confusing when we deal with diagrams.

For example, let L and $\mathfrak{A}$ be as above, with $\emptyset \neq M \subseteq |\mathfrak{A}|$. We next form $L(M)$. Now, this language has new constants, $\bar{i}$, for each $i \in M$. Thus, when we proceed to interpret it into $\mathfrak{A}_M$, we *already have formal counterparts* of all $i \in M$ in $L(M)$, and we need only import the constants (as per I.5.4) of $|\mathfrak{A}| - M$.

But is there any harm done if we re-import *all* $i \in |\mathfrak{A}|$, as $\tilde{i}$, to form $L(M)(\mathfrak{A})$? Not really. Let us focus on an $i \in M$ which has been imported as $\bar{i}$ when forming $L(M)$ and then again as $\tilde{i}$ to do what I.5.4 prescribes towards interpreting $L(M)$ in $\mathfrak{A}_M$. Thus, $\bar{i}$ is a closed term t of $L(M)(\mathfrak{A})$ for which $t^{\mathfrak{A}_M} = i$. Since $\tilde{i}$ is the imported name for i, Lemma I.5.11 yields that, for any term s and formula $\mathscr{A}$ over $L(M)(\mathfrak{A})$ of at most one free variable x,

$$(s[x \leftarrow \tilde{i}])^{\mathfrak{A}_M} = (s[x \leftarrow \bar{i}])^{\mathfrak{A}_M}$$

and

$$(\mathscr{A}[x \leftarrow \tilde{i}])^{\mathfrak{A}_M} = (\mathscr{A}[x \leftarrow \bar{i}])^{\mathfrak{A}_M}$$

In short, meaning does not depend on which *formal alias* of $i \in M$ we use. We will be free to assume then that the $\tilde{i}$ were not new (for $i \in M$); they were just the $\bar{i}$.

In particular, the statement "the sentence $\mathscr{A}$ over $L(\mathfrak{A})$ is true in $\mathfrak{A}_{|\mathfrak{A}|}$" is equivalent to the statement "the $\mathfrak{A}$-instance[†] $\mathscr{A}$ of a formula over L is true in $\mathfrak{A}$", since a closed formula of $L(\mathfrak{A})$ is an $\mathfrak{A}$-instance of one over L and conversely $(\tilde{i} \equiv \bar{i})$.

Diagrams offer more than one would expect from just nomenclature, as a number of applications in the balance of this section will readily demonstrate. First, we translate I.6.13, I.6.14, and I.6.15 into the diagram terminology.

I.6.23 Lemma (Main Diagram Lemma). *Let $\mathfrak{M}$ and $\mathfrak{K}$ be two structures for L, and $\phi : |\mathfrak{M}| \to |\mathfrak{K}|$ be a total function. Then*

(1) $\phi : \mathfrak{M} \to \mathfrak{K}$ *(embedding) iff* $\models_{\mathfrak{K}_{\phi[|\mathfrak{M}|]}} D(\mathfrak{M})$;

(2) $\phi : \mathfrak{M} \to_{\prec} \mathfrak{K}$ *(elementary embedding) iff* $\models_{\mathfrak{K}_{\phi[|\mathfrak{M}|]}} \mathrm{Th}(\mathfrak{M}_{|\mathfrak{M}|})$.

Moreover, if $|\mathfrak{M}| \subseteq |\mathfrak{K}|$, then

[†] Cf. I.5.7, p. 55.

(3) $\mathfrak{M} \subseteq \mathfrak{K}$ *(substructure) iff* $\models_{\mathfrak{K}_{|\mathfrak{M}|}} D(\mathfrak{M})$;

(4) $\mathfrak{M} \prec \mathfrak{K}$ *(elementary substructure) iff* $\models_{\mathfrak{K}_{|\mathfrak{M}|}} \mathrm{Th}(\mathfrak{M}_{|\mathfrak{M}|})$.

Proof. Direct translation of the facts quoted prior to the lemma statement. For example, the hypothesis of the if part in item (1) above is the same as the displayed formula in the statement of I.6.13 if we recall that, e.g., an atomic sentence over $L(M)$ is an $\mathfrak{M}$-instance $\mathscr{A}(\bar{i}_1, \dots, \bar{i}_n)$ of an atomic formula $\mathscr{A}$ over L and that $\models_{\mathfrak{M}} \mathscr{A}[\![\bar{i}_n]\!]$ is *argot* for $\left(\mathscr{A}(\bar{i}_1, \dots, \bar{i}_n)\right)^{\mathfrak{M}} = \mathbf{t}$ (cf. I.5.17).

The only-if part in (1)–(4) is due to the remark following I.6.14, p. 82. □

We present a number of applications of diagrams. First, we revisit the upward Löwenheim-Skolem theorem (I.5.43).

I.6.24 Theorem (Upward Löwenheim-Skolem Theorem, Version 2). *Let* $\mathfrak{A}$ *be an infinite structure for* L. *For every cardinal* $\mathfrak{n}$ *that is not smaller than the cardinal of* $\mathfrak{A}$ *(that is, of* $|\mathfrak{A}|$*) and of* L, *there is a structure* $\mathfrak{B}$ *for* L, *of cardinality exactly* $\mathfrak{n}$, *such that* $\mathfrak{A} \prec \mathfrak{B}$.

Proof. As in the proof of I.5.43, pick a set N of cardinality $\mathfrak{n}$. The set of sentences over $L(\mathfrak{A})(N)$

$$\mathcal{Q} = \mathrm{Th}(\mathfrak{A}_{|\mathfrak{A}|}) \cup \{\neg\bar{c} = \bar{d} : c \neq d \text{ on } N\}$$

where $\bar{c}$ are distinct new constants, one for each $c \in N$, is finitely satisfiable. This is because every finite subset of $\mathcal{Q}$ involves only finitely many of the sentences $\neg\bar{c} = \bar{d}$, and thus there is capacity in $\mathfrak{A} = (A, \mathscr{I})$ (as A is infinite) to extend $\mathscr{I}$ into $\mathscr{I}'$ (keeping A the same, but defining distinct $\bar{c}^{\mathscr{I}'}, \bar{d}^{\mathscr{I}'}$, etc.) to satisfy these sentences in an expanded structure $\mathfrak{A}' = (A, \mathscr{I}')$.

By compactness, there is a model $\mathfrak{D} = (D, \mathscr{I}_D)$ for $\mathcal{Q}$ (1)

We will look at first into its reduct on $L(\mathfrak{A})$, $\mathfrak{C} = (D, \mathscr{I}_C)$ – that is, $\mathscr{I}_C = \mathscr{I}_D \restriction L(\mathfrak{A})$. $\mathfrak{C}$ is a model of $\mathrm{Th}(\mathfrak{A}_{|\mathfrak{A}|})$.

We define next a function $f : A \to D$ by $f(\bar{n}^{\mathscr{I}}) = \bar{n}^{\mathscr{I}_C}$. The function f is total because all $n \in A$ have been imported, as $\bar{n}$, in the language $L(\mathfrak{A})$ of $\mathrm{Th}(\mathfrak{A}_{|\mathfrak{A}|})$.

Thus, if $\mathfrak{C}' = (D, \mathscr{I}_{C'})$ is the reduct of $\mathfrak{C}$ on L – that is, $\mathscr{I}_{C'} = \mathscr{I}_C \restriction L$ – then the expansion $\mathfrak{C}'_{f[A]}$ of $\mathfrak{C}'$ is just $\mathfrak{C}$, hence, $\models_{\mathfrak{C}'_{f[A]}} \mathrm{Th}(\mathfrak{A}_A)$. By the main diagram lemma above,

$$f : \mathfrak{A} \to_{\prec} \mathfrak{C}'   (2)$$

Without loss of generality, we may assume in (2) that $A \subseteq D$ (see Remark I.6.17) and that f is the inclusion map $(\overline{n}^{\mathscr{I}} = \overline{n}^{\mathscr{I}c})$. That is,

$$\mathfrak{A} \prec \mathfrak{C}'$$

Thus, it remains to settle the cardinality claims. First off, since $c \neq d$ in N implies that $\neg \overline{c} = \overline{d}$ is true in $\mathfrak{D}$ by (1), it follows that the cardinality of D is $\geq n$.[†] That is, the cardinality of $\mathfrak{C}'$ (that is, of its domain D) is $\geq n$.

By the downward Löwenheim-Skolem theorem (in the form I.6.21) there is a structure $\mathfrak{B}$ of cardinality exactly n such that $\mathfrak{A} \prec \mathfrak{B} \prec \mathfrak{C}'$. ☐

 Theorem I.6.24 provides different information than its counterpart, I.5.43. The former ignores axioms and works with a language and its structures. In the process it gives a strong type of extension between the given structure and the constructed structure. The latter downplays structures and is about a language, a set of axioms, and the size of the models that we can build for those axioms.

Version I.5.43 has the important consequence that theories over countable languages that have any infinite model at all – such as ZFC – must also have models of any infinite cardinality. In particular they must have countable models.

We conclude the section with a further sampling of applications.

I.6.25 Definition. We call a theory $\mathscr{T}$ *open* iff it is the set of theorems of a set of axioms Γ that consists entirely of *open formulas*, i.e., quantifier-free formulas. Such is, for example, group theory and ROB arithmetic (I.9.32). Such theories are also called *universal*, since one can generate them with a different Γ'. The latter's formulas are all the universal closures of open formulas, those in Γ. ☐

 Thus, by the completeness theorem, a theory $\mathscr{T}$ is open iff there is a set of open formulas, Γ, such that $\mathscr{T}$ and Γ have the same models.

We have the following model-theoretic result for open theories:

I.6.26 Theorem (Łoś-Tarski). *A theory $\mathscr{T}$ (in the sense of p. 38) is open iff every substructure of every model of $\mathscr{T}$ is also a model.*

Proof. Only-if part: Exercise I.53.

If part: All this is over some language L. Let Γ be the set of all open theorems of $\mathscr{T}$ (over L).[‡] We need to show that $\mathscr{T} = \textbf{Thm}_\Gamma$, or, for every structure $\mathfrak{M}$

[†] The function $N \ni n \mapsto \overline{n}^{\mathscr{I}D} \in D$ is total and 1-1.

[‡] According to the definition of a theory on p. 38, in particular the displayed formula (1) there, "open theorems of $\mathscr{T}$" is synonymous with "open formulas *in* $\mathscr{T}$".

for L,

$$\models_{\mathfrak{M}} \Gamma \quad \text{implies} \quad \models_{\mathfrak{M}} \mathcal{T} \tag{1}$$

To this end, we assume the hypothesis in (1) and consider $D(\mathfrak{M}) \cup \mathcal{T}$. We claim that this is a consistent set of formulas over $L(\mathfrak{M})$. Otherwise $\mathcal{T} \cup \{\mathscr{A}_1, \ldots, \mathscr{A}_n\} \vdash \neg x = x$, where $\mathscr{A}_1, \ldots, \mathscr{A}_n$ is a finite set of sentences of $D(\mathfrak{M})$. By the deduction theorem

$$\mathcal{T} \vdash \neg x = x \vee \neg \mathscr{A}_1 \vee \cdots \vee \neg \mathscr{A}_n$$

Hence (tautological implication and logical axiom $x = x$)

$$\mathcal{T} \vdash \neg \mathscr{A}_1 \vee \cdots \vee \neg \mathscr{A}_n \tag{2}$$

By the theorem on constants, (I.4.15) – since $\mathcal{T}$ is over L – (2) yields

$$\mathcal{T} \vdash \neg \overline{\mathscr{A}}_1 \vee \cdots \vee \neg \overline{\mathscr{A}}_n$$

where $\neg \overline{\mathscr{A}}_1 \vee \cdots \vee \neg \overline{\mathscr{A}}_n$ is obtained from $\neg \mathscr{A}_1 \vee \cdots \vee \neg \mathscr{A}_n$ by replacing all the imported constants $\bar{i}, \bar{j}, \ldots$ which appear in the latter formula by distinct *new* variables. Thus, $\neg \overline{\mathscr{A}}_1 \vee \cdots \vee \neg \overline{\mathscr{A}}_n$ is in Γ hence is valid in $\mathfrak{M}$. That is, assuming the formula's variables are among $\vec{y}_r$, for all $\vec{a}_r$ in $|\mathfrak{M}|$,

$$\models_{\mathfrak{M}} (\neg \overline{\mathscr{A}}_1 \vee \cdots \vee \neg \overline{\mathscr{A}}_n)[\![\vec{a}_r]\!] \tag{3}$$

We can thus choose the $\vec{a}_r$ so that $\mathscr{A}_j \equiv (\overline{\mathscr{A}}_j)[\![\vec{a}_r]\!]$ for $j = 1, \ldots, n$. Since (3) entails that $(\overline{\mathscr{A}}_j)[\![\vec{a}_r]\!]$ is false[†] in $\mathfrak{M}_{|\mathfrak{M}|}$ for at least one j, this contradicts that all the $\mathscr{A}_j$ are in $D(\mathfrak{M})$.

Thus, by the consistency theorem, there is a model $\mathfrak{K}$ for $D(\mathfrak{M}) \cup \mathcal{T}$. Without loss of generality (cf. Remark I.6.17), the mapping $\bar{i}^{\mathfrak{M}} \mapsto \bar{i}^{\mathfrak{K}}$ is the identity. If we now call $\mathfrak{A}$ the reduct of $\mathfrak{K}$ on L, then the expansion $\mathfrak{A}_{|\mathfrak{M}|}$ of $\mathfrak{A}$ is $\mathfrak{K}$; therefore $\models_{\mathfrak{A}_{|\mathfrak{M}|}} D(\mathfrak{M})$. The diagram lemma implies that $\mathfrak{M} \subseteq \mathfrak{A}$. Since $\mathfrak{A}$ is a model of $\mathcal{T}$, then so is $\mathfrak{M}$ by hypothesis. This settles (1). □

I.6.27 Example. Group theory is usually formulated in a language that employs the nonlogical symbols "∘" (binary function), "$^{-1}$" (unary function) and "1" (constant). Its axioms are open, the following:

(1) $x \circ (y \circ z) = (x \circ y) \circ z$[‡];
(2) $x \circ 1 = 1 \circ x = x$ (using "=" conjunctionally);
(3) $x \circ x^{-1} = x^{-1} \circ x = 1$.

---

† Cf. discussion on p. 89.
‡ Where we are using the habitual infix notation, "$x \circ y$", rather than $\circ xy$ or $\circ (x, y)$.

The models of the above axioms are called *groups*. Over this language, every substructure of a group, that is, every subset that includes (the interpretation of) 1 and is *closed under* ∘ and $^{-1}$, is itself a group, as the only-if direction of the above theorem guarantees.

It is possible to formulate group theory in a language that has only "∘" and "1" as nonlogical symbols. Its axioms are not open:

(i) $x \circ (y \circ z) = (x \circ y) \circ z$;
(ii) $x \circ 1 = 1 \circ x = x$;
(iii) $(\exists y)x \circ y = 1$;
(iv) $(\exists y)y \circ x = 1$.

Since the Łoś-Tarski theorem is an "iff" theorem, it must be that group theory so formulated has models (still called groups) that have substructures that are not groups (they are *semigroups with unit*, however). □

I.6.28 Definition. Constructions in model theory often result in increasing sequences of structures, either of the type

$$\mathfrak{M}_0 \subseteq \mathfrak{M}_1 \subseteq \mathfrak{M}_2 \subseteq \cdots \tag{1}$$

or of the type

$$\mathfrak{M}_0 \prec \mathfrak{M}_1 \prec \mathfrak{M}_2 \prec \cdots \tag{2}$$

Sequences of type (1) are called *increasing chains* or *ascending chains* of structures. Those of type (2) are called *elementary chains*. □

Chains are related to theories that can be axiomatized using exclusively *existential axioms*, that is, axioms of the form $(\exists x_1) \ldots (\exists x_n)\mathscr{A}$, for $n \geq 0,$[†] where $\mathscr{A}$ is open. Such theories are called *inductive*. By taking the universal closures of existential axioms we obtain formulas of the form $(\forall y_1) \ldots (\forall y_m)(\exists x_1) \ldots (\exists x_n)\mathscr{A}$. For this reason, inductive theories are also called ∀∃-theories.

The following result is easy to prove, and is left as an exercise.

I.6.29 Theorem (Tarski's Lemma). *The union $\mathfrak{M}$ of an elementary chain (2) above is an elementary extension of every $\mathfrak{M}_i$.*

[†] $n = 0$ means that the prefix $(\exists x_1) \ldots (\exists x_n)$ is missing. Thus an open formula is a special case of an existential one. As a matter of fact, we can force a nonempty existential prefix on an open formula: If z does not occur in $\mathscr{A}$, then we can be prove $\mathscr{A} \leftrightarrow (\exists z)\mathscr{A}$ without nonlogical axioms.

Proof. First off, a chain of type (2) is also of type (1); thus, $\mathfrak{M}_i \subseteq \mathfrak{M}_{i+1}$ for all $i \geq 0$. Let M and M_i denote the universes of $\mathfrak{M}$ and $\mathfrak{M}_i$ respectively. Then by "union of the $\mathfrak{M}_i$" we understand that we have taken $M = \bigcup_{i \geq 0} M_i$ and

(i) $P^{\mathfrak{M}} = \bigcup_{i \geq 0} P^{\mathfrak{M}_i}$, for each predicate P,

(ii) $f^{\mathfrak{M}} = \bigcup_{i \geq 0} f^{\mathfrak{M}_i}$ for each function f (note that $f^{\mathfrak{M}_i} \subseteq f^{\mathfrak{M}_{i+1}}$ for all $i \geq 0$), and

(iii) $a^{\mathfrak{M}} = a^{\mathfrak{M}_0}$, for each constant.

The real work is delegated to Exercise I.54.   □

I.6.30 Theorem (**Chang-Łoś-Suszko**). *The union of every type-(1) chain (I.6.28) of models of a theory $\mathscr{T}$ (the latter in the sense of p. 38) is a model of $\mathscr{T}$ iff the theory is inductive.*

Proof. The proof is standard (e.g., Chang and Keisler (1973), Keisler (1978), Shoenfied (1967)).

If part: Delegated to Exercise I.55.

Only-if part: Assume the hypothesis for a theory $\mathscr{T}$ over L. We will prove that it is inductive. To this end, let Γ by the set of all existential consequences of $\mathscr{T}$ – i.e., formulas $(\exists x_1) \ldots (\exists x_n) \mathscr{A} - n \geq 0$ – with $\mathscr{A}$ open. As in the proof of I.6.26, we endeavour to prove

$$\models_{\mathfrak{M}} \Gamma \quad \text{implies} \quad \models_{\mathfrak{M}} \mathscr{T} \tag{1}$$

Assume the hypothesis in (1), and let $D_{\forall}(\mathfrak{M})$ denote the set of all *universal sentences*[†] over $L(\mathfrak{M})$ that are true in $\mathfrak{M}_M$, where we have written M for $|\mathfrak{M}|$.

We argue that $D_{\forall}(\mathfrak{M}) \cup \mathscr{T}$ is consistent. If not, $\mathscr{T} \vdash \neg x = x \vee \neg \mathscr{A}_1 \vee \cdots \vee \neg \mathscr{A}_n$, for some $\mathscr{A}_i \in D_{\forall}(\mathfrak{M})$; hence (notation as in the proof of I.6.26)

$$\mathscr{T} \vdash \neg(\overline{\mathscr{A}}_1 \wedge \cdots \wedge \overline{\mathscr{A}}_n)$$

Since $\vdash \mathscr{A} \leftrightarrow (\forall z)\mathscr{A}$ if z is not free in $\mathscr{A}$, and $\forall$ distributes over $\wedge$ (see Exercise I.23), we can rewrite the above as

$$\mathscr{T} \vdash \neg(\forall \vec{x}_r)(\mathscr{B}_1 \wedge \cdots \wedge \mathscr{B}_n) \tag{2}$$

where the $\mathscr{B}_i$ are open over L, and $(\forall \vec{x}_r)$ is short for $(\forall x_1) \ldots (\forall x_r)$. The formula in (2) is (logically equivalent to one that is) existential. Thus, it is in Γ, and hence is true in $\mathfrak{M}$, an impossibility, since no $\mathscr{A}_i \in D_{\forall}(\mathfrak{M})$ can be false in $\mathfrak{M}_M$.

[†] A universal formula has the form $(\forall x_1) \ldots (\forall x_n) \mathscr{A}$, $n \geq 0$, where $\mathscr{A}$ is open. Thus $(n = 0)$ every open formula is also universal, as well as existential.

We can now have a model, $\mathfrak{K}' = (K, \ldots, (a)_{a \in M})$, of $D_\forall(\mathfrak{M}) \cup \mathcal{T}$, where "$(a)_{a \in M}$" indicates the special status of the constants $a \in M \subseteq K$: These have been added to L, as $\bar{a}$, to form $L(\mathfrak{M})$. If $\mathfrak{K} = (K, \ldots)$ is the reduct of $\mathfrak{K}'$ to L, then $\mathfrak{K}' = \mathfrak{K}_M$; hence

$$\models_{\mathfrak{K}_M} D_\forall(\mathfrak{M}) \tag{3}$$

Since $D(\mathfrak{M}) \subseteq D_\forall(\mathfrak{M})$, we obtain

$$\mathfrak{M} \subseteq \mathfrak{K} \quad (\text{I.6.23}), \quad \text{and} \quad \models_{\mathfrak{K}} \mathcal{T} \quad (\mathcal{T} \text{ is over } L) \tag{4}$$

Remark I.6.17 was invoked without notice in assuming that the $\bar{a}$ of $L(\mathfrak{M})$ are interpreted in $\mathfrak{K}'$ exactly as in $\mathfrak{M}_M$: as a.

We next argue that the theory $\mathcal{S} = D(\mathfrak{K}) \cup \text{Th}(\mathfrak{M}_M)$ over[†] $L(K)$ is consistent, where $D(\mathfrak{K})$ is as in I.6.22, i.e.,the set of all atomic and negated atomic sentences over $L(K)$ that are true in $\mathfrak{K}_K = (K, \ldots, (a)_{a \in M}, (a)_{a \in K-M})$. Note that $L(K)$ is obtained by *adding* to $L(M)$ the constants $(\bar{a})_{a \in K - M}$, each $\bar{a}$ interpreted as a in $\mathfrak{K}_K$. That is, $a \in M$ were not re-imported as something other than the original $\bar{a}$ of $L(M)$.

Now, if $\mathcal{S}$ is not consistent, then some $\mathcal{A}_i \in D(\mathfrak{K})$ ($i = 1, \ldots, n$) jointly with $\text{Th}(\mathfrak{M}_M)$ prove $\neg x = x$. Since the $\mathcal{A}_i$ are sentences (over $L(K)$), the usual technique yields

$$\text{Th}(\mathfrak{M}_M) \vdash \neg \mathcal{A}_1 \vee \cdots \vee \neg \mathcal{A}_n$$

Let $\neg \mathcal{A}'_1 \vee \cdots \vee \neg \mathcal{A}'_n$ be obtained from $\neg \mathcal{A}_1 \vee \cdots \vee \neg \mathcal{A}_n$ by replacing all its constants $\bar{a}, \bar{b}, \ldots$ – where $a, b, \ldots$ are in $K - M$ – by distinct new variables $\vec{x}_r$. Thus (I.4.15)

$$\text{Th}(\mathfrak{M}_M) \vdash (\forall \vec{x}_r)(\neg \mathcal{A}'_1 \vee \cdots \vee \neg \mathcal{A}'_n) \tag{5}$$

Since $(\forall \vec{x}_r)(\neg \mathcal{A}'_1 \vee \cdots \vee \neg \mathcal{A}'_n)$ is a sentence over $L(M)$, (5) implies that it is in $\text{Th}(\mathfrak{M}_M)$, i.e., true in $\mathfrak{M}_M$. Thus, being universal, it is in $D_\forall(\mathfrak{M})$. By (3) it is true in $\mathfrak{K}_M$. In particular, its $\mathfrak{K}_M$-instance (cf. I.5.7) $\neg \mathcal{A}_1 \vee \cdots \vee \neg \mathcal{A}_n$ – a sentence of $L(K)$ – is true in $\mathfrak{K}_M$ in the sense of I.5.7, and hence also in $\mathfrak{K}_K$. This is impossible, since $\mathcal{A}_i \in D(\mathfrak{K})$ implies that no sentence $\mathcal{A}_i$ can be false in $\mathfrak{K}_K$.

Thus, there is a model $\mathfrak{A} = (A, \ldots)$ of $\mathcal{S}$ over $L(K)$. Now, on one hand we have

$$\models_{\mathfrak{A}} D(\mathfrak{K}) \tag{6}$$

[†] The part $\text{Th}(\mathfrak{M}_M)$ is over $L(M)$, of course.

and we may assume, without loss of generality (cf. I.6.17), that

$$\text{for } a \in K, \quad \bar{a}^{\mathfrak{K}} \mapsto \bar{a}^{\mathfrak{A}} \text{ is the inclusion map} \tag{7}$$

Therefore, $K \subseteq A$, and we may write $\mathfrak{A}$ a bit more explicitly as

$$\mathfrak{A} = (A, \ldots, (a)_{a \in M}, (a)_{a \in K - M}) \tag{8}$$

Let us call $\mathfrak{M}' = (A, \ldots)$ the reduct of $\mathfrak{A}$ on L. By (8), $\mathfrak{A} = \mathfrak{M}'_K$, so that (6) reads $\models_{\mathfrak{M}'_K} D(\mathfrak{K})$; hence (by I.6.23)

$$\mathfrak{K} \subseteq \mathfrak{M}' \tag{9}$$

On the other hand, every sentence of $\mathrm{Th}(\mathfrak{M}_M)$ – a subtheory of $\mathscr{S}$ – is true in $\mathfrak{A}$. The relevant part of $\mathfrak{A}$ is its reduct on $L(M)$, namely, $\mathfrak{M}'_M$. Thus, $\models_{\mathfrak{M}'_M} \mathrm{Th}(\mathfrak{M}_M)$, and therefore

$$\mathfrak{M} \prec \mathfrak{M}' \tag{10}$$

After all this work we concentrate on the core of what we have obtained: (4), (9), and (10). By (10) and remarks in I.6.16 we have that $\models_{\mathfrak{M}'} \Gamma$. Thus we can repeat our construction all over again, using $\mathfrak{M}'$ in the place of $\mathfrak{M}$, to obtain $\mathfrak{K}' \supseteq \mathfrak{M}'$ with $\models_{\mathfrak{K}'} \mathscr{S}$, and also a $\mathfrak{M}''$ that is an elementary extension of $\mathfrak{M}'$.

In short, we can build, by induction on n, an *alternating chain*

$$\mathfrak{M}_0 \subseteq \mathfrak{K}_0 \subseteq \mathfrak{M}_1 \subseteq \mathfrak{K}_1 \subseteq \cdots \tag{11}$$

such that

$$\mathfrak{M}_0 \prec \mathfrak{M}_1 \prec \cdots \tag{12}$$

and

$$\models_{\mathfrak{K}_n} \mathscr{S} \quad \text{for} \quad n \geq 0 \tag{13}$$

where $\mathfrak{M}_0 = \mathfrak{M}$, and, for $n \geq 0$, $\mathfrak{K}_n$ and $\mathfrak{M}_{n+1}$ are obtained as "$\mathfrak{K}$" and "$\mathfrak{M}'$" respectively from $\mathfrak{M}_n$, the latter posing as the "$\mathfrak{M}$" in the above construction. By assumption and (13), the union of the alternating chain (11) is a model, $\mathfrak{B}$, of $\mathscr{S}$. Since $\mathfrak{B}$ also equals the union of the chain (12), Tarski's lemma (I.6.29) yields $\mathfrak{M} \prec \mathfrak{B}$. Hence $\models_{\mathfrak{M}} \mathscr{S}$, proving (1). $\qquad\square$

Some issues on the cardinality of models are addressed by the upward and downward Löwenheim-Skolem theorems. Here we sample the connection between possible uniqueness of models of a given cardinality and completeness.

As before, a theory $\mathscr{T}$ over some language L is (simply) *complete* iff for all sentences $\mathscr{A}$, one of $\mathscr{T} \vdash \mathscr{A}$ or $\mathscr{T} \vdash \neg \mathscr{A}$ holds. If $\mathscr{T}$ is in the sense of p. 38, then it is complete iff for all sentences $\mathscr{A}$, one of $\mathscr{A} \in \mathscr{T}$ or $(\neg \mathscr{A}) \in \mathscr{T}$ holds.

I.6.31 Definition. A theory $\mathscr{T}$ over a language L is κ-*categorical* (where $\kappa \geq$ cardinality of L) iff any two models of $\mathscr{T}$ of cardinality κ are isomorphic. In essence, there is only one model of cardinality κ. $\square$

I.6.32 Theorem (Łoś-Vaught Test). *If $\mathscr{T}$ over L has no finite models and is κ-categorical for some κ, then it is complete.*

Proof. The hypothesis is vacuously satisfied by an inconsistent theory. If $\mathscr{T}$ is inconsistent, then it is complete. Let it be consistent then, and assume that there is some sentence $\mathscr{A}$ that is neither provable nor refutable. Then both $\mathscr{T} \cup \{\mathscr{A}\}$ and $\mathscr{T} \cup \{\neg \mathscr{A}\}$ are consistent. Let $\mathfrak{A}$ and $\mathfrak{B}$ be respective models. By the upward Löwenheim-Skolem theorem there are structures $\mathfrak{A}'$ and $\mathfrak{B}'$ of cardinality κ each such that $\mathfrak{A} \prec \mathfrak{A}'$ and $\mathfrak{B} \prec \mathfrak{B}'$; hence (by assumption and remarks in I.6.16)

$$\mathfrak{A} \equiv \mathfrak{A}' \cong \mathfrak{B}' \equiv \mathfrak{B}$$

Thus $\mathscr{A}$ is true in $\mathfrak{B}$ (since it is so in $\mathfrak{A}$), which is absurd. $\square$

Our last application is perhaps the most startling. It is the rigorous introduction of Leibniz's *infinitesimals*, that is, "quantities" that are non-zero,[†] yet their absolute values are less than every positive real number. Infinitesimals form the core of a user-friendly (sans ε-δ, that is) introduction to limits and the differential and integral calculus – in the manner that Leibniz conceived calculus (but he was never able to prove that infinitesimals "existed"). Such approaches to learning calculus (sometimes under the title *non-standard analysis*) have been accessible to the undergraduate student for quite some time now (e.g., Henle and Kleinberg (1980); see also Keisler (1976)). The legitimization of infinitesimals was first announced in 1961 by Abraham Robinson (1961, 1966). Besides making calculus more "natural", non-standard analysis has also been responsible for the discovery of new results.

All that we need to do now is to extend the standard structure for the reals, $\mathfrak{R} = (\mathbb{R}, \dots)$ – where "$\dots$" includes "$+$", "$\times$", "$<$", "0", "1", etc. – so that it is enriched with "new" numbers, including infinitesimals. Given the tools at our disposal, this will be surprisingly easy:

[†] Actually, 0 *is* an infinitesimal, as it will turn out. But there are uncountably many that are non-zero.

Thus, we start by extending the first order language for the reals, by importing all constants, all functions, and all relations over $\mathbb{R}$ into the language as formal constants, functions, and predicates respectively.

 There are two issues we want to be clear about.

(1) In the exposition that follows we will almost exclusively use (an *argot* of) the *formal* notation for the extended language of the reals – and, infrequently, informal notations for concrete objects (e.g., reals). We are thus best advised to keep the *formal* language notation uncomplicated, and complicate – if we must – only the notation for the *concrete* objects. Thus, e.g., we will denote by "$<$" the formal, and by "$^{\circ}<$" the concrete, predicate on the reals; by "$\sqrt{2}$" the formal, and by "$^{\circ}\sqrt{2}$" the concrete constant.[†] In short, we import a, f, P for all informal $^{\circ}a \in \mathbb{R}$, for all functions $^{\circ}f$ with inputs and outputs in $\mathbb{R}$, and for all relations $^{\circ}P$ on $\mathbb{R}$. Having done that, we interpret as follows: $a^{\mathfrak{R}} = {}^{\circ}a$, $f^{\mathfrak{R}} = {}^{\circ}f$, and $P^{\mathfrak{R}} = {}^{\circ}P$. We will call the language so obtained, simply, L. Due to lack of imagination, we will call the expansion of the original $\mathfrak{R}$ still $\mathfrak{R}$.[‡] Thus

$$\mathfrak{R} = (\mathbb{R}, \ldots, (^{\circ}a : {}^{\circ}a \in \mathbb{R}), (^{\circ}f : {}^{\circ}f \text{ on } \mathbb{R}), (^{\circ}P : {}^{\circ}P \text{ on } \mathbb{R})) \qquad (i)$$

(2) Not all functions on $\mathbb{R}$ are *total* (i.e., everywhere defined). For example, the function $x \mapsto \sqrt{x}$ is undefined for $x < 0$. This creates a minor annoyance because functions of a first order language are supposed to be interpreted as (i.e., are supposed to name) total functions. We get around this by simply *not* importing (names for) nontotal functions. However, we do import their *graphs*,[§] for these are just relations. All we have to do is to invent appropriate notation for the formal name of such graphs. For example, suppose we want to formally name "$y = {}^{\circ}\sqrt{x}$". We just use $y = \sqrt{x}$. Thus, the formal (binary) predicate name is the compound symbol "$= \sqrt{}$", which we employ in infix notation when building an atomic formula from it. See also I.6.37 later in this section (p. 101).

With the above out of the way, we fix a cardinal $\mathfrak{m}$ that is bigger than the maximum of the cardinal of L and $\mathbb{R}$. By the upward Löwenheim-Skolem

---

[†] There is nothing odd about this. Both "$\sqrt{2}$" and "$^{\circ}\sqrt{2}$" are names. Unlike what we do in normal use, we pretend here that the former is the formal name, but the latter is the informal name for the object called the "square root of two".

[‡] No confusion should arise from this, for neither the original bare bones $\mathfrak{R}$ nor its language – which we haven't named – will be of any future use.

[§] The graph of an n-ary function $^{\circ}f$ is the relation $\{\langle y, \vec{x}_n \rangle : y = {}^{\circ}f(\vec{x}_n) \text{ is true in } \mathbb{R}\}$, or simply, $y = {}^{\circ}f(\vec{x}_n)$.

theorem there is a structure

$$^*\mathfrak{R} = (^*\mathbb{R}, \dots, (^*a : {}^\circ a \in \mathbb{R}), (^*f : {}^\circ f \text{ on } \mathbb{R}), (^*P : {}^\circ P \text{ on } \mathbb{R})) \qquad (ii)$$

for L such that

$$\mathfrak{R} \prec {}^*\mathfrak{R} \qquad (1)$$

and

$$(\text{cardinality of } {}^*\mathbb{R}) = \mathfrak{m} \qquad (2)$$

In particular (from (1) and (2)),

$$\mathbb{R} \subset {}^*\mathbb{R} \qquad (3)$$

I.6.33 Definition. The members of $^*\mathbb{R}$ are called the *hyperreals* or *hyperreal numbers*. By (3), all reals are also hyperreals. The members of $^*\mathbb{R} - \mathbb{R}$ – a nonempty set by (3) – are the *non-standard numbers* or *non-standard hyperreals*. The members of $\mathbb{R}$ are the *standard numbers* or *standard hyperreals*.   □

(1) says that a first order sentence is true in $\mathfrak{R}$ iff it is true in $^*\mathfrak{R}$. This is neat. It gives us a practical *transfer principle* (as it is called in Keisler (1982), although in a different formulation): To verify a first order sentence in $^*\mathfrak{R}$ (respectively, in $\mathfrak{R}$) one only needs to verify it in $\mathfrak{R}$ (respectively, in $^*\mathfrak{R}$).

By (1) (which implies $\mathfrak{R} \subseteq {}^*\mathfrak{R}$) we also have

$$^\circ a = {}^*a \qquad \text{for all} \quad {}^\circ a \in \mathbb{R} \qquad (4)$$

$$^\circ f \subseteq {}^*f \qquad \text{for all} \quad {}^\circ f \text{ on } \mathbb{R} \qquad (5)$$

and

$$^\circ P \subseteq {}^*P \qquad \text{for all} \quad {}^\circ P \text{ on } \mathbb{R} \qquad (6)$$

I.6.34 Example. Here is an example of how (4) and (6) apply:

$$(\forall x)(x < 0 \rightarrow \neg(\exists y)(y = \sqrt{x})) \qquad (7)$$

is true in $\mathfrak{R}$. Hence it is true in $^*\mathfrak{R}$. Moreover, $(-6) < 0$ is true in $^*\mathfrak{R}$, since it is true in $\mathfrak{R}$.

Pause. We have used *formal* notation in writing the two previous formulas.

By (7) and specialization, $(-6) < 0 \rightarrow \neg(\exists y)(y = \sqrt{(-6)})$. Thus,

$$\neg(\exists y)(y = \sqrt{(-6)})$$

is true in *$\mathfrak{R}$. *Concretely* this says that if we are working in the set *$\mathbb{R}$, then (using $x \mapsto {}^*\sqrt{x}$ for the concrete function on *$\mathbb{R}$, and also using (4))

$$^*\sqrt{^\circ(-6)} \text{ is undefined}$$

So, nontotal functions continue (i.e., their extensions on *$\mathbb{R}$ do) being undefined on the *same reals* on which they failed to be defined on $\mathbb{R}$.

The reader can similarly verify that *$\mathbb{R}$ will *not* forgive division by zero, using, for example, the formal sentence $\neg(\exists y)(y \cdot 0 = 1)$.   □

I.6.35 Example. Now start with the sentence (written formally!) $\sin(\pi) = 0$. This is true in $\mathfrak{R}$, and hence in *$\mathfrak{R}$. A more messy way to say this is to say that by (5) (note that $^\circ\sin$ is total; hence the name sin was imported), $^\circ\sin \subseteq {}^*\sin$, and hence (cf. (4)) $^*\sin(^*\pi) = {}^*\sin(^\circ\pi) = {}^\circ\sin(^\circ\pi) = {}^\circ 0 = {}^*0$. Thus, $\sin(\pi) = 0$, the formal counterpart of $^*\sin(^*\pi) = {}^*0$, is true in *$\mathfrak{R}$.   □

Elementary arithmetic on $\mathbb{R}$ carries over on *$\mathbb{R}$ without change. Below is a representative sample.

I.6.36 Example. First off, $^* <$ is a total order on *$\mathbb{R}$. That is, the following three sentences are true in *$\mathfrak{R}$ (because they are so in $\mathfrak{R}$):

$$(\forall x)(\neg x < x)$$

$$(\forall x)(\forall y)(x < y \lor y < x \lor x = y)$$

and

$$(\forall x)(\forall y)(\forall z)(x < y \land y < z \rightarrow x < z)$$

The formal symbol $\leq$, as usual, is introduced by a definition

$$x \leq y \leftrightarrow x = y \lor x < y$$

Can we add inequalities, term by term, in *$\mathbb{R}$? Of course, since

$$(\forall x)(\forall y)(\forall z)(\forall w)(x < y \land z < w \rightarrow x + z < y + w)$$

is true in $\mathfrak{R}$ therefore in *$\mathfrak{R}$ as well.

The above has all sorts of obvious variations using, selectively, $\leq$ instead of $<$. Multiplication, term by term, goes through in *$\mathfrak{R}$ under the usual caveat:

$$(\forall x)(\forall y)(\forall z)(\forall w)(0 \leq x < y \land 0 \leq z < w \rightarrow xz < yw)$$

Two more useful inequalities are

$$(\forall x)(\forall y)(\forall z)(z < 0 \wedge x < y \rightarrow zx > zy)$$

and

$$(\forall x)(\forall y)((0 < x \wedge x < 1/y) \leftrightarrow (0 < y \wedge y < 1/x))$$

The formal function $|\ldots|$ (absolute value) enjoys the usual properties in $^*\mathfrak{R}$ because it does so in $\mathfrak{R}$. Here is a useful sample:

$$(\forall x)(\forall y)(|x + y| \leq |x| + |y|)$$

$$(\forall x)(\forall y)(|x - y| \leq |x| + |y|)$$

$$(\forall x)(\forall y)(|x| \leq y \leftrightarrow -y \leq x \leq y)$$

Back to equalities: $+$ and $\times$ are, of course, commutative on $^*\mathbb{R}$, for $(\forall x)(\forall y)(x + y = y + x)$ is true in $\mathfrak{R}$. Moreover,

$$(\forall x)(\forall y)(|xy| = |x||y|)$$

where we have used "implied $\times$" above. Finally, the following are true in $^*\mathfrak{R}$: $(\forall x)(0x = 0)$, $(\forall x)(1x = x)$, and $(\forall x)((-1)x = -x)$. $\square$

I.6.37 Remark (Importing Functions: the Last Word). We can now conclude our discussion, on p. 98, about importing nontotal functions. Let then $^\circ f$ be an n-ary nontotal function on $\mathbb{R}$. The graph, $y = {}^\circ f(\vec{x}_n)$ is imported as the formal predicate "$(= f)$" – we write, formally, "$y(= f)\vec{x}_n$" rather than "$y = f\vec{x}_n$" or "$y = f(\vec{x}_n)$".[†] Let $y = {}^*f(\vec{x}_n)$ be the extension of $y = {}^\circ f(\vec{x}_n)$ on $^*\mathbb{R}$ (cf. (6), p. 99). We note two things:

One, for any $b, \vec{a}_n$ in $\mathbb{R}$, if $b = {}^\circ f(\vec{a}_n)$, then $b = {}^*f(\vec{a}_n)$ by (6) and (4) (p. 99).

Two, since

$$(\forall x_1)\ldots(\forall x_n)(\forall y)(\forall z)(y(= f)\vec{x}_n \wedge z(= f)\vec{x}_n \rightarrow y = z)$$

is true in $\mathfrak{R}$ by the assumption that $^\circ f$ is a function – i.e., that the relation $y = {}^\circ f(\vec{x}_n)$ is single-valued in y – it is true in $^*\mathfrak{R}$ as well, so that the concrete relation $y = {}^*f(\vec{x}_n)$ is single-valued in y. That is, $^*f$ is a function. Along with remark one, above, this yields $^\circ f \subseteq {}^*f$. This is the counterpart of (5) (p. 99)

[†] This avoids a possible misunderstanding that, in something like $y = f\vec{x}_n$, f is an n-ary function letter and hence $f\vec{x}_n$ is a term. f is *part* of a language symbol, namely, of the predicate "$(= f)$". It is not a symbol itself.

for nontotal functions, and we are pleased to note that it has exactly the *form* of (5), without any caveats.

Pause. Is $^*f$ total on $^*\mathbb{R}$?

We can now pretend, in practice (i.e., in *argot*) that *all* functions of $\mathbb{R}$, total or not, have been imported and thus have extensions in $^*\mathbb{R}$.   □

We next explore $^*\mathfrak{R}$ looking for strange numbers, such as infinitesimals and infinite numbers. To ensure that we know what we are looking for, we define:

I.6.38 Definition (Infinitesimals). An *infinitesimal h* is a member of $^*\mathbb{R}$ such that $|h| < x$ is true in $^*\mathfrak{R}$ for *all* positive $x \in \mathbb{R}$. A member h of $^*\mathbb{R}$ is a *finite hyperreal* iff $|h| \leq x$ is true in $^*\mathfrak{R}$ for *some* positive $x \in \mathbb{R}$. A member h of $^*\mathbb{R}$ is an *infinite hyperreal* iff it is not finite. That is, $|h| > x$ is true in $^*\mathfrak{R}$ for all positive $x \in \mathbb{R}$.   □

I.6.39 Remark. Thus, 0 is an infinitesimal, and every infinitesimal is finite.

The reader has surely noticed that we have dropped the annoying left-superscripting of members of $^*\mathbb{R}$ (or of $\mathbb{R}$). This is partly because of (4) (p. 99). Since $^{\circ}a = {}^*a$ for all $^{\circ}a \in \mathbb{R}$, it is smart to name both by the simpler *formal name*, a. Moreover, since the left-superscript notation originates in the structure $\mathfrak{R}$, it is not applicable to objects of $^*\mathfrak{R}$ *unless* these are extensions (functions or predicates) of objects of $\mathfrak{R}$. It is thus pointless to write "$^*h$".

Even in the cases of functions and predicates, we can usually get away without superscripts, letting the context indicate whether we are in $\mathbb{R}$, in $^*\mathbb{R}$, or in the formal language. For example we have used "$|h|$" rather than "$^*|h|$", but then we are clear that it is the latter that we are talking about, as the absolute value here is applicable to non-standard inputs.   □

I.6.40 Proposition. $h \in {}^*\mathbb{R}$ *is a nonzero infinitesimal iff* $1/|h|$ *is infinite.*

Proof. *If part:* Suppose that $1/|h|$ is infinite, and let $0 < r \in \mathbb{R}$ be arbitrary. Then $1/r < 1/|h|$. Specialization of the appropriate inequality in I.6.36 (a true fact in $^*\mathfrak{R}$) yields $|h| < r$.

Only-if part: Let $0 < r \in \mathbb{R}$ be arbitrary. By hypothesis, $|h| < 1/r$. By the fact invoked above, $r < 1/|h|$.   □

I.6.41 Proposition. *Let h and h$'$ be infinitesimals. Then so are*

(1) $h + h'$,
(2) hh',
(3) hr *for any* $r \in \mathbb{R}$.

Proof. (1): Let $0 < r \in \mathbb{R}$ be arbitrary. Then $|h + h'| \leq |h| + |h'| < r/2 + r/2 < r$ is true in *$\mathfrak{R}$.

(2): Let $0 < r \in \mathbb{R}$ be arbitrary. Then $|hh'| = |h||h'| < r1 = r$ is true in *$\mathfrak{R}$.

(3): If $r = 0$, then $h0 = 0$ (cf. I.6.36), an infinitesimal. Otherwise, let $0 < s \in \mathbb{R}$ be arbitrary. $|h| < s/|r|$ by hypothesis; hence $|hr| = |h||r| < s$ (the reader can easily verify that we have used legitimate *$\mathbb{R}$-arithmetic). □

I.6.42 Proposition. *Let h and h' be infinite hyperreals. Then so are*

(1) hh',
(2) hr for any $0 \neq r \in \mathbb{R}$,
(3) $h + r$ for any $r \in \mathbb{R}$.

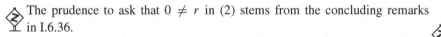

The prudence to ask that $0 \neq r$ in (2) stems from the concluding remarks in I.6.36.

Proof. (1): Let $0 < r \in \mathbb{R}$ be arbitrary. Then $|hh'| = |h||h'| > r1 = r$ is true in *$\mathfrak{R}$.

(2): Let $0 < s \in \mathbb{R}$ be arbitrary. $|h| > s/|r|$ by hypothesis; hence $|hr| = |h||r| > s$.

(3): Let $0 < s \in \mathbb{R}$. Now, $|h + r| \geq |h| - |r|$.

Pause. Do you believe this?

Hence, $s < |h + r|$, since $s + |r| < |h|$. □

I.6.43 Example. We observe the phenomenon of indeterminate forms familiar from elementary calculus. There we use the following symbols:

(i) *Form $\infty - \infty$.* This translates into "there is no fixed rule for what $h - h'$ will yield" (both positive infinite). For example, if $h = 2h'$ (certainly infinite, by (2)), then $h - h'$ is infinite. If a is an infinitesimal, then $h' + a$ is infinite.

   Pause. Is that true?

   Hence, if $h = h' + a$, then $h - h' = a$ is an infinitesimal. In particular, it is finite.

(ii) *Form ∞/∞.* There are three different outcomes for h/h', according as $h = h'$, $h = (h')^2$ (see (1) above), or $h' = h^2$.

(iii) *Form $0/0$.* This translates to the question "what is the result of h/h', if both are infinitesimal?" It depends: Typical cases are $h = h'$, $h = (h')^2$, and $h' = h^2$. □

The following terminology and notation are useful. They are at the heart of the limiting processes.

I.6.44 Definition. We say that two hyperreals a and b are *infinitely close*, in symbols $a \approx b$, iff $a - b$ is an infinitesimal.

Thus, in particular (since $a - 0 = a$ is true), $a \approx 0$ says that a is an infinitesimal.   □

I.6.45 Proposition. $\approx$ *is an equivalence relation on* $^*\mathbb{R}$*. That is, for all* x*,* y*,* z*, it satisfies*

(1) $x \approx x$,
(2) $x \approx y \to y \approx x$,
(3) $x \approx y \approx z \to x \approx z$.

Proof. Exercise I.59.   □

I.6.46 Proposition. *If* $r \approx s$ *and* r *and* s *are real, then* $r = s$.

Proof. $r - s \approx 0$ and $r - s \in \mathbb{R}$. But, trivially, 0 is the only real infinitesimal.   □

I.6.47 Theorem (Main Theorem). *For any finite non-standard number* h *there is a unique real* r *such that* $h \approx r$.

Throughout the following proof we use superscriptless notation. In each case it is clear where we are: In L, in $\mathfrak{R}$, or in $^*\mathfrak{R}$.

Proof. Uniqueness is by I.6.45 and I.6.46. For the existence part let $|h| \leq b$, $0 < b \in \mathbb{R}$, and define

$$H = \{x \in \mathbb{R} : x < h\} \tag{1}$$

H, a subset of $\mathbb{R}$, is bounded above by b: Indeed, in $^*\mathbb{R}$, and for any fixed $x \in H$, $x < h \leq b$ holds. Hence (cf. I.6.36), still in $^*\mathbb{R}$, $x < b$ holds. Then it is true in $\mathbb{R}$ as well (cf. p. 99, (4) and (6)). Let $r \in \mathbb{R}$ be the least upper bound of H (over the reals, least upper bounds of sets bounded above exist). We now argue that

$$h \approx r \tag{2}$$

Suppose that (2) is false. Then there is a $s > 0$, in $\mathbb{R}$, such that[†]

$$s \leq |h - r| \text{ is true in } {}^*\mathbb{R} \tag{3}$$

There are two cases:

Case $|h - r| = h - r$. Then (3) implies $s + r \leq h$. Since $s + r$ is standard, but h is not, we have $s + r < h$ hence $s + r \in H$, and thus $s + r \leq r$ from the choice of r (upper bound), that is, $s \leq 0$, contrary to the choice of s.

Case $|h - r| = r - h$. Then (3) implies $h \leq r - s$. Thus $r - s$ is an upper bound of H (in $\mathbb{R}$), contrary to choice of r (*least* upper bound, yet $r - s < r$). □

 It is unreasonable to expect that an infinite number h is infinitely close to a real r, for if that were possible, then $h - r = a \approx 0$. But $r + a$ is finite (right?).

I.6.48 Definition (Standard Parts). Let h be a finite hyperreal. The unique real r such that $h \approx r$ is called *the standard part* of h. We write $\text{st}(h) = r$. □

 I.6.49 Example. For any real r, $\text{st}(r) = r$. This is by $r \approx r$ and uniqueness of standard parts. Also, since h is an infinitesimal iff $h \approx 0$, then h is an infinitesimal iff $\text{st}(h) = 0$. □

We can now prove that infinitesimals and infinite numbers exist in great abundance.

I.6.50 Theorem. *There is at least one, and hence there are uncountably many, non-standard infinitesimals.*

Proof. Pick an $h \in {}^*\mathbb{R} - \mathbb{R}$ (cf. (3) preceding Definition I.6.33). If it is infinite, then $1/|h|$ is an infinitesimal (by I.6.40). It is also nonzero, and hence non-standard (0 is the only standard infinitesimal).

Pause. $\neg(\exists x)(1 = x0)$ is true in $\mathfrak{R}$.

If $|h|$ is finite, then $\text{st}(h)$ exists and $a = h - \text{st}(h) \approx 0$. If $a \in \mathbb{R}$, so is $h = \text{st}(h) + a$.

Why uncountably many? Well, fix a non-standard infinitesimal h. The function $r \mapsto rh$ is a 1-1, total function on the reals.[‡] Thus, its range, $\{rh : r \in \mathbb{R}\}$ is in 1-1 correspondence with $\mathbb{R}$, and hence is uncountable. □

[†] Let us settle an inconsistency in terminology: We use "true in $\mathbb{R}$" (with or without the $*$ superscript) synonymously with "true in $\mathfrak{R}$"

[‡] 1-1 because $rh = r'h \rightarrow r = r'$, since $h \neq 0$ – being non-standard – and the sentence $(\forall x)(\forall y)(\forall z)(z \neq 0 \rightarrow xz = yz \rightarrow x = y)$ is true in $\mathfrak{R}$.

I.6.51 Corollary. *There is at least one, and hence there are uncountably many, infinite numbers.*

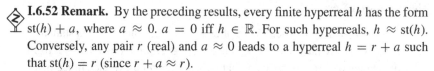

I.6.52 Remark. By the preceding results, every finite hyperreal h has the form $\mathrm{st}(h) + a$, where $a \approx 0$. $a = 0$ iff $h \in \mathbb{R}$. For such hyperreals, $h \approx \mathrm{st}(h)$. Conversely, any pair r (real) and $a \approx 0$ leads to a hyperreal $h = r + a$ such that $\mathrm{st}(h) = r$ (since $r + a \approx r$).

Thus, if we were to depict the set $*\mathbb{R}$ on a line, in analogy with the $\mathbb{R}$-line, we could start with the latter, then stretch it and insert nonstandard numbers (respecting order, $*<$) so that each real r lies in a "cloud" of nonstandard numbers that are infinitely close to r. Then each such cloud[†] contains only one real; for $r \approx h \approx r'$ with r, r' real implies $r \approx r'$ and hence $r = r'$. The cloud in which 0 lies is the set of all infinitesimals.

We then add all the positive infinite numbers to the right end of the line (again, respecting order, $*<$) and all the negative infinite numbers to the left end of the line.   □

The definition that a function f is continuous that we will eventually give essentially requires that the *standard part function* st, commute with f. As a prelude towards that we present a proposition below that deals with the special cases of addition, multiplication, and a few other elementary functions.

But first, the non-standard counterpart to the *pinching lemma* of calculus.[‡]

I.6.53 Lemma (Pinching Lemma). *If* $0 < h < h'$ *and* $h' \approx 0$, *then* $h \approx 0$. *Either or both of "$<$" can be replaced by "$\leq$".*

Proof. Exercise I.60.   □

I.6.54 Corollary. *If* $a < b < c$ *and* $a \approx c$, *then* $a \approx b$ *and* $b \approx c$. *Moreover, this remains true if we replace one or both of "$<$" by "$\leq$".*

Proof. $0 < b - a < c - a$.   □

I.6.55 Proposition (Elementary Algebra of st). *Throughout, a and b are fixed finite hyperreals.*

(1) $a \geq 0$ *implies* $\mathrm{st}(a) \geq 0$.
(2) $\mathrm{st}(a + b) = \mathrm{st}(a) + \mathrm{st}(b)$.
(3) $\mathrm{st}(a - b) = \mathrm{st}(a) - \mathrm{st}(b)$.

[†] The cloud for $r \in \mathbb{R}$ is $\{r + a : a \approx 0\}$. Of course, if $a < 0$, then $r + a < r$, while if $a > 0$, then $r + a > r$.

[‡] If $0 < g(x) < f(x)$ for all x in an open interval (a, b), $c \in (a, b)$, and $\lim_{x \to c} f(x) = 0$, then also $\lim_{x \to c} g(x) = 0$.

(4) $\text{st}(a \cdot b) = \text{st}(a) \cdot \text{st}(b)$.

(5) *If* $\text{st}(b) \neq 0$, *then* $\text{st}(a/b) = \text{st}(a)/\text{st}(b)$.

(6) $\text{st}(a^n) = \text{st}(a)^n$ *for all* $n \in \mathbb{N}$.

(7) *If* $\text{st}(a) \neq 0$, *then* $\text{st}(a^{-n}) = \text{st}(a)^{-n}$ *for all* $0 < n \in \mathbb{N}$.

(8) $\text{st}(a^{1/n}) = \text{st}(a)^{1/n}$ *for all* $0 < n \in \mathbb{N}$ $(a \geq 0$ *assumed for n even*).

Proof. We sample a few cases and leave the rest as an exercise (Exercise I.61).

(1): Assume the hypothesis, yet $\text{st}(a) < 0$. By I.6.54, $\text{st}(a) \approx 0$; hence $\text{st}(a) = 0$, a contradiction.

(5): $\text{st}(b) \neq 0$ implies that $b \not\approx 0$,[†] in particular, $b \neq 0$. The formula to prove thus makes sense. Now, $a = a(a/b)$; hence, by (4), $\text{st}(a) = \text{st}(a(a/b)) = \text{st}(a)\,\text{st}(a/b)$.

(7): Having proved (6) by induction on n, we note that $\text{st}(a^n) = \text{st}(a)^n \neq 0$. Moreover, $a^{-n} = 1/a^n$. Hence $\text{st}(a^{-n}) = \text{st}(1/a^n) = \text{st}(1)/st(a^n) = 1/\text{st}(a)^n$, since $\text{st}(1) = 1$.

(8): For odd n, it is true in $\mathfrak{R}$ that

$$(\forall x)(\exists y)x = y^n$$

or, more colloquially,

$$(\forall x)(\exists y)y = x^{1/n} \tag{i}$$

(i) is also true in $^*\mathfrak{R}$, so it makes sense, for any $a \in {}^*\mathbb{R}$ and odd n, to form $a^{1/n}$. Similarly, if n is even, then

$$(\forall x)(x \geq 0 \rightarrow (\exists y)y = x^{1/n}) \tag{ii}$$

is true in $^*\mathfrak{R}$, so it makes sense, for any $0 \leq a \in {}^*\mathbb{R}$ and even n, to form $a^{1/n}$.

For any such $a^{1/n}$ that makes sense, so does $\text{st}(a)^{1/n}$, by (1).

Thus, noting that $a = (a^{1/n})^n$ we get $\text{st}(a) = \text{st}((a^{1/n})^n) = \text{st}(a^{1/n})^n$, the second "=" by (6). Hence $\text{st}(a^{1/n}) = \text{st}(a)^{1/n}$. $\qquad\square$

A corollary to (1) above, insignificant (and easy) as it may sound, is the non-standard counterpart to the statement that a real closed interval $[a, b]$ is *compact*, that is, every sequence of members of $[a, b]$ that converges, converges to a number in $[a, b]$.

I.6.56 Corollary. $h \leq h'$ *implies* $\text{st}(h) \leq \text{st}(h')$. *In particular, if a and b are real and the closed interval in* $^*\mathbb{R}$ *is*

$$[a, b] \overset{\text{def}}{=} \{x \in {}^*\mathbb{R} : a \leq x \leq b\}$$

then whenever $h \in [a, b]$, *it also follows that* $\text{st}(h) \in [a, b]$.

[†] $b \not\approx 0$ means, of course, $\neg(b \approx 0)$.

The notion of the limit of a *real function*[†] of one variable f at a point a depends on what the function does when its inputs are in a *neighbourhood* of a – that is, an open interval (c, b) that contains a – but not on what it does on a itself. For this reason the limit is defined in terms of a *punctured neighbourhood* of a, which means a set like $(c, a) \cup (a, b)$, where $c < a < b$. We are interested in calculating better and better approximations to the values $f(x)$ as x gets very close to a (but never becomes equal to a – for all we care, f might not even be defined on a).

We can define limits *à la* Leibniz now, replacing "very close" by "infinitely close":

I.6.57 Definition (Limits). Let $^\circ f$ be a real function of one variable defined in some punctured *real* neighbourhood of the real a. Let b be also real.[‡] The notation $\lim_{x \to a} {}^\circ f(x) = b$ abbreviates (i.e., is defined to mean)

$$\text{for all non-standard } h \approx 0, \qquad {}^*f(a + h) \approx b \qquad (1)$$

In practice (*argot*) we will let the context fend for itself and simply write (1) as "for all non-standard $h \approx 0$, $f(a + h) \approx b$" and, similarly, $\lim_{x \to a} f(x) = b$ for its abbreviation, that is, dropping the $*$ and $\circ$ left superscripts. We have just defined the so-called *two-sided finite limit*.

Similarly one defines a whole variety of other limits. We give two examples of such definitions (in simplified notation):

Suppose f is defined in the open interval (a, c) and let b be real. Then the symbol $\lim_{x \to a^+} f(x) = b$ abbreviates

$$\text{for all positive } h \approx 0, \qquad f(a + h) \approx b \qquad (2)$$

We have just defined the so-called *right finite limit*.

Finally, let f be defined in the open interval (a, c). Then $\lim_{x \to a^+} f(x) = +\infty$ abbreviates

$$\text{for all positive } h \approx 0, \qquad f(a + h) \text{ is positive infinite} \qquad (3)$$

We have just defined the so-called *right positive infinite limit*. □

I.6.58 Remark.

(A) In (1) in the above definition, $h \in {}^*\mathbb{R} - \mathbb{R}$ guarantees that $h \neq 0$. This in turn guarantees that we are unconcerned with what f wants to be on a – as

---

[†] That is, a function $f : \mathbb{R} \to \mathbb{R}$ – not necessarily a total one.
[‡] Recall that, because $^\circ a =^* a$ by (4) on p. 99, we have already decided to use the formal name "a" for either $^\circ a$ or $^*a$.

we should be – since $a + h \neq a$. Furthermore, $^*f(a + h)$ is defined for all such h, so that (1) makes sense. This is easy: To fix ideas, let $^\circ f$ be defined in the punctured neighbourhood $(d, a) \cup (a, c)$. First off, $0 < |h| < \min(a - d, c - a)$, since $0 \neq h \approx 0$. Hence $h > 0 \rightarrow a < a + h < c$, while $h < 0 \rightarrow d < a + h < a$. Secondly, the sentence $(\forall x)(d < x < a \vee a < x < c \rightarrow (\exists y)y = f(x))$ is true in $\mathfrak{R}$ by the assumption on $^\circ f$, hence also in $^*\mathfrak{R}$. Thus, $^*f$ is defined in the non-standard (punctured) neighbourhood $\{x \in {}^*\mathbb{R} : d < x < a \vee a < x < c\}$.

 This neighbourhood we are going to denote by $(d, a) \cup (a, c)$ as well, and let the context indicate whether or not this symbol denotes a standard or a non-standard neighbourhood.

(B) Another way to state (1) above is

$$\text{for all nonzero } h \approx 0, \qquad \text{st}(f(a + h)) = b \qquad (4)$$

That is, the standard part above *is independent of the choice of h.*   □

I.6.59 Example. We compute $\lim_{x \to 2}(3x^3 - x^2 + 1)$: Let $0 \neq h \approx 0$ be arbitrary. Then

$$
\begin{aligned}
\text{st}(3(2 + h)^3 - (2 + h)^2 + 1) &= \text{st}(3(2^3 + 12h + 6h^2 + h^3) \\
&\quad -(4 + 4h + h^2) + 1) \\
&= \text{st}(21 + 32h + 17h^2 + 3h^3) \\
&= 21, \qquad \text{by I.6.55}
\end{aligned}
$$

We compute $\lim_{x \to 0}(x/|x|)$. We want $\text{st}(h/|h|)$ for $0 \neq h \approx 0$. In order to remove the absolute value sign we consider cases:

$$
\text{st}\left(\frac{h}{|h|}\right) = \begin{cases} 1 & \text{if } h > 0 \\ -1 & \text{if } h < 0 \end{cases}
$$

According to the definition (cf. (B), previous remark), the limit $\lim_{x \to 0}(x/|x|)$ does not exist. The calculation above shows however that $\lim_{x \to 0^+}(x/|x|) = 1$ and $\lim_{x \to 0^-}(x/|x|) = -1$.

We see that calculating limits within non-standard calculus is easy because we calculate with equalities, rather than with inequalities as in standard calculus.   □

We show next that the non-standard definition of limit is equivalent to Weierstrass's ε-δ definition:

I.6.60 Theorem. *Let* $^{\circ}f$ *be a real function, defined in some punctured neighbourhood of (the real) a. Let b be also real. The following statements are equivalent:*

(i) (1) *in Definition* I.6.57
(ii) $(\forall 0 < \varepsilon \in \mathbb{R})(\exists 0 < \delta \in \mathbb{R})(\forall x \in \mathbb{R})(0 < |x - a| < \delta \to |^{\circ}f(x) - b| < \varepsilon).$[†]

Proof. $(i) \to (ii)$: Let then (1) in I.6.57 hold. To prove (ii), fix an $0 < \varepsilon \in \mathbb{R}$.[‡] It now suffices to show the truth of the *formal* sentence (iii) below:

$$(\exists \delta > 0)(\forall x)(0 < |x - a| < \delta \to |f(x) - b| < \varepsilon) \qquad (iii)$$

One way to prove an existential sentence such as (iii) is to exhibit a δ that works. Since it is a sentence over L, it suffices to verify it in $^{*}\mathfrak{R}$. It will then be true in $\mathfrak{R}$ – which is what we want.

Thus, we take $\delta = h$ where $0 \neq h \approx 0$ and show that it works. Let now $x \in {}^{*}\mathbb{R}$ be arbitrary such that $0 < |x - a| < h$. Thus (I.6.53), $|x - a| \approx 0$; hence $x - a \approx 0$ from $-|x - a| \leq x - a \leq |x - a|$ and I.6.54 (via I.6.55). We can now write $x = a + h'$, with $h' \approx 0$ and $h' \neq 0$. By hypothesis – i.e., (i) – $^{*}f(a + h') \approx b$ is true in $^{*}\mathfrak{R}$, i.e., $^{*}f(x) \approx b$ is true. Hence $^{*}f(x) - b \approx 0$, and therefore $|^{*}f(x) - b|$ is less than any positive real. In particular, it is less than ε.

$(ii) \to (i)$: We assume (ii), pick an arbitrary h such that $0 \neq h \approx 0$, and prove $^{*}f(a + h) \approx b$. This requires

$$\text{for all real } \varepsilon > 0, \qquad |^{*}f(a + h) - b| < \varepsilon \qquad (iv)$$

So fix an arbitrary real $\varepsilon > 0$. Assumption (ii) translates into the assumption that *the sentence (iii) is true in* $\mathfrak{R}$. Let then $\delta > 0$ be real, so that the L-sentence below is true in $\mathfrak{R}$ (δ and ε below are formal constants):

$$(\forall x)(0 < |x - a| < \delta \to |f(x) - b| < \varepsilon) \qquad (v)$$

(v) is also true in $^{*}\mathfrak{R}$. By specialization in the metalanguage, take $x = a + h$. Now, $0 < |x - a| = |h|$ by choice of h. Also, $|x - a| = |h| < \delta$ is also true, since $\delta > 0$ and real, and $h \approx 0$. Thus, by (v), translated into $^{*}\mathfrak{R}$, we have $|^{*}f(x) - b| < \varepsilon$. This proves (iv). $\qquad \square$

[†] This *argot* is a bit awkward, but not unusual. "$(\forall 0 < \varepsilon \in \mathbb{R})\ldots$" stands for "$(\forall \varepsilon)(0 < \varepsilon \wedge \varepsilon \in \mathbb{R} \to \ldots)$".

[‡] We have fixed a real ε. Recall that the name "ε" is also used for the *formal constant* that denotes this real ε.

Worth repeating: The part $(i) \to (ii)$ of the above proof was an instance where we were able to prove a first order fact in $*\mathfrak{R}$ and then transfer it back to $\mathfrak{R}$. This is the essential use we get from the elementary extension $\mathfrak{R} \prec *\mathfrak{R}$, for if all the facts we needed could easily be proved in $\mathfrak{R}$, the whole fuss of obtaining an extension that contains weird numbers would be pointless.

We conclude with the definition of continuity and with one more elementary application of transferring facts from $*\mathfrak{R}$ back to $\mathfrak{R}$. Some more techniques and facts will be discovered by the reader in the Exercises section.

I.6.61 Definition. Let f be a real function of one real variable, defined at least on an open real interval (a, b). We say that f is *continuous* at $c \in (a, b)$ (a real point) iff $\lim_{x \to c} f(x) = f(c)$.

If f is also defined at a, then we say that it is continuous at the *left endpoint* a of $[a, b)$, meaning that $\lim_{x \to a^+} f(x) = f(a)$. In a similar situation at the *right endpoint* b, we require that $\lim_{x \to b^-} f(x) = f(b)$.

We say that f is continuous on $[a, b]$ iff it is so at every real $x \in [a, b]$. $\square$

I.6.62 Remark. The above is the standard definition. Since it involves the concept of limit, we may translate it to a corresponding non-standard definition. Let then f be defined on the real *closed interval* $[a, b]$. Then for any (real) $c \in (a, b)$ continuity requires (using h as a free variable over $*\mathbb{R}$)

$$0 \neq h \approx 0 \to \text{st}(*f(c + h)) = f(c) \tag{1}$$

Continuity at the endpoints reads

$$0 < h \approx 0 \to \text{st}(*f(a + h)) = f(a) \tag{2}$$

and

$$0 > h \approx 0 \to \text{st}(*f(b + h)) = f(b) \tag{3}$$

Does it matter if we take the $0 \neq$ part away? No, since the limit is equal to the function value.

Suppose now that f is *continuous on the real interval* $[a, b]$. We now extend $[a, b]$ to include non-standard numbers as in I.6.56. Then, whenever $x \in [a, b]$, where x is a hyperreal, we also have $\text{st}(x) \in [a, b]$ by I.6.56. Thus, $x \in [a, b]$ implies that $x = r + h$ where r is real – $a \leq r \leq b$ – and $h \approx 0$. We can now capture (1)–(3) by the single statement

$$x \in [a, b] \to \text{st}(*f(x)) = *f(\text{st}(x)) \tag{4}$$

Thus, continuity is the state of affairs where st commutes with the function letter. By the way, since st(x) is real, so is $^*f(\text{st}(x))$; indeed, it is the same as $^\circ f(\text{st}(x))$ (cf. (5), p. 99), which we write, more simply, $f(\text{st}(x))$. In practice one writes (4) above as

$$x \in [a, b] \to \text{st}(f(x)) = f(\text{st}(x))$$

I.6.63 Example. The function $x \mapsto \sqrt{x}$ is continuous on any $[a, b]$ where $0 \leq a$. Indeed, by I.6.55, $0 \leq x$ implies $\text{st}(\sqrt{x}) = \sqrt{\text{st}(x)}$. Now invoke (4) in I.6.62.   □

I.6.64 Theorem. *Suppose that f is continuous on the real interval $[a, b]$. Then f is* bounded *on $[a, b]$, that is, there is a real $B > 0$ such that*

$$x \in [a, b] \cap \mathbb{R} \to |f(x)| < B \tag{1}$$

Proof. We translate the theorem conclusion into a sentence over L. "f", as usual, plays a dual role: name of the real and name of the formal object. The translation is

$$(\exists y)(\forall x)(a \leq x \leq b \to |f(x)| < y) \tag{1'}$$

Now (1') is true in $^*\mathfrak{R}$ *under the assumption that $^\circ f$, which we still call f, is continuous.*

Here is why: Take $y = H$, where $H \in {}^*\mathbb{R}$ is some positive infinite hyperreal. Pick any hyperreal x in $[a, b]$ (extended interval). Now, the assumption on continuity, in the form (4) of I.6.62, has the side effect that

$$\text{st}(f(x)) \text{ is defined}$$

Hence $f(x)$ is finite. Let then $0 < r_x \in \mathbb{R}$ such that $|f(x)| < r_x$. This r_x depends on the picked x. But $r_x < H$; thus $|f(x)| < H$ for the arbitrary hyperreal x in $[a, b]$, establishing the truth of (1') in $^*\mathfrak{R}$. So it is true in $\mathfrak{R}$ too.   □

I.7. Defined Symbols

We have already mentioned that the language lives, and it is being constantly enriched by new nonlogical symbols through definitions. The reason we do this is to abbreviate undecipherably long formal texts, thus making them humanly understandable.

There are three possible kinds of formal abbreviations, namely, abbreviations of *formulas*, abbreviations of *variable terms* (i.e., "objects" that depend on free variables), and abbreviations of *constant* terms (i.e., "objects" that do not depend on free variables). Correspondingly, we introduce a new nonlogical symbol for a *predicate*, a *function*, or a *constant* in order to accomplish such abbreviations.

Here are three simple examples, representative of each case.

We introduce a new *predicate* (symbol), "⊆", in set theory by a definition[†]

$$A \subseteq B \leftrightarrow (\forall x)(x \in A \to x \in B)$$

An introduction of a function symbol by definition is familiar from elementary mathematics. There is a theorem that says

> "for every non-negative real number x there is a unique
> non-negative real number y such that $x = y \cdot y$" (1)

This justifies the introduction of a 1-ary function symbol f that, for each such x, produces the corresponding y. Instead of using the generic "$f(x)$", we normally adopt one of the notations "$\sqrt{x}$" or "$x^{1/2}$". Thus, we enrich the language (of, say, algebra or real analysis) by the function symbol $\sqrt{}$ and add as an axiom the definition of its behaviour. This would be

$$x = \sqrt{x}\sqrt{x}$$

or

$$y = \sqrt{x} \leftrightarrow x = y \cdot y$$

where the restriction $x \geq 0$ is implied by the context.

The "enabling formula" (1) – stated in *argot* above – is crucial in order that we be allowed to introduce $\sqrt{}$ and its defining axiom. That is, before we introduce an abbreviation of a (variable or constant) term – i.e., an *object* – we *must have a proof in our theory* of an *existential* formula, i.e., one of the type $(\exists!y)\mathscr{A}$, that asserts that (if applicable, for each value of the free variables) a *unique* such object exists.

 The symbol "$(\exists!y)$" is read "there is a unique y". It is a "logical" abbreviation (defined logical symbol, just like ∀) given (in least-parenthesized form) by

$$(\exists x)(\mathscr{A} \wedge \neg(\exists z)(\mathscr{A} \wedge \neg x = z))$$

Finally, an example of introducing a new constant symbol, from set theory, is the introduction of the symbol ∅ into the language, as the name of *the unique*

[†] In practice we state the above definition in *argot*, probably as "$A \subseteq B$ means that, for all x, $x \in A \to x \in B$".

I. Basic Logic

object[†] y that satisfies $\neg U(y) \wedge (\forall x)x \notin y$, read "*y is a set,[‡]* and it has no members". Thus, $\emptyset$ is defined by

$$\neg U(\emptyset) \wedge (\forall x)x \notin \emptyset$$

or, equivalently, by

$$y = \emptyset \leftrightarrow \neg U(y) \wedge (\forall x)x \notin y$$

The general situation is this: We start with a theory Γ, spoken in some *basic[§] formal language L*. As the development of Γ proceeds, gradually and continuously we *extend L* into languages L_n, for $n \geq 0$ (we have set $L_0 = L$). Thus the symbol L_{n+1} stands for some arbitrary extension of L_n effected at *stage n + 1*. The theory itself is being extended by stages, as a sequence Γ_n, $n \geq 0$.

A *stage* is marked by the event of introducing a *single* new symbol into the language via a definition of a new predicate, function or constant symbol. At that same stage we also add to Γ_n the *defining nonlogical axiom* of the new symbol in question, thus *extending* the theory Γ_n into Γ_{n+1}. We set $\Gamma_0 = \Gamma$.

Specifically, if[¶] $\mathcal{Q}(\vec{x}_n)$ is some formula, we then can introduce a new predicate symbol "P"[#] that stands for $\mathcal{Q}$.

In the present description, $\mathcal{Q}$ is a syntactic (meta-)variable, while P is a new *formal* predicate symbol.

This entails adding P to L_k (i.e., to its alphabet $\mathcal{V}_k$) as a new n-ary predicate symbol, and adding

$$P\vec{x}_n \leftrightarrow \mathcal{Q}(\vec{x}_n) \tag{i}$$

to Γ_k as the defining axiom for P. "$\subseteq$" is such a defined (2-ary) predicate in set theory.

Similarly, a new n-ary function symbol f is added into L_k (to form L_{k+1}) by a definition of its behaviour. That is, we add f to L_k and also add the following

---

[†] Uniqueness follows from *extensionality*, while existence follows from *separation*. These facts – and the italicized terminology – are found in volume 2, Chapter III.

[‡] U is 1-ary (unary) predicate. It is one of the two primitive nonlogical symbols of formal set theory. With the help of this predicate we can "test" an object for set or atom status. "$U(y)$" asserts that y is an atom, thus "$\neg U(y)$" asserts that y is a set – since we accept that sets or atoms are the only *types of objects that the formal system axiomatically characterizes*.

[§] "Basic" means here the language given originally, before any *new* symbols were added.

[¶] Recall that (see Remark I.1.11, p. 18) the notation $\mathcal{Q}(\vec{x}_n)$ asserts that $\vec{x}_n$, i.e., $x_1, \ldots, x_n$, is the *complete list* of the free variables of $\mathcal{Q}$.

[#] Recall that predicate letters are denoted by *non*-calligraphic capital letters P, Q, R with or without subscripts or primes.

formula (ii) to Γ_k as a new nonlogical axiom:

$$y = fy_1 \ldots y_n \leftrightarrow \mathscr{Q}(y, y_1, \ldots, y_n) \qquad (ii)$$

provided we have a proof in Γ_k of the formula

$$(\exists! y)\mathscr{Q}(y, y_1, \ldots, y_n) \qquad (iii)$$

 Of course, the variables y, $\vec{y}_n$ are distinct.

Depending on the theory, and the number of free variables ($n \geq 0$), "f" may take theory-specific *names* such as $\emptyset$, ω, $\sqrt{}$, etc. (in this illustration, for the sake of economy of effort, we have thought of defined constants, e.g., $\emptyset$ and ω, as 0-ary functions – something that we do not normally do).

In effecting these definitions, we want to be assured of two things:

1. Whatever we can state in the richer language L_k (for any $k > 0$) we can also state in the original ("basic") language $L = L_0$ (although awkwardly, which justifies our doing all this). "Can state" means that we can "translate" any formula $\mathscr{F}$ over L_k (hopefully in a *natural* way) into a formula $\mathscr{F}^*$ over L so that the extended theory Γ_k can *prove* that $\mathscr{F}$ and $\mathscr{F}^*$ are equivalent.[†]
2. We also want to be assured that the new symbols offer *no more than convenience*, in the sense that any formula $\mathscr{F}$, *over the basic language L*, that Γ_k ($k > 0$) is able to prove, one way or another (perhaps with the help of defined symbols), Γ can also prove.[‡]

These assurances will become available shortly, as Metatheorems I.7.1 and I.7.3. Here are the "natural" translation rules, that take us from a language stage L_{k+1} back to the previous, L_k (so that, iterating the process, we are back to L):

Rule (1). Suppose that $\mathscr{F}$ is a formula over L_{k+1}, and that the predicate P (whose definition took us from L_k to L_{k+1}, and hence is a symbol of L_{k+1} but not of L_k) occurs in $\mathscr{F}$ zero or more times. Assume that P has been defined by the axiom (i) above (included in Γ_{k+1}), where $\mathscr{Q}$ is a formula over L_k.

We eliminate P from $\mathscr{F}$ by replacing *all its occurrences* by $\mathscr{Q}$. By this we mean that whenever $P\vec{t}_n$ is a subformula of $\mathscr{F}$, all its occurrences are replaced by $\mathscr{Q}(\vec{t}_n)$. We can always arrange by I.4.13 that the simultaneous substitution $\mathscr{Q}[\vec{x}_n \leftarrow \vec{t}_n]$ is defined.

This results to a formula $\mathscr{F}^*$ over L_k.

[†] Γ, spoken over L, can have no opinion, of course, since it cannot see the new symbols, nor does it have their "definitions" among its "knowledge".

[‡] Trivially, any $\mathscr{F}$ over L that Γ can prove, any Γ_k ($k > 0$) can prove as well, since the latter understands the language (L) *and* contains all the axioms of Γ. Thus Γ_k extends the theory Γ. That it cannot have more theorems *over L* than Γ makes this extension *conservative*.

Rule (2). If f is a defined n-ary function symbol as in (ii) above, introduced into L_{k+1}, and if it occurs in $\mathscr{F}$ as $\mathscr{F}[ft_1 \ldots t_n],^\dagger$ then this formula is logically equivalent to[‡]

$$(\exists y)(y = ft_1 \ldots t_n \wedge \mathscr{F}[y]) \qquad\qquad (iv)$$

provided that y is not free in $\mathscr{F}[ft_1 \ldots t_n]$.

Using the definition of f given by (ii) and I.4.13 to ensure that $\mathcal{Q}(y, \vec{t}_n)$ is defined, we eliminate this occurrence of f, writing (iv) as

$$(\exists y)(\mathcal{Q}(y, t_1, \ldots, t_n) \wedge \mathscr{F}[y]) \qquad\qquad (v)$$

which says the same thing as (iv) in any theory that thinks that (ii) is true (this observation is made precise in the proof of Metatheorem I.7.1). Of course, f may occur many times in $\mathscr{F}$, even "within itself", as in $ffz_1 \ldots z_n y_2 \ldots y_n,^{\S}$ or even in more complicated configurations. Indeed, it may occur within the scope of a quantifier. So the rule becomes: *Apply the transformation taking every atomic subformula $\mathscr{A}[f\vec{t}_n]$ of $\mathscr{F}$ into form (v) by stages, eliminating at each stage the leftmost innermost[¶] occurrence of f (in the atomic formula we are transforming at this stage), until all occurrences of f are eliminated.*

We now have a formula $\mathscr{F}^*$ over L_k.

I.7.1 Metatheorem (Elimination of Defined Symbols: I). *Let Γ be any theory over some formal language L.*

(a) *Let the formula $\mathcal{Q}$ be over L, and P be a new predicate symbol that extends L into L' and Γ into Γ' via the axiom $P\vec{x}_n \leftrightarrow \mathcal{Q}(\vec{x}_n)$. Then, for any formula $\mathscr{F}$ over L', the P-elimination as in Rule (1) above yields a $\mathscr{F}^*$ over L such that*

$$\Gamma' \vdash \mathscr{F} \leftrightarrow \mathscr{F}^*$$

(b) *Let $\mathscr{F}[x]$ be over L, and let t stand for $ft_1, \ldots, t_n$, where f is introduced by (ii) above as an axiom that extends Γ into Γ'. Assume that no t_i contains the letter f and that y is not free in $\mathscr{F}[t]$. Then[#]*

$$\Gamma' \vdash \mathscr{F}[t] \leftrightarrow (\exists y)(\mathcal{Q}(y, \vec{t}_n) \wedge \mathscr{F}[y])$$

[†] This notation allows for the possibility that $ft_1, \ldots, t_n$ does not occur at all in $\mathscr{F}$ (see the convention on brackets, p. 18).

[‡] See (C) in the proof of Metatheorem I.7.1 below.

[§] Or $f(f(z_1, \ldots, z_n), y_2, \ldots, y_n))$, using brackets and commas to facilitate reading.

[¶] A term $ft_1, \ldots, t_n$ is "innermost" iff none of the t_i contains "f".

[#] As we already have remarked, in view of I.4.13, it is unnecessary pedantry to make assumptions on substitutability explicit.

Here "L'''" is "L_{k+1}" (for some k) and "L" is "L_k".

Proof. First observe that this metatheorem indeed gives the assurance that, after applying the transformations (1) and (2) to obtain $\mathscr{F}^*$ from $\mathscr{F}$, Γ' thinks that the two are equivalent.

(a): This follows immediately from the Leibniz rule (I.4.25).

(b): Start with

$$\vdash \mathscr{F}[t] \to t = t \wedge \mathscr{F}[t] \text{ (By } \vdash t = t \text{ and } \models_{\text{Taut}}\text{-implication)} \qquad (A)$$

Now, by **Ax2**, substitutability, and non-freedom of y in $\mathscr{F}[t]$,

$$\vdash t = t \wedge \mathscr{F}[t] \to (\exists y)(y = t \wedge \mathscr{F}[y])$$

Hence

$$\vdash \mathscr{F}[t] \to (\exists y)(y = t \wedge \mathscr{F}[y]) \qquad (B)$$

by (A) and $\models_{\text{Taut}}$-implication.[†]

Conversely,

$$\vdash y = t \to (\mathscr{F}[y] \leftrightarrow \mathscr{F}[t]) \qquad (\textbf{Ax4}; \text{ substitutability was used here})$$

Hence (by $\models_{\text{Taut}}$)

$$\vdash y = t \wedge \mathscr{F}[y] \to \mathscr{F}[t]$$

Therefore, by $\exists$-introduction (allowed, by our assumption on y),

$$\vdash (\exists y)(y = t \wedge \mathscr{F}[y]) \to \mathscr{F}[t]$$

which, along with (B), establishes

$$\vdash \mathscr{F}[t] \leftrightarrow (\exists y)(y = t \wedge \mathscr{F}[y]) \qquad (C)$$

Finally, by (ii) (which introduces Γ' to the left of $\vdash$), (C), and the Leibniz rule,

$$\Gamma' \vdash \mathscr{F}[t] \leftrightarrow (\exists y)(\mathcal{Q}(y, \vec{t}_n) \wedge \mathscr{F}[y]) \qquad (D)$$

$\square$

The import of Metatheorem I.7.1 is that if we transform a formula $\mathscr{F}$ – written over some *arbitrary* extension by definitions, L_{k+1}, of the basic language L – into a formula $\mathscr{F}^*$ over L, then Γ_{k+1} (the theory over L_{k+1} that has the benefit of all the added axioms) thinks that $\mathscr{F} \leftrightarrow \mathscr{F}^*$. The reason for this is that we can

---

[†] We will often write just "by $\models_{\text{Taut}}$" meaning to say "by $\models_{\text{Taut}}$-implication".

imagine that we eliminate *one new symbol at a time*, repeatedly applying the metatheorem above – part (b) to atomic subformulas – forming a sequence of increasingly more "basic" formulas $\mathscr{F}_{k+1}, \mathscr{F}_k, \mathscr{F}_{k-1}, \ldots, \mathscr{F}_0$, where $\mathscr{F}_0$ is the same string as $\mathscr{F}^*$ and $\mathscr{F}_{k+1}$ is the same string as $\mathscr{F}$.

Now, $\Gamma_{i+1} \vdash \mathscr{F}_{i+1} \leftrightarrow \mathscr{F}_i$ for $i = k, \ldots, 0$, where, if a defined function letter was eliminated at step $i + 1 \to i$, we invoke (D) above and the Leibniz rule. Hence, since $\Gamma_0 \subseteq \Gamma_1 \subseteq \cdots \subseteq \Gamma_{k+1}$, $\Gamma_{k+1} \vdash \mathscr{F}_{i+1} \leftrightarrow \mathscr{F}_i$ for $i = k, \ldots, 0$, therefore $\Gamma_{k+1} \vdash \mathscr{F}_{k+1} \leftrightarrow \mathscr{F}_0$.

I.7.2 Remark (One Point Rule). The absolutely provable formula in (C) above is sometimes called the *one point rule* (Gries and Schneider (1994), Tourlakis (2000a, 2001b)). Its dual

$$\mathscr{F}[t] \leftrightarrow (\forall y)(y = t \to \mathscr{F}[y])$$

is also given the same nickname and is easily (absolutely) provable using (C) by eliminating $\exists$. □

I.7.3 Metatheorem (Elimination of Defined Symbols: II). *Let Γ be a theory over a language L.*

(a) *If L' denotes the extension of L by the new predicate symbol P, and Γ' denotes the extension of Γ by the addition of the axiom $P\vec{x}_n \leftrightarrow \mathcal{Q}(\vec{x}_n)$, where $\mathcal{Q}$ is a formula over L, then $\Gamma \vdash \mathscr{F}$ for any formula $\mathscr{F}$ over L such that $\Gamma' \vdash \mathscr{F}$.*

(b) *Assume that*

$$\Gamma \vdash (\exists! y)\mathscr{R}(y, x_1, \ldots, x_n) \qquad (*)$$

pursuant to which we have defined the new function symbol f by the axiom

$$y = f x_1 \ldots x_n \leftrightarrow \mathscr{R}(y, x_1, \ldots, x_n) \qquad (**)$$

and thus extended L to L' and Γ to Γ'. Then $\Gamma \vdash \mathscr{F}$ for any formula $\mathscr{F}$ over L such that $\Gamma' \vdash \mathscr{F}$.

Proof. This metatheorem assures that extensions of theories by definitions are conservative in that they produce convenience but no additional power (the same old theorems over the original language are the only ones provable).

(*a*): By the completeness theorem, we show instead that

$$\Gamma \models \mathscr{F} \qquad (1)$$

So let $\mathfrak{M} = (M, \mathscr{I})$ be an arbitrary model of Γ, i.e., let

$$\models_{\mathfrak{M}} \Gamma \tag{2}$$

We now *expand* the structure $\mathfrak{M}$ into $\mathfrak{M}' = (M, \mathscr{I}')$ – *without adding any new individuals to its domain M* – by adding an interpretation $P^{\mathscr{I}'}$ for the new symbol P. We define for *every* $a_1, \ldots, a_n$ in M

$$P^{\mathscr{I}'}(a_1, \ldots, a_n) = \mathbf{t} \text{ iff } \models_{\mathfrak{M}'} \mathscr{Q}(\bar{a}_1, \ldots, \bar{a}_n) \qquad \text{[i.e., iff}$$
$$\models_{\mathfrak{M}} \mathscr{Q}(\bar{a}_1, \ldots, \bar{a}_n)]$$

Clearly then, $\mathfrak{M}'$ is a model of the new axiom, since, for all $\mathfrak{M}'$-instances of the axiom – such as $P(\bar{a}_1, \ldots, \bar{a}_n) \leftrightarrow \mathscr{Q}(\bar{a}_1, \ldots, \bar{a}_n)$ – we have

$$(P(\bar{a}_1, \ldots, \bar{a}_n) \leftrightarrow \mathscr{Q}(\bar{a}_1, \ldots, \bar{a}_n))^{\mathscr{I}'} = \mathbf{t}$$

It follows that $\models_{\mathfrak{M}'} \Gamma'$, since we have $\models_{\mathfrak{M}'} \Gamma$, the latter by (2), due to having made no changes to $\mathfrak{M}$ that affect the symbols of L. Thus, $\Gamma' \vdash \mathscr{F}$ yields $\models_{\mathfrak{M}'} \mathscr{F}$; hence, since $\mathscr{F}$ is over L, we obtain $\models_{\mathfrak{M}} \mathscr{F}$. Along with (2), this proves (1).

(*b*): As in (*a*), assume (2) in an attempt to prove (1). By (∗),

$$\models_{\mathfrak{M}} (\exists!y)\mathscr{R}(y, x_1, \ldots, x_n)$$

Thus, there is a concrete (i.e., in the metatheory) function $\widehat{f}$ of n arguments that takes its inputs from M and gives its outputs to M, the input-output relation being given by (3) below ($\vec{b}_n$ in, a out). To be specific, the semantics of "$\exists!$" implies that for all $b_1, \ldots, b_n$ in M there is a unique $a \in M$ such that

$$(\mathscr{R}(\bar{a}, \bar{b}_1, \ldots, \bar{b}_n))^{\mathscr{I}} = \mathbf{t} \tag{3}$$

We now expand the structure $\mathfrak{M}$ into $\mathfrak{M}' = (M, \mathscr{I}')$,[†] so that all we add to it is an interpretation for the new function symbol f. We let $f^{\mathscr{I}'} = \widehat{f}$. From (2) it follows that

$$\models_{\mathfrak{M}'} \Gamma \tag{2'}$$

since we made no changes to $\mathfrak{M}$ other than adding an interpretation of f, and since no formula in Γ contains f. By (3), if $a, b_1, \ldots, b_n$ are any members of M, then we have

$$\models_{\mathfrak{M}'} \bar{a} = f\bar{b}_1\ldots\bar{b}_n \text{ iff } a = \widehat{f}(b_1, \ldots, b_n)$$
$$\text{iff } \models_{\mathfrak{M}} \mathscr{R}(\bar{a}, \bar{b}_1, \ldots, \bar{b}_n) \qquad \text{by the definition of } \widehat{f}$$
$$\text{iff } \models_{\mathfrak{M}'} \mathscr{R}(\bar{a}, \bar{b}_1, \ldots, \bar{b}_n)$$

[†] This part is independent of part (a); hence this is a different $\mathscr{I}'$ in general.

the last "iff" being because $\mathscr{R}$ (over L) means the same thing in $\mathfrak{M}$ and $\mathfrak{M}'$.
Thus,

$$\models_{\mathfrak{M}'} y = f x_1 \ldots x_n \leftrightarrow \mathscr{R}(y, x_1, \ldots, x_n) \qquad (4)$$

Now (∗∗), (2′), and (4) yield $\models_{\mathfrak{M}'} \Gamma'$ which implies $\models_{\mathfrak{M}'} \mathscr{F}$ (from $\Gamma' \vdash \mathscr{F}$).
Finally, since $\mathscr{F}$ contains no f, we have $\models_{\mathfrak{M}} \mathscr{F}$. This last fact, and (2) give (1).

$\square$

I.7.4 Remark.

(a) We note that translation rules (1) and (2) – the latter applied to atomic
subformulas – preserve the syntactic structure of quantifier prefixes. For
example, suppose that we have introduced f by

$$y = f x_1 \ldots x_n \leftrightarrow \mathcal{Q}(y, x_1, \ldots, x_n) \qquad (5)$$

in set theory. Now, an application of the *collection axiom* of set theory has
a hypothesis of the form

$$\text{"}(\forall x \in Z)(\exists w)(\ldots \mathscr{A}[f t_1 \ldots t_n] \ldots)\text{"} \qquad (6)$$

where, say, $\mathscr{A}$ is atomic and the displayed f is innermost. Eliminating this
f, we have the translation

$$\text{"}(\forall x \in Z)(\exists w)(\ldots (\exists y)(\mathscr{A}[y] \wedge \mathcal{Q}(y, t_1, \ldots, t_n)) \ldots)\text{"} \qquad (7)$$

which still has the $\forall\exists$-prefix and still looks exactly like a collection axiom
hypothesis.

(b) Rather than worrying about the "ontology" of the function *symbol* formally
introduced by (5) above – i.e., the question of the exact nature of the symbol
that we *named* "f" – in practice we shrug this off and resort to *metalinguistic*
devices to name the function symbol, or the *term* that naturally arises from
it. For example, one can use the notation "$f_{\mathcal{Q}}$" for the function – where the
subscript "$\mathcal{Q}$" is the exact string over the language that "$\mathcal{Q}$" denotes – or,
for the corresponding term, the notation of Whitehead and Russell (1912),

$$(\iota z)\mathcal{Q}(z, x_1, \ldots, x_n) \qquad (8)$$

The "z" in (8) above is a bound variable.[†] This new type of term is read
"*the* unique z such that …". *This* "ι" *is* not *one of our primitive symbols.*[‡]

[†] That it must be distinct from the x_i is obvious.

[‡] It is however possible to enlarge our alphabet to include "ι", and then add definitions of the
syntax of "ι-terms" and axioms for the behaviour of "ι-terms". At the end of all this one gets a

It is just meant to lead to the friendly shorthand (8) above that avoids the "ontology" issue. Thus, once one proves

$$(\exists! z)\mathcal{Q}(z, x_1, \ldots, x_n) \tag{9}$$

one can then introduce (8) by the axiom

$$y = (\iota z)\mathcal{Q}(z, x_1, \ldots, x_n) \leftrightarrow \mathcal{Q}(y, x_1, \ldots, x_n) \tag{5'}$$

which, of course, is an alias for the axiom (5), using more suggestive notation for the term $f x_1, \ldots, x_n$. By (9), the axioms (5) or (5') can be replaced by

$$\mathcal{Q}(f x_1, \ldots, x_n, x_1, \ldots, x_n)$$

and

$$\mathcal{Q}((\iota z)\mathcal{Q}(z, x_1, \ldots, x_n), x_1, \ldots, x_n) \tag{10}$$

respectively. For example, from (5') we get (10) by substitution. Now, **Ax4** (with some help from $\models_{\text{Taut}}$) yields

$$\mathcal{Q}((\iota z)\mathcal{Q}(z, x_1, \ldots, x_n), \ x_1, \ldots, x_n) \rightarrow$$
$$y = (\iota z)\mathcal{Q}(z, x_1, \ldots, x_n) \rightarrow \mathcal{Q}(y, x_1, \ldots, x_n)$$

Hence, assuming (10),

$$y = (\iota z)\mathcal{Q}(z, x_1, \ldots, x_n) \rightarrow \mathcal{Q}(y, x_1, \ldots, x_n) \tag{11}$$

Finally, deploying (9), we get

$$\mathcal{Q}((\iota z)\mathcal{Q}(z, x_1, \ldots, x_n), \ x_1, \ldots, x_n) \rightarrow$$
$$\mathcal{Q}(y, x_1, \ldots, x_n) \rightarrow y = (\iota z)\mathcal{Q}(z, x_1, \ldots, x_n)$$

Hence

$$\mathcal{Q}(y, x_1, \ldots, x_n) \rightarrow y = (\iota z)\mathcal{Q}(z, x_1, \ldots, x_n)$$

by (10). This, along with (11), yields (5'). □

The Indefinite Article. We often have the following situation: We have proved a statement like

$$(\exists x)\mathcal{A}[x] \tag{1}$$

conservative extension of the original theory, i.e., any ι-free formula provable in the new theory can be also proved in the old (Hilbert and Bernays (1968)).

and we want next to derive a statement $\mathcal{B}$. To this end, we start by picking a symbol c not in $\mathcal{B}$ and say "let c be such that $\mathcal{A}[c]$ is true".[†] That is, we add $\mathcal{A}[c]$ as a nonlogical axiom, treating c as a new constant.

From all these assumptions we then manage to prove $\mathcal{B}$, hopefully treating all the free variables of $\mathcal{A}[c]$ as constants during the argument. We then conclude that $\mathcal{B}$ has been derived *without the help of $\mathcal{A}[c]$ or c* (see I.4.27).

Two things are noteworthy in this technique: One, c does not occur in the conclusion, and, two, c is not uniquely determined by (1). So we have $\underline{a}$ (rather than *the*) c that makes $\mathcal{A}[c]$ true.

Now the suggestion that the free variables of the latter be frozen during the derivation of $\mathcal{B}$ is unnecessarily restrictive, and we have a more general result: Suppose that

$$\Gamma \vdash (\exists x)\mathcal{A}(x, y_1, \ldots, y_n) \tag{2}$$

Add a *new* function symbol f to the language L of Γ (thus obtaining L') via the axiom

$$\mathcal{A}(fy_1, \ldots, y_n, y_1, \ldots, y_n) \tag{3}$$

This says, intuitively, "for any $y_1, \ldots, y_n$, let $x = f\vec{y}_n$ make $\mathcal{A}(x, \vec{y}_n)$ true". Again, this x is not uniquely determined by (2).

Finally, suppose that we have a proof

$$\Gamma + \mathcal{A}(f\vec{y}_n, \vec{y}_n) \vdash \mathcal{B} \tag{4}$$

such that f, the new function symbol, occurs nowhere in $\mathcal{B}$, i.e., the latter formula is over L. We can conclude then that

$$\Gamma \vdash \mathcal{B} \tag{5}$$

that is, the extension $\Gamma + \mathcal{A}(f\vec{y}_n, \vec{y}_n)$ of Γ is conservative.

A proof of the legitimacy of this technique, based on the completeness theorem, is easy. Let

$$\models_{\mathfrak{M}} \Gamma \tag{6}$$

and show

$$\models_{\mathfrak{M}} \mathcal{B} \tag{7}$$

Expand the model $\mathfrak{M} = (M, \mathcal{I})$ to $\mathfrak{M}' = (M, \mathcal{I}')$, so that $\mathcal{I}'$ interprets the new symbol f. The interpretation is chosen as follows: (2) guarantees

[†] Cf. I.5.44.

that, for all choices of $i_1, \ldots, i_n$ in M, the set $S(i_1, \ldots, i_n) = \{a \in M : \models_{\mathfrak{M}} \mathscr{A}(\bar{a}, \bar{i_1}, \ldots, \bar{i_n})\}$ is not empty.

By the axiom of choice (of *informal* set theory), we can pick an $a(i_1, \ldots, i_n)^{\dagger}$ in each $S(i_1, \ldots, i_n)$. Thus, we define a function $\widehat{f} : M^n \to M$ by letting, for each $i_1, \ldots, i_n$ in M, $\widehat{f}(i_1, \ldots, i_n) = a(i_1, \ldots, i_n)$.

The next step is to set

$$f^{\mathscr{T}'} = \widehat{f}$$

Therefore, for all $i_1, \ldots, i_n$ in M,

$$(f\bar{i_1} \ldots \bar{i_n})^{\mathscr{T}'} = \widehat{f}(i_1, \ldots, i_n) = a(i_1, \ldots, i_n)$$

It is now clear that $\models_{\mathfrak{M}'} \mathscr{A}(fy_1 \ldots y_n, y_1, \ldots, y_n)$, for, by I.5.11,

$$(\mathscr{A}(f\bar{i_1} \ldots \bar{i_n}, \bar{i_1}, \ldots, \bar{i_n}))^{\mathscr{T}'} = \mathbf{t} \leftrightarrow (\mathscr{A}(\overline{a(i_1, \ldots, i_n)}, \bar{i_1}, \ldots, \bar{i_n}))^{\mathscr{T}'} = \mathbf{t}$$

and the right hand side of the above is true by the choice of $a(i_1, \ldots, i_n)$.

Thus, $\models_{\mathfrak{M}'} \Gamma + \mathscr{A}(fy_1 \ldots y_n, y_1, \ldots, y_n)$; hence $\models_{\mathfrak{M}'} \mathscr{B}$, by (4). Since $\mathscr{B}$ contains no f, we also have $\models_{\mathfrak{M}} \mathscr{B}$; thus we have established (7) from (6). We now have (5).

One can give a number of names to a function like f: a *Skolem function*, an *ε-term* (Hilbert (1968)), or a *τ-term* (Bourbaki (1966b)). In the first case one may ornament the symbol f, e.g., $f_{\exists \mathscr{A}}$, to show where it is coming from, although such mnemonic naming is not, of course, mandatory.

The last two terminologies actually apply to the *term* $fy_1 \ldots y_n$, rather than to the function *symbol* f. Hilbert would have written

$$(\varepsilon x)\mathscr{A}(x, y_1 \ldots, y_n) \tag{8}$$

and Bourbaki

$$(\tau x)\mathscr{A}(x, y_1 \ldots, y_n) \tag{9}$$

– each denoting $fy_1 \ldots y_n$. The "x" in each of (8) and (9) is a bound variable (different from each y_i).

I.8. Computability and Uncomputability

Computability (or "recursion theory") is nowadays classified as an area of logic (e.g., it is one of the areas represented in the *Handbook of Mathematical Logic*, Barwise (1978)). It has its origins in the work of several logicians in the 1930s

---

† The "$(i_1, \ldots, i_n)$" part indicates that "a" depends on $i_1, \ldots, i_n$.

(Gödel, Turing, Kleene, Church, Post, *et al.*). Motivation for this research was partly provided by Hilbert's program to found all mathematics on formalism. This was a formalism that one ought to be able to certify by *finitary* means (for each particular formalized theory) to be free of contradiction. Moreover, it was a formalism, for which – Hilbert expected – a "method" ought to exist to solve the *Entscheidungsproblem* (decision problem), that is, the question "is this arbitrary formula a theorem, or not?"

What *was* a "method" supposed to be, exactly, mathematically speaking? Was the expectation that the *Entscheidungsproblem* of any theory is amenable to algorithmic solution realistic? Work of Church (lack of a decision algorithm for certain theories (1936)) showed that it was not, nor for that matter was the expectation of certifying freedom of contradiction of all formal theories by finitary means (Gödel's second incompleteness theorem).

One of these two negative answers (Church's) built on an emerging theory of computable (or algorithmic) functions and the mathematical formulation of the concepts of *algorithm* or method. The other one, Gödel's, while it used existing (pre-Turing and pre-Kleene) rudiments of computability (*primitive recursive functions* of Dedekind), can be recast, in hindsight, in the framework of modern computability. This recasting shows the intimate connection between the phenomena of *incompletableness* of certain theories and *uncomputability*, and thus it enhances our understanding of both phenomena.

With the advent of computers and the development of computer science, computability gained a new set of practitioners and researchers: theoretical computer scientists. This group approaches the area from two (main) standpoints: to study the power and limitations of mathematical models of computing devices (after all, computer programs are algorithms), and also to understand why some problems have "easy" while others have "hard" algorithmic solutions (complexity theory) – in the process devising several "practical" (or *efficient*) solutions, and techniques, for a plethora of practical problems.

We develop the basics of computability here *informally*, that is, within "real mathematics" (in the metatheory of pure and applied first order logic).

Computability, generally speaking, formalizes the concept of a "computable function" $f : \mathbb{N}^k \to \mathbb{N}$. That is, it concerns itself with the issue of separating the set of all so-called *number-theoretic functions* – that is,[†] functions with inputs in $\mathbb{N}$ and outputs in $\mathbb{N}$ – into computable and uncomputable.

Because we want the theory to be as inclusive as possible, we allow it to study both total and nontotal functions $f : \mathbb{N}^k \to \mathbb{N}$.

[†] More precisely, this is what *ordinary computability* or *ordinary recursion theory* studies. *Higher* recursion theory, invented by Kleene, also looks into functions that have higher order inputs such as number-theoretic functions.

The trivial reason is that in everyday computing we do encounter both total and nontotal functions. There are computer programs which (whether or not according to the programmer's intent) do *not* stop to yield an answer for all possible inputs. We do want to have formal counterparts of those in our theory, if we are hoping to have a theory that is inclusive.

A less trivial reason is that unless we allow nontotal functions in the theory, an obvious diagonalization can show the existence of total (intuitively) computable functions that are not in the theory.

I.8.1 Definition. Any number-theoretic function $f : \mathbb{N}^k \to \mathbb{N}$ is a *partial function*. If its domain, $\text{dom}(f)$, equals $\mathbb{N}^k$ – the set of all potential inputs, or *left field* – then we say that f is *total*. If it does not, then f is *nontotal*.

That $a \in \text{dom}(f)$ is also denoted by $f(a) \downarrow$, and we say that f is defined at a or that $f(a)$ *converges*.[†] In the opposite case we write $f(a) \uparrow$ and say that f is undefined at a or that $f(a)$ *diverges*.

A number-theoretic relation is a subset of $\mathbb{N}^k$. We usually write such relations in *relational notation*. That is, we write $R(a_1, \ldots, a_n)$ for $\langle a_1, \ldots, a_n \rangle \in R$. Thus our notation of relations parallels that of formulas of a first order language, and we use the logical connectives ($\exists, \forall, \neg, \vee$, etc.) *informally* to combine relations. We carry that parallel to the next natural step, and use the phrases "... a relation R ..." and "... a relation $R(y_1, \ldots, y_n)$ ..." interchangeably, the latter to convey that the *full list* of the relation's variables is exactly $y_1, \ldots, y_n$ (cf. p. 18).

We occasionally use λ-*notation* to modify a given relation $R(y_1, \ldots, y_n)$. This notation is employed as in $\lambda z_1 \ldots z_r.R$, or even $\lambda z_1 \ldots z_r.R(y_1, \ldots, y_n)$. The part "$\lambda z_1 \ldots z_r.$" denotes that "$z_1, \ldots, z_r$" is the *active* variables list and *supersedes* the list "$y_1, \ldots, y_n$". Any y_i that is not in the list $z_1, \ldots, z_r$ is treated as a constant (or "parameter" – i.e., it is "frozen"). The list $z_1, \ldots, z_r$ may contain additional variables not in the list $y_1, \ldots, y_n$.

Thus, e.g., $\lambda xy.x < 2 = \{0, 1\} \times \mathbb{N}$, while $\lambda yx.x < 2 = \mathbb{N} \times \{0, 1\}$. On the other hand, $\lambda x.x < y = \{x : x < y\}$, which denotes a different relation for different values of the parameter y.

Finally, as before, $\vec{z}_r$ or just $\vec{z}$ (if r is understood) denotes $z_1, \ldots, z_r$, so that we may write $\lambda \vec{z}_r.R(y_1, \ldots, y_n)$. □

I.8.2 Definition (Bounded Quantification). For any relation R, the symbols $(\exists x)_{<z}R$, $(\forall x)_{<z}R$, $(\exists x)_{\leq z}R$, $(\forall x)_{\leq z}R$ stand for $(\exists x)(x < z \wedge R)$, $(\forall x)(x < z \to R)$, $(\exists x)(x \leq z \wedge R)$, $(\forall x)(x \leq z \to R)$, respectively. We say that they denote *bounded quantification*. □

[†] This nomenclature parallels that of "convergent" or "halting" computations.

I.8.3 Definition. If $R \subseteq \mathbb{N}^n$ is a relation and $f : \mathbb{N}^k \to \mathbb{N}$ a function, then $R(w_1, \ldots, w_m, f(\vec{x}_k), w_{m+1}, \ldots, w_n)$ means[†]

$$(\exists z)(R(w_1, \ldots, w_m, z, w_{m+1}, \ldots, w_n) \wedge z = f(\vec{x}_k))$$

We have just one important exception to this rule: If Q is $g(\vec{y}) = w$, then $g(\vec{y}) = f(\vec{x}_k)$ means

$$g(\vec{y}) \uparrow \wedge f(\vec{x}_k) \uparrow \vee (\exists z)(z = g(\vec{y}) \wedge z = f(\vec{x}_k))$$

One often writes $g(\vec{y}) \simeq f(\vec{x}_k)$ for the above to alert the reader that "weak equality" (a notion due to Kleene) applies, but we will rather use "=" throughout and let the context determine the meaning. □

Clearly, weak equality restores reflexivity of "=" (which fails if the general understanding of substitution above applied to "=" as well).

λ-notation comes in handy in denoting number-theoretic functions. Instead of saying "consider the function g obtained from $f : \mathbb{N}^k \to \mathbb{N}$, by setting, for all $\vec{w}_m$,

$$g(\vec{w}_m) = f(\vec{x}_k)$$

where if an x_i is not among the w_j it is understood to be an (unspecified) constant", we simply say "consider $g = \lambda \vec{w}_m . f(\vec{x}_k)$".

I.8.4 Example. Turning to inequalities, $f(x) > 0$ means (is equivalent to) $(\exists y)(y = f(x) \wedge y > 0)$. In particular, it implies that $f(x) \downarrow$. □

I.8.5 Example. In the presence of partial functions, $\neg A = B$ and $A \neq B$ are not interchangeable. For example, $f(a) \neq b$ says (by I.8.3) that $(\exists y)(f(a) = y \wedge y \neq b)$. In particular, this entails that $f(a) \downarrow$. On the other hand, $\neg f(a) = b$ holds iff $f(a) \uparrow \vee (\exists y)(f(a) = y \wedge y \neq b)$.

We are not changing the rules of logic here, but are just amending our understanding of the semantics of the metanotation "$\neq$", to make it correct in the presence of partial functions. □

There are many approaches to defining computable functions, and they are all equivalent, that is, they define exactly the same set of functions. All except two of them begin by defining a notion of "computation model", that is, a set

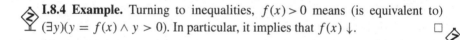

[†] Cf. I.7.2.

of string-manipulation algorithms (e.g., Turing machines, Markov algorithms, Kleene's equation manipulation processes), and then they define a computable function as one whose input-output relationship – coded as a relation *on strings* – can be verified by an algorithm belonging to the computation model.

There are two number-theoretic approaches, both due to Kleene, one using so-called *Kleene schemata*[†] and one that inductively defines the set of computable functions, bypassing the concepts of "algorithm" or "computation".[‡]

We follow the latter approach in this section. According to this, the set of computable functions is the smallest set of functions that includes some "indisputably computable" functions, and is closed under some "indisputably algorithmic" operations.[§]

The following are operations (on number-theoretic functions) that are centrally important:

I.8.6 Definition (Composition). Let $\lambda \vec{x}.g_i(\vec{x})$ $(i = 1, \ldots, n)$ and $\lambda \vec{y}_n.f(\vec{y}_n)$ be given functions.[¶] Then $h = \lambda \vec{x}.f(g_1(\vec{x}), \ldots, g_n(\vec{x}))$ is the result of their *composition*. □

 Note the requirement that *all* the variables of the "outermost" function, f, be substituted, and that each substitution (a function application, $g_i(\vec{x})$) apply to the same variable list $\vec{x}$. With additional tools, we can eventually relax this very rigid requirement.

I.8.7 Definition (Primitive Recursion). Let $\lambda x \vec{y}_n z.g(x, \vec{y}_n, z)$ and $\lambda \vec{y}_n.h(\vec{y}_n)$ be given. We say that $\lambda x \vec{y}_n.f(x, \vec{y}_n)$ is obtained by *primitive recursion* from h and g just in case it satisfies, for all x and $\vec{y}_n$, the following equations (the so-called *primitive recursive schema*):

$$f(0, \vec{y}_n) = h(\vec{y}_n)$$
$$f(x + 1, \vec{y}_n) = g(x, \vec{y}_n, f(x, \vec{y}_n))$$ □

I.8.8 Definition (Unbounded Search). Given $\lambda \vec{x} y_n.g(x, \vec{y}_n)$. f is defined from g by *unbounded search on the variable x* just in case, for all $\vec{y}_n$, the following

---

[†] These characterize inductively the set of all number-tuples $\langle z, \vec{x}, y \rangle$ which are intuitively understood to "code" the statement that the machine, or algorithm, z, when presented with input $\vec{x}$, will eventually output y.

[‡] Work on this originated with Dedekind, who characterized in this manner a proper subset of computable functions, that of *primitive recursive functions*.

[§] The reader will agree, once all the details are in hand, that the qualification "indisputably" is apt.

[¶] A function in this section, unless otherwise explicitly stated, is a number-theoretic partial function.

holds:

$$f(\vec{y}_n) = \begin{cases} \min\{x : g(x, \vec{y}_n) = 0 \wedge (\forall z)_{<x} g(z, \vec{y}_n) \downarrow\} \\ \uparrow \text{ if the minimum above does not exist} \end{cases} \quad (1)$$

In (1) above, the case "$\uparrow$" is short for "$f(\vec{y}_n)$ is undefined". We write $f(\vec{y}_n) = (\mu x)g(x, \vec{y}_n)$ as a short form of (1).   □

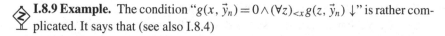 **I.8.9 Example.** The condition "$g(x, \vec{y}_n) = 0 \wedge (\forall z)_{<x} g(z, \vec{y}_n) \downarrow$" is rather complicated. It says that (see also I.8.4)

$$g(0, \vec{y}_n) > 0, \qquad g(1, \vec{y}_n) > 0, \ldots, \qquad g(x - 1, \vec{y}_n) > 0$$

but $g(x, \vec{y}_n) = 0$. For example, suppose that

$$g(x, y) = \begin{cases} 0 & \text{if } x = y = 1 \\ \uparrow & \text{otherwise} \end{cases}$$

Then, while the smallest x such that $g(x, 1) = 0$ holds is $x = 1$, this is not what (1) "computes". The definition (1) yields *undefined* in this case, since $g(0, 1) \uparrow$. Of course, the part "$(\forall z)_{<x} g(z, \vec{y}_n) \downarrow$" in (1) is superfluous if g is total.   □

The following functions are intuitively computable. They form the basis of an inductive definition of all computable functions.

I.8.10 Definition (Initial Functions).

Zero: z $(\lambda x.0)$
Successor: s $(\lambda x.x + 1)$
Identities or *projections*: u_i^n, for $n \geq 1$ and $1 \leq i \leq n$ $(\lambda \vec{x}_n.x_i)$.   □

I.8.11 Definition. The set of *partial computable* or *partial recursive* functions, $\mathfrak{P}$, is the closure of the initial functions above, under the operations composition, primitive recursion, and unbounded search.

The set of *computable* or *recursive* functions, $\mathfrak{R}$, is the set of all total functions of $\mathfrak{P}$.   □

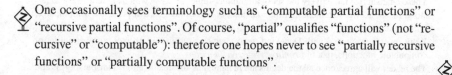 One occasionally sees terminology such as "computable partial functions" or "recursive partial functions". Of course, "partial" qualifies "functions" (not "recursive" or "computable"): therefore one hopes never to see "partially recursive functions" or "partially computable functions".

I.8.12 Definition. The set of *primitive recursive* functions, $\mathfrak{PR}$, is the closure of the initial functions above under the operations *composition* and *primitive recursion*. ☐

 The primitive recursive functions were defined by Dedekind and were called "recursive" until the recursive functions of I.8.11 were defined. Then the name of the functions of Dedekind was qualified to be "primitive".

Why are the functions in $\mathfrak{P}$ "computable"?[†] Well, an (informal) induction on the definition (I.8.11) shows why this is "correct".

The initial functions are clearly intuitively computable (e.g., by pencil and paper, by anyone who knows how to add 1 to an arbitrary natural number).

Suppose that each of $\lambda \vec{x}.g_i(\vec{x})$ $(i = 1, \ldots, n)$ and $\lambda \vec{y}_n.f(\vec{y}_n)$ are intuitively computable (i.e., we know how to compute the output, given the input). To compute $f(g_1(\vec{a}), \ldots, g_n(\vec{a}))$, given $\vec{a}$, we compute each of the $g_i(\vec{a})$, and then use the results as inputs to f.

To see why f (defined by a primitive recursive schema from h and g) is computable if h and g are, let us first introduce the notation $z := x$, which we understand to say "copy the value of x into z".

Then we can write an "algorithm" for the computation of $f(a, \vec{b}_n)$:

$$(1) \quad z := h(\vec{b}_n)$$

Repeat (2) below for $i = 0, 1, 2, \ldots, a - 1$:

$$(2) \quad z := g(i, \vec{b}_n, z)$$

Since (I.H.) the computations $h(\vec{b}_n)$ and $g(i, \vec{b}_n, z)$ can be carried out – regardless of the input values $\vec{b}_n$, i, and z – at the end of the "computation" indicated above, z holds the value $f(a, \vec{b}_n)$.

Finally, let $\lambda x \vec{y}_n.g(x, \vec{y}_n)$ be intuitively computable. We show how to compute $\lambda \vec{y}_n.(\mu x)g(x, \vec{y}_n)$:

(1) $x := 0$.
(2) if $g(x, \vec{b}_n) = 0$, go to step (5).
(3) $x := x + 1$.
(4) go back to step (2).
(5) Done! x holds the result.

The above algorithm justifies the term "unbounded search". We are searching by letting $x = 0, 1, 2, \ldots$ in turn. It is "unbounded" since we have no *a priori*

[†] We have "*computable*" and *computable*. The former connotes our intuitive understanding of the term. It means "intuitively computable". The latter has an exact definition (I.8.11).

knowledge of how far the search will have to go. It is also clear that the algorithm satisfies the definition of (μx):[†] We will hit step (5) iff progress was never blocked at step (2) (i.e., iff all along $g(i, \vec{b}_n) > 0$ (see I.8.4) until the first (smallest) i came along for which $g(i, \vec{b}_n) = 0$).

We have our first few simple results:

I.8.13 Proposition. $\mathfrak{R} \subset \mathfrak{P}$.

Proof. The $\subseteq$-part is by definition. The $\neq$-part follows from the fact that $e \in \mathfrak{P}$ but $e \notin \mathfrak{R}$, where we have denoted by "e" the totally undefined (empty) function $\lambda y.(\mu x)s(u_1^2(x, y))$ (in short, $e(y)$, for any y, is the smallest x such that $x + 1 = 0$; but such an x does not exist). $\qquad\square$

I.8.14 Proposition. $\mathfrak{R}$ *is closed under composition and primitive recursion.*

Proof. These two operations preserve total functions (why?). $\qquad\square$

I.8.15 Corollary. $\mathfrak{PR} \subseteq \mathfrak{R}$.

Proof. By induction on $\mathfrak{PR}$, since the initial functions (common to $\mathfrak{PR}$ and $\mathfrak{P}$) are total and hence are in $\mathfrak{R}$. $\qquad\square$

Thus all primitive recursive functions are total.

It can be shown that the inclusion $\mathfrak{PR} \subseteq \mathfrak{R}$ is proper, but we will not need this result (see, e.g., Tourlakis (1984)).

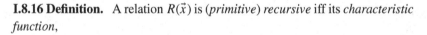

I.8.16 Definition. A relation $R(\vec{x})$ is (*primitive*) *recursive* iff its *characteristic function*,

$$\chi_R = \lambda \vec{x}. \begin{cases} 0 & \text{if } R(\vec{x}) \\ 1 & \text{if } \neg R(\vec{x}) \end{cases}$$

is (primitive) recursive.

The set of all primitive recursive (recursive) *relations*, or *predicates*,[‡] is denoted by $\mathfrak{PR}_*$ $(\mathfrak{R}_*)$. $\qquad\square$

[†] By the way, in modern Greek, one pronounces "μ" exactly like the English word "me".
[‡] Relations are often called "predicates" by computability practitioners.

Since we are to stay within $\mathbb{N}$, we need a special kind of subtraction, *proper subtraction:*[†]

$$x \overset{\cdot}{-} y \overset{\text{def}}{=} \begin{cases} x - y & \text{if } x \geq y \\ 0 & \text{otherwise} \end{cases}$$

I.8.17 Example. This example illustrates some important techniques used to circumvent the rigidity of our definitions.

We prove that $\lambda xy.x \overset{\cdot}{-} y \in \mathfrak{PR}$. First, we look at a special case. Let $p = \lambda x.x \overset{\cdot}{-} 1$ and $\widehat{p} = \lambda xy.p(x)$. Now $\widehat{p}$ is primitive recursive, since

$$\begin{aligned} \widehat{p}(0, y) &= z(y) \\ \widehat{p}(x + 1, y) &= u_1^3(x, y, \widehat{p}(x, y)) \end{aligned} \tag{1}$$

Thus, so is

$$p = \lambda x.\widehat{p}\big(u_1^1(x), z(x)\big) \tag{2}$$

Finally, let $d = \lambda xy.y \overset{\cdot}{-} x$. This is in $\mathfrak{PR}$, since

$$\begin{aligned} d(0, y) &= u_1^1(y) \\ d(x + 1, y) &= p\big(u_3^3(x, y, d(x, y))\big) \end{aligned} \tag{3}$$

Thus, $\lambda xy.x \overset{\cdot}{-} y$ is primitive recursive, since

$$\lambda xy.x \overset{\cdot}{-} y = \lambda xy.d\big(u_2^2(x, y), u_1^2(x, y)\big) \tag{4}$$

Our acrobatics here have worked around the following formal difficulties:

(i) Our number-theoretic functions have at least one argument. Thus, any instance of the primitive recursive schema must define a function of at least two arguments. This explains the introduction of $\widehat{p}$ in the schema (1).
(ii) A more user-friendly way to write (1) (in the *argot* of recursion theory) is

$$\begin{aligned} p(0) &= 0 \\ p(x + 1) &= x \end{aligned}$$

Indeed, "$u_1^3(x, y, \widehat{p}(x, y))$" is a fancy way (respecting the *form* of the primitive recursive schema) to just say "x". Moreover, one simply writes $p = \lambda x.\widehat{p}(x, 0)$ instead of (2) above.
(iii) Finally, (3) and (4) get around the fact that the primitive recursion schema iterates via the first variable. As this example shows, this is not cast in stone, for we can swap variables (with the help of the u_i^n).

[†] Some authors pronounce proper subtraction *monus*.

One must be careful not to gloss over this last hurdle by shrugging it off: "What's in a name?". It is not a matter of changing names *everywhere* to go from $\lambda xy.x \overset{\cdot}{-} y$ to $\lambda yx.y \overset{\cdot}{-} x$. We actually needed to work with the *first variable in the λ-list*, but (because of the nature of "$\overset{\cdot}{-}$") this variable should be *after* "$\overset{\cdot}{-}$". That is, we did need $d = \lambda xy.y \overset{\cdot}{-} x$.

In *argot*, (3) takes the simple form

$$x \overset{\cdot}{-} 0 = x$$
$$x \overset{\cdot}{-} (y + 1) = (x \overset{\cdot}{-} y) \overset{\cdot}{-} 1$$

The reader must have concluded (correctly) that the *argot* operations of permuting variables, identifying variables, augmenting the variable list with new variables (also, replacing a single variable with a function application or a constant) are not *argot* at all, but are derived "legal" *operations of substitution* (due to Grzegorczyk (1953) – see Exercise I.68).

Therefore, from now on we will relax our notational rigidity and benefit from the presence of these operations of substitution. □

I.8.18 Example. $\lambda xy.x + y$, $\lambda xy.x \times y$ (or, in implied multiplication notation, $\lambda xy.xy$), and $\lambda xy.x^y$ are in $\mathfrak{PR}$. Let us leave the first two as an easy exercise, and deal with the third one, since it entails an important point:

$$x^0 = 1$$
$$x^{y+1} = x \times x^y$$

The "important point" is regarding the basis case, $x^0 = 1$. We learn in "ordinary math" that 0^0 is undefined. If we sustain this point of view, then $\lambda xy.x^y$ cannot possibly be in $\mathfrak{PR}$ (why?). So we *re-define* 0^0 to be 1.

One does this kind of re-definition a lot in recursion theory (it is akin to removing removable discontinuities in calculus) when a function threatens *not* to be, say, primitive recursive for trivial reasons.

A trivial corollary is that $\lambda x.0^x \in \mathfrak{PR}$ (why?). This is a useful function, normally denoted by $\overline{sg}$. Clearly,

$$\overline{sg}(x) = \begin{cases} 1 & \text{if } x = 0 \\ 0 & \text{otherwise} \end{cases}$$

We also see that $\overline{sg}(x) = 1 \overset{\cdot}{-} x$, which provides an alternative proof that $\lambda x.0^x \in \mathfrak{PR}$. □

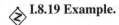

I.8.19 Example.

$$\lambda xyz. \begin{cases} y & \text{if } x = 0 \\ z & \text{if } x \neq 0 \end{cases}$$

is in $\mathfrak{PR}$. This function is often called the "switch" or "if-then-else", and is sometimes denoted by the name "sw".

We rest our case, since

$$sw(0, y, z) = y$$
$$sw(x + 1, y, z) = z$$

We see immediately that $sw(x, 1, 0) = 0^x = 1 \mathbin{\dot{-}} x$. The function $\lambda x.sw(x, 0, 1) = \lambda x.1 \mathbin{\dot{-}} (1 \mathbin{\dot{-}} x)$ has a special symbol: "sg". It is often called the *signum*, since it gives the sign of its argument. □

I.8.20 Lemma. $R(\vec{x})$ *is in* $\mathfrak{PR}_*$ *(respectively,* $\mathfrak{R}_*$*) iff, for some* $f \in \mathfrak{PR}$ *(respectively,* $f \in \mathfrak{R}$*),* $R(\vec{x}) \leftrightarrow f(\vec{x}) = 0$.

Proof. Only-if part: Take $f = \chi_R$. *If part:* $\chi_R = \lambda \vec{x}.sg(f(\vec{x}))$. □

I.8.21 Theorem. $\mathfrak{PR}_*$ *(respectively,* $\mathfrak{R}_*$*) is closed under replacement of variables by primitive recursive (respectively, recursive) functions.*

Proof. If χ_R is the characteristic function of $R(\vec{x}, y, \vec{z})$ and f is a total function, then $\lambda \vec{x}\vec{w}\vec{z}.\chi_R(\vec{x}, f(\vec{w}), \vec{z})$ is the characteristic function of $R(\vec{x}, f(\vec{w}), \vec{z})$. (See also Exercise I.68.) □

I.8.22 Theorem. $\mathfrak{PR}_*$ *and* $\mathfrak{R}_*$ *are closed under Boolean connectives ("Boolean operations") and bounded quantification.*

Proof. It suffices to cover $\neg$, $\vee$, $(\exists y)_{<z}$. We are given $R(\vec{x})$, $Q(\vec{y})$, and $P(y, \vec{x})$, all in $\mathfrak{PR}_*$ (or $\mathfrak{R}_*$; the argument is the same for both cases).

Case for $\neg$: $\chi_{\neg R} = \lambda \vec{x}.\overline{sg}(\chi_R(\vec{x}))$.
Case for $\vee$: $\chi_{R \vee Q} = \lambda \vec{x}\vec{y}.\chi_R(\vec{x})\chi_Q(\vec{y})$ (where we have used implied multiplication notation).
Case for $(\exists y)_{<z}$: To unclutter the notation, let us denote by $\chi_{\exists P}$ the characteristic function of $(\exists y)_{<z}P(y, \vec{x})$. Then

$$\chi_{\exists P}(0, \vec{x}) = 1$$
$$\chi_{\exists P}(z + 1, \vec{x}) = \chi_P(z, \vec{x})\chi_{\exists P}(z, \vec{x})$$ □

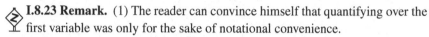

I.8.23 Remark. (1) The reader can convince himself that quantifying over the first variable was only for the sake of notational convenience.

(2) The case for $(\exists y)_{\leq z}$ (and therefore for $(\forall y)_{\leq z}$) can be easily made:

$$(\exists y)_{\leq z} Q(y, \vec{x}) \leftrightarrow (\exists y)_{<z+1} Q(y, \vec{x})$$

(here we have used I.8.21). □

I.8.24 Example (Bounded Summation and Multiplication). We are collecting tools that will be useful in our arithmetization of $\mathfrak{P}$. Two such tools are the operations $\sum_{y<z} f(y, \vec{x})$ and $\prod_{y<z} f(y, \vec{x})$. Both $\mathfrak{PR}$ and $\mathfrak{R}$ are closed under these operations. For example, here is the reason for $\prod$:

$$\prod_{y<0} f(y, \vec{x}) = 1$$
$$\prod_{y<z+1} f(y, \vec{x}) = f(z, \vec{x}) \prod_{y<z} f(y, \vec{x})$$

□

I.8.25 Definition (Bounded Search). For a total function $\lambda y\vec{x}.g(y, \vec{x})$ we define for all $\vec{x}$

$$(\mu y)_{<z} g(y, \vec{x}) \stackrel{\text{def}}{=} \begin{cases} \min\{y : y < z \wedge g(y, \vec{x}) = 0\} \\ z \text{ if the minimum does not exist} \end{cases}$$

The symbol $(\mu y)_{\leq z} g(y, \vec{x})$ is defined to mean $(\mu y)_{\leq z+1} g(y, \vec{x})$. □

Bounded search, $(\mu y)_{<z}$, searches a predetermined *domain*, $0, 1, \ldots, z - 1$. If unsuccessful, it returns the first number to the right of the domain.

We extend the use of search on predicates:

I.8.26 Definition. For a predicate $R(y, \vec{x})$, the symbols $(\mu y)R(y, \vec{x})$, $(\mu y)_{<z}$ $R(y, \vec{x})$, and $(\mu y)_{\leq z} R(y, \vec{x})$ mean $(\mu y)\chi_R(y, \vec{x})$, $(\mu y)_{<z}\chi_R(y, \vec{x})$, and $(\mu y)_{\leq z}\chi_R(y, \vec{x})$ respectively. □

I.8.27 Theorem (Definition by Cases). $\mathfrak{R}$ *and* $\mathfrak{PR}$ *are closed under the schema of definition by cases (below), where it is understood that the relations* R_i *are mutually exclusive:*

$$f(\vec{x}) = \begin{cases} g_1(\vec{x}) & \text{if } R_1(\vec{x}) \\ g_2(\vec{x}) & \text{if } R_2(\vec{x}) \\ \vdots \\ g_k(\vec{x}) & \text{if } R_k(\vec{x}) \\ g_{k+1}(\vec{x}) & \text{otherwise} \end{cases}$$

Proof. Perhaps the simplest proof is to observe that

$$f(\vec{x}) = g_1(\vec{x})\overline{sg}(\chi_{R_1}(\vec{x})) + \cdots + g_k(\vec{x})\overline{sg}(\chi_{R_k}(\vec{x})) + g_{k+1}(\vec{x})\overline{sg}(\chi_Q(\vec{x}))$$

a fixed-length sum, where $Q \leftrightarrow \neg(R_1 \vee \cdots \vee R_k)$. □

I.8.28 Theorem. $\mathfrak{R}$ *and* $\mathfrak{PR}$ *are closed under bounded search.*

Proof. Let $g(z, \vec{x}) = (\mu y)_{<z} f(y, \vec{x})$. Then

$$g(0, \vec{x}) = 0$$
$$g(z + 1, \vec{x}) = \textbf{if } g(z, \vec{x}) \neq z \textbf{ then } g(z, \vec{x})$$
$$\textbf{else if } f(z, \vec{x}) = 0 \textbf{ then } z$$
$$\textbf{else } z + 1$$

The second equation above is $g(z + 1, \vec{x}) = k(z, \vec{x}, g(z, \vec{x}))$, where

$$k(z, \vec{x}, w) = \textbf{if } w \neq z \textbf{ then } w$$
$$\textbf{else if } f(z, \vec{x}) = 0 \textbf{ then } z$$
$$\textbf{else } z + 1$$

Clearly k is wherever f is (in $\mathfrak{R}$ or $\mathfrak{PR}$), since (see I.8.19) $sw \in \mathfrak{PR}$. □

I.8.29 Proposition. *The following are all primitive recursive:*

(i) $\lambda x y. \left\lfloor \dfrac{x}{y} \right\rfloor \left(\text{the quotient of the division } \dfrac{x}{y}\right)$

(ii) $\lambda x y. rem(x, y) \left(\text{the remainder of the division } \dfrac{x}{y}\right)$

(iii) $\lambda x y. x | y$ ("*x divides y*")

(iv) $Pr(x)$ (*x is a prime*)

(v) $\lambda n. p_n$ (*the n th prime*)

(vi) $\lambda n x. \exp(n, x)$ (*the exponent of p_n in the prime factorization of x*)

(vii) $Seq(x)$ ("*x's prime number factorization contains at least one prime, but no gaps*")

Proof. (i):

$$\left\lfloor \dfrac{x}{y} \right\rfloor = (\mu z)_{\leq x}((z + 1)y > x) \tag{1}$$

(1) is correct for all $y \neq 0$. Since we do not want the quotient function to fail primitive recursiveness for a trivial reason (we have a "removable nontotalness" – see also I.8.18), we define $\lfloor x/y \rfloor$ to equal the right hand side of (1) at all times (of course, the right hand side *is* total).

(ii): $rem(x, y) = x \mathbin{\dot-} \lfloor x/y \rfloor y$.
(iii): $x|y \leftrightarrow rem(y, x) = 0$.
(iv): $Pr(x) \leftrightarrow x > 1 \wedge (\forall y)_{\leq x}(y|x \rightarrow y = 1 \vee y = x)$.
(v):

$$p_0 = 2$$
$$p_{n+1} = (\mu y)_{\leq 2^{2^{n+1}}}(Pr(y) \wedge y > p_n)$$

The above is based on Euclid's proof that there are infinitely many primes
($p_0 p_1 \cdots p_n + 1$ is either a prime, $q \geq p_{n+1}$, or it has a prime divisor $q \geq p_{n+1}$)
and an induction on n that shows $p_n \leq 2^{2^n}$.

(vi): $\exp(n, x) = (\mu y)_{\leq x}\left(\neg(p_n^{y+1}|x)\right)$.
(vii): $Seq(x) \leftrightarrow x > 1 \wedge (\forall y)_{\leq x}(\forall z)_{\leq x}(y|x \wedge Pr(y) \wedge Pr(z) \wedge z < y \rightarrow z|x)$.
$\square$

I.8.30 Definition (Coding and Decoding Number Sequences). An arbitrary
(finite) sequence of natural numbers $a_0, a_1, \ldots, a_{n-1}$ will be coded as

$$p_0^{a_0+1} p_1^{a_1+1} \cdots p_{n-1}^{a_{n-1}+1}$$

We use the notation

$$\langle a_0, a_1, \ldots, a_{n-1} \rangle \stackrel{\text{def}}{=} \prod_{y<n} p_y^{a_y+1} \tag{1}$$

$\square$

In set theory one likes to denote tuples by $\langle a_0, \ldots, a_{n-1} \rangle$ as well, a practice
that we have been following (cf. Section I.2). To avoid notational confusion, in
those rare cases where we want to write down both a code $\langle a_0, \ldots, a_{n-1} \rangle$ of a
sequence $a_0, \ldots, a_{n-1}$ and an n-tuple in set theory's sense, we write the latter
in the "old" notation, with round brackets, that is, $(a_0, \ldots, a_{n-1})$.

Why "$+ 1$" in the exponent? Without that, all three sequences "2", "2, 0",
and "2, 0, 0" get the same code, namely 2^2. This is a drawback, for if we are
given the code 2^2 but do not know the length of the coded sequence, then we
cannot decode 2^2 back into the original sequence correctly. Contrast this with
the schema (1) above, where these three examples are coded as 2^3, $2^3 \cdot 3$ and
$2^3 \cdot 3 \cdot 5$ respectively. We see that the coding (1) above codes the length n of
the sequence $a_0, a_1, \ldots, a_{n-1}$ into the code $z = \langle a_0, a_1, \ldots, a_{n-1} \rangle$. This length
is the number of primes in the decomposition of z (of course, $Seq(z)$ is true),
and it is useful to have a function for it, called "lh". There are many ways to
define a primitive recursive length function, lh, that does the job. The simplest
definition allows lh to give nonsensical answers for all inputs that do *not* code

sequences. Examples of such inputs are 0, 1, 10;[†] in short, any number z such that $\neg Seq(z)$. We let

$$lh(z) = (\mu y)_{\leq z}(\neg p_y | z)$$

Clearly, $lh \in \mathfrak{PR}$.

From all the above, we get that $Seq(z)$ iff, for some $a_0, a_1, \ldots, a_{n-1}, z = \langle a_0, a_1, \ldots, a_{n-1} \rangle$ (this justifies the mnemonic "Seq" for "sequence").

Clearly, $lh(z) = n$ in this case, and[‡] $\exp(i, z) \overset{\cdot}{-} 1 = a_i$ for $i = 0, 1, \ldots, n-1$ ($\exp(i, z) \overset{\cdot}{-} 1 = 0$ if $i \geq n$).

It is customary to use the more compact symbol

$$(z)_i \overset{\text{def}}{=} \exp(i, z) \overset{\cdot}{-} 1$$

Thus, if $Seq(z)$, then the sequence $(z)_i$, for $i = 0, \ldots, lh(z) \overset{\cdot}{-} 1$, decodes z.

We will also need to express sequence *concatenation* primitive recursively. We define concatenation, "$*$", by

$$\langle a_0, \ldots, a_{n-1} \rangle * \langle b_0, \ldots, b_{m-1} \rangle \overset{\text{def}}{=} \langle a_0, \ldots, a_{n-1}, b_0, \ldots, b_{m-1} \rangle \qquad (2)$$

Of course, for $\lambda xy.x * y$ to be in $\mathfrak{PR}$ we must have a total function to begin with, so that "$*$" must be defined on all natural numbers, not on just those satisfying Seq.

The following definition is at once seen to satisfy all our requirements:

$$x * y \overset{\text{def}}{=} x \cdot \prod_{i < lh(y)} p_{i+lh(x)}^{\exp(i, y)} \qquad (3)$$

I.8.31 Theorem (Course-of-Values (Primitive) Recursion). *Let* $H(x, \vec{y}_n)$, *the history function of* f, *stand for* $\langle f(0, \vec{y}_n), \ldots, f(x, \vec{y}_n) \rangle$ *for* $x \geq 0$.

Then $\mathfrak{PR}$ *and* $\mathfrak{R}$ *are closed under the following schema of* course-of-values *recursion:*

$$f(0, \vec{y}_n) = h(\vec{y}_n)$$
$$f(x + 1, \vec{y}_n) = g(x, \vec{y}_n, H(x, \vec{y}_n))$$

Proof. It follows from the (ordinary) primitive recursion

$$H(0, \vec{y}_n) = \langle h(\vec{y}_n) \rangle$$
$$H(x + 1, \vec{y}_n) = H(x, \vec{y}_n) * \langle f(x + 1, \vec{y}_n) \rangle$$
$$= H(x, \vec{y}_n) * \langle g(x, \vec{y}_n, H(x, \vec{y}_n)) \rangle$$

and $f(x, \vec{y}_n) = (H(x, \vec{y}_n))_x$. $\qquad \square$

[†] Our definition gives $lh(0) = 1$, $lh(1) = 0$, $lh(10) = 1$.
[‡] Since $Seq(z)$, we have $\exp(i, z) \overset{\cdot}{-} 1 = \exp(i, z) - 1$ for $i = 0, 1, \ldots, n-1$.

We next *arithmetize* $\mathfrak{P}$-functions and their computations. We will assign "*program codes*" to each function. A program code – called in the literature a *Gödel number*, or a *ϕ-index*, or just an *index* – is, intuitively, a number in $\mathbb{N}$ that codes the "instructions" necessary to compute a $\mathfrak{P}$-function.

 If $i \in \mathbb{N}$ is $a^\dagger$ code for $f \in \mathfrak{P}$, then we write

$$f = \{i\} \qquad \text{(Kleene's notation)}$$

or[‡]

$$f = \phi_i \qquad \text{(Rogers's (1967) notation)}$$

Thus, either $\{i\}$ or ϕ_i denotes the function with code i.

The following table indicates how to assign Gödel numbers (middle column) to all partial recursive functions by following the definition of $\mathfrak{P}$. In the table, $\widehat{f}$ indicates a code of f:

Function	Code	Comment
$\lambda x.0$	$\langle 0, 1, 0 \rangle$	
$\lambda x.x + 1$	$\langle 0, 1, 1 \rangle$	
$\lambda \vec{x}_n.x_i$	$\langle 0, n, i, 2 \rangle$	$1 \leq i \leq n$
Composition:	$\langle 1, m, \widehat{f}, \widehat{g}_1, \ldots, \widehat{g}_n \rangle$	f must be n-ary
$\quad f(g_1(\vec{y}_m), \ldots, g_n(\vec{y}_m))$		g_i must be m-ary
Primitive recursion from		
$\quad$ *basis h* and *iterated* part *g*	$\langle 2, n+1, \widehat{h}, \widehat{g} \rangle$	h must be n-ary
		g must be $(n+2)$-ary
Unbounded search:	$\langle 3, n, \widehat{f} \rangle$	f must be $(n+1)$-ary
$\quad (\mu y) f(y, \vec{x}_n)$		and $n > 0$

 We have been somewhat loose in our description above. "The following table indicates how to assign Gödel numbers (middle column) to all partial recursive functions by following the definition of $\mathfrak{P}$", we have said, perhaps leading the reader to think that we are defining the codes by recursion on $\mathfrak{P}$. Not so. After all, each function has infinitely many codes.

What was really involved in the table – see also below – was arguing backwards: a specification of how we would like our ϕ-indices behave once we

---

[†] The indefinite article is appropriate here. Just as in real life a computable function has infinitely many different programs that compute it, a partial recursive function f has infinitely many different codes (see I.8.34 later on).

[‡] That is where the name "ϕ-index" comes from.

obtained them. We now turn to showing how to actually obtain them by *directly defining* the set of all ϕ-indices, Φ, as an inductively defined subset of $\{z : Seq(z)\}$:

$$\Phi = Cl(\mathscr{I}, \mathscr{R})$$

where $\mathscr{I} = \{\langle 0, 1, 0 \rangle, \langle 0, 1, 1 \rangle\} \cup \{\langle 0, n, i, 2 \rangle : n > 0 \wedge 1 \le i \le n\}$, and the rule set $\mathscr{R}$ consists of the following three operations:

(i) *Coding composition*: Input a and b_i ($i = 1, \ldots, n$) causes output

$$\langle 1, m, a, b_1, \ldots, b_n \rangle$$

provided $(a)_1 = n$ and $(b_i)_1 = m$, for $i = 1, \ldots, n$.

(ii) *Coding primitive recursion*: Input a and b causes output

$$\langle 2, n + 1, a, b \rangle$$

provided $(a)_1 = n$ and $(b)_1 = n + 2$.

(iii) *Coding unbounded search*: Input a causes output

$$\langle 3, n, a \rangle$$

provided $(a)_1 = n + 1$ and $n > 0$.[†]

By the uniqueness of prime number decomposition, the pair $(\mathscr{I}, \mathscr{R})$ is unambiguous (see I.2.10, p. 24). Therefore we define by recursion on Φ (cf. I.2.13) a *total* function $\lambda a.\{a\}$ (or $\lambda a.\phi_a$)[‡] for each $a \in \Phi$:

$$\{\langle 0, 1, 0 \rangle\} = \lambda x.0$$
$$\{\langle 0, 1, 1 \rangle\} = \lambda x.x + 1$$
$$\{\langle 0, n, i, 2 \rangle\} = \lambda \vec{x}_n.x_i$$
$$\{\langle 1, m, a, b_1, \ldots, b_n \rangle\} = \lambda \vec{y}_m.\{a\}(\{b_1\}(\vec{y}_m), \ldots, \{b_n\}(\vec{y}_m))$$
$$\{\langle 2, n + 1, a, b \rangle\} = \lambda x \vec{y}_n.Prec(\{a\}, \{b\})$$
$$\{\langle 3, n, a \rangle\} = \lambda \vec{x}_n.(\mu y)\{a\}(y, \vec{x}_n)$$

In the above recursive definition we have used the abbreviation $Prec(\{a\}, \{b\})$ for the function given (for all x, $\vec{y}_n$) by the primitive recursive schema (I.8.7) with h-part $\{a\}$ and g-part $\{b\}$.

We can now make the intentions implied in the above table official:

[†] By an obvious I.H. the other cases can fend for themselves, but, here, reducing the number of arguments must not result in 0 arguments, as we have decided not to allow 0-ary functions.

[‡] The input, a, is a code; the output, $\{a\}$ or ϕ_a, is a function.

I.8.32 Theorem. $\mathfrak{P} = \{\{a\} : a \in \Phi\}$.

Proof. $\subseteq$-part: Induction on $\mathfrak{P}$. The table encapsulates the argument diagrammatically.

$\supseteq$-part: Induction on Φ. It follows trivially from the recursive definition of $\{a\}$ and the fact that $\mathfrak{P}$ contains the initial functions and is closed under composition, primitive recursion, and unbounded search. $\qquad\square$

 I.8.33 Remark. (Important.) Thus, $f \in \mathfrak{P}$ iff for some $a \in \Phi$, $f = \{a\}$. $\quad\square$

I.8.34 Example. Every function $f \in \mathfrak{P}$ has infinitely many ϕ-indices. Indeed, let $f = \{\widehat{f}\}$. Since $f = \lambda \vec{x}_n . u_1^1(f(\vec{x}_n))$, we obtain $f = \{\langle 1, n, \langle 0, 1, 1, 2\rangle, \widehat{f}\rangle\}$ as well. Since $\langle 1, n, \langle 0, 1, 1, 2\rangle, \widehat{f}\rangle > \widehat{f}$, the claim follows. $\qquad\square$

I.8.35 Theorem. *The relation $x \in \Phi$ is primitive recursive.*

Proof. Let χ denote the characteristic function of the relation "$x \in \Phi$". Then

$$\chi(0) = 1$$
$$\chi(x+1) = 0 \text{ if } \quad x + 1 = \langle 0, 1, 0\rangle \vee x + 1 = \langle 0, 1, 1\rangle \vee$$
$$(\exists n, i)_{\leq x}(n > 0 \wedge 0 < i \leq n \wedge x + 1 = \langle 0, n, i, 2\rangle) \vee$$
$$(\exists a, b, m, n)_{\leq x}(\chi(a) = 0 \wedge (a)_1 = n \wedge Seq(b) \wedge$$
$$lh(b) = n \wedge (\forall i)_{<n}(\chi((b)_i) = 0 \wedge$$
$$((b)_i)_1 = m) \wedge x + 1 = \langle 1, m, a\rangle * b) \vee$$
$$(\exists a, b, n)_{\leq x}(\chi(a) = 0 \wedge (a)_1 = n \wedge \chi(b) = 0 \wedge$$
$$(b)_1 = n + 2 \wedge x + 1 = \langle 2, n + 1, a, b\rangle) \vee$$
$$(\exists a, n)_{\leq x}(\chi(a) = 0 \wedge (a)_1 = n + 1 \wedge n > 0 \wedge$$
$$x + 1 = \langle 3, n, a\rangle)$$
$$= 1 \text{ otherwise}$$

The above can easily be seen to be a course-of-values recursion. For example, if $H(x) = \langle \chi(0), \ldots, \chi(x)\rangle$, then an occurrence of "$\chi(a) = 0$" above can be replaced by "$(H(x))_a = 0$", since $a \leq x$. $\qquad\square$

 We think of[†] a *computation* as a *sequence of equations* like $\{e\}(\vec{a}) = b$. Such an equation is intuitively read as "the program e, when it runs on input $\vec{a}$, produces

---

[†] "We think of" indicates our determination to avoid a rigorous definition. The integrity of our exposition will not suffer from this.

output *b*". An equation will be *legitimate* iff

(i) it states an input-output relation of some initial function (i.e., $(e)_0 = 0$), or
(ii) it states an input-output relation according to ϕ-indices e such that $(e)_0 \in \{1, 2, 3\}$, using results (i.e., equations) that have *already appeared in the sequence.*

For example, in order to state $(\mu y)\{e\}(y, \vec{a}_n) = b$ – that is, $\{\langle 3, n, e \rangle\}(\vec{a}_n) = b$ – one must ensure that all the equations,

$$\{e\}(b, \vec{a}_n) = 0, \{e\}(0, \vec{a}_n) = r_0, \ldots, \{e\}(b - 1, \vec{a}_n) = r_{b-1}$$

where the r_i's are all non-zero, have already appeared in the sequence. In our coding, every equation $\{a\}(\vec{a}_n) = b$ will be denoted by a triple $\langle e, \vec{a}_n, b \rangle$ that codes, in that order, the ϕ-index, the input, and the output. We will collect (code) all these triples into a single code, $u = \langle \ldots, \langle e, \vec{a}_n, b \rangle, \ldots \rangle$.

Before proceeding, let us define the primitive recursive predicates

(1) $\lambda uv. u \in v$ ("v is a term in the (coded) sequence u"),
(2) $\lambda uvw. v <_u w$ ("v occurs *before* w in the (coded) sequence u").

Primitive recursiveness follows from the equivalences

$$v \in u \leftrightarrow Seq(u) \wedge (\exists i)_{<lh(u)}(u)_i = v$$

$$v <_u w \leftrightarrow v \in u \wedge w \in u \wedge (\exists i, j)_{<lh(u)}((u)_i = v \wedge (u)_j = w \wedge i < j)$$

We are now ready to define the relation "*Computation(u)*" which holds iff u codes a computation according to the previous understanding. This involves a lengthy formula. In the interest of readability, comments enclosed in { }-brackets are included in the left margin, to indicate the case under consideration.

$Computation(u) \leftrightarrow Seq(u) \wedge (\forall v)_{\leq u}[v \in u \rightarrow$
$\{\lambda x.0\}$ $\quad (\exists x)_{\leq u} v = \langle \langle 0, 1, 0 \rangle, x, 0 \rangle \vee$
$\{\lambda x.x + 1\}$ $\quad (\exists x)_{\leq u} v = \langle \langle 0, 1, 1 \rangle, x, x + 1 \rangle \vee$
$\{\lambda \vec{x}_n.x_i\}$ $\quad (\exists x, n, i)_{\leq u}\{Seq(x) \wedge n = lh(x) \wedge i < n \wedge$
$\qquad v = \langle \langle 0, n, i + 1, 2 \rangle \rangle * x * \langle (x)_i \rangle \} \vee$
$\{\text{composition}\}$ $\quad (\exists x, y, \widehat{f}, \widehat{g}, m, n, z)_{\leq u}\{Seq(x) \wedge Seq(y) \wedge Seq(\widehat{f}) \wedge$
$\qquad Seq(\widehat{g}) \wedge n = lh(x) \wedge n = lh(\widehat{g}) \wedge m = lh(y) \wedge$
$\qquad (\widehat{f})_1 = n \wedge (\forall i)_{<n}(Seq((\widehat{g})_i) \wedge ((\widehat{g})_i)_1 = m) \wedge$
$\qquad v = \langle \langle 1, m, \widehat{f} \rangle * \widehat{g} \rangle * y * \langle z \rangle \wedge$
$\qquad \langle \widehat{f} \rangle * x * \langle z \rangle <_u v \wedge$
$\qquad (\forall i)_{<n} \langle (\widehat{g})_i \rangle * y * \langle (x)_i \rangle <_u v \} \vee$

{prim. recursion} $(\exists x, y, \widehat{h}, \widehat{g}, n, c)_{\leq u}\{Seq(\widehat{h}) \wedge (\widehat{h})_1 = n \wedge Seq(\widehat{g})\wedge$
$(\widehat{g})_1 = n + 2 \wedge Seq(y) \wedge lh(y) = n \wedge Seq(c)\wedge$
$lh(c) = x + 1 \wedge v = \langle\langle 2, n+1, \widehat{h}, \widehat{g}\rangle, x\rangle * y * \langle(c)_x\rangle\wedge$
$\langle\widehat{h}\rangle * y * \langle(c)_0\rangle <_u v \wedge (\forall i)_{<x}\langle\widehat{g}, i\rangle * y * \langle(c)_i,$
$(c)_{i+1}\rangle <_u v\}\vee$

$\{(\mu y)f(y, \vec{x}_n)\}$ $(\exists \widehat{f}, y, x, n, r)_{\leq u}\{Seq(\widehat{f}) \wedge (\widehat{f})_1 = n + 1 \wedge n > 0\wedge$
$Seq(x) \wedge lh(x) = n \wedge Seq(r) \wedge lh(r) = y\wedge$
$v = \langle\langle 3, n, \widehat{f}\rangle\rangle * x * \langle y\rangle\wedge$
$\langle\widehat{f}, y\rangle * x * \langle 0\rangle <_u v\wedge$
$(\forall i)_{<y}(\langle\widehat{f}, i\rangle * x * \langle(r)_i\rangle <_u v \wedge (r)_i > 0)\}]$

Clearly, the formula to the right of "$\leftrightarrow$" above we see that *Computation(u)* is primitive recursive.

I.8.36 Definition (The Kleene T-Predicate). For each $n \in \mathbb{N}$, $T^{(n)}(a, \vec{x}_n, z)$ stands for *Computation*$((z)_1) \wedge \langle a, \vec{x}_n, (z)_0\rangle \in (z)_1$. $\square$

The above discussion yields immediately:

I.8.37 Theorem (Kleene Normal Form Theorem).

(1) $y = \{a\}(\vec{x}_n) \equiv (\exists z)(T^{(n)}(a, \vec{x}_n, z) \wedge (z)_0 = y)$.
(2) $\{a\}(\vec{x}_n) = ((\mu z)T^{(n)}(a, \vec{x}_n, z))_0$.
(3) $\{a\}(\vec{x}_n) \downarrow\equiv (\exists z)T^{(n)}(a, \vec{x}_n, z)$.

I.8.38 Remark. (Very important.) The right hand side of I.8.37(2) above is meaningful for all $a \in \mathbb{N}$, while the left hand side is only meaningful for $a \in \Phi$.

We now extend the symbols $\{a\}$ and ϕ_a to be meaningful for *all* $a \in \mathbb{N}$. In all cases, the meaning is given by the right hand side of (2).

Of course, if $a \notin \Phi$, then $(\mu z)T^{(n)}(a, \vec{x}_n, z) \uparrow$ *for all* $\vec{x}_n$, since $T^{(n)}(a, \vec{x}_n, z)$ will be false under the circumstances. Hence also $\{a\}(\vec{x}_n) \uparrow$, as it should be intuitively: In computer programmer's jargon, "if the program a is syntactically incorrect, then it will not run, for it will not even compile. Thus, it will define the everywhere undefined function".

By the above, I.8.33 now is strengthened to read "thus, $f \in \mathfrak{P}$ iff for some $a \in \mathbb{N}$, $f = \{a\}$". $\square$

We can now define a $\mathfrak{P}$-counterpart to $\mathfrak{R}_*$ and $\mathfrak{P}\mathfrak{R}_*$ and consider its closure properties.

I.8.39 Definition (Semi-recursive Relations or Predicates). A relation $P(\vec{x})$ is *semi-recursive* iff for some $f \in \mathfrak{P}$, the equivalence

$$P(\vec{x}) \leftrightarrow f(\vec{x}) \downarrow \qquad (1)$$

holds (for all $\vec{x}$, of course). Equivalently, we can state that $P = dom(f)$.

The set of all semi-recursive relations is denoted by $\mathfrak{P}_*$.[†]

If $f = \{a\}$ in (1) above, then we say that a is *a semi-recursive index* of P.

If P has one argument (i.e., $P \subseteq \mathbb{N}$) and a is one of its semi-recursive indices, then we write $P = W_a$ (Rogers (1967)).

We have at once

I.8.40 Corollary (Normal Form Theorem for Semi-recursive Relations). $P(\vec{x}_n) \in \mathfrak{P}_*$ *iff, for some* $a \in \mathbb{N}$,

$$P(\vec{x}_n) \leftrightarrow (\exists z)T^{(n)}(a, \vec{x}_n, z)$$

Proof. By definition (and Theorem I.8.32 along with Remark I.8.38), $P(\vec{x}_n) \in \mathfrak{P}_*$ iff, for some $a \in \mathbb{N}$, $P(\vec{x}_n) \leftrightarrow \{a\}(\vec{x}_n) \downarrow$. Now invoke I.8.37(3). $\quad\square$

Rephrasing the above (hiding the "a", and remembering that $\mathfrak{PR}_* \subseteq \mathfrak{R}_*$) we have

I.8.41 Corollary (Strong Projection Theorem). $P(\vec{x}_n) \in \mathfrak{P}_*$ *iff, for some* $Q(\vec{x}_n, z) \in \mathfrak{R}_*$,

$$P(\vec{x}_n) \leftrightarrow (\exists z)Q(\vec{x}_n, z)$$

Proof. For the only-if part take $Q(\vec{x}_n, z)$ to be $\lambda\vec{x}_n z.T^{(n)}(a, \vec{x}_n, z)$ for appropriate $a \in \mathbb{N}$. For the if part take $f = \lambda\vec{x}_n.(\mu z)Q(\vec{x}_n, z)$. Then $f \in \mathfrak{P}$ and $P(\vec{x}_n) \leftrightarrow f(\vec{x}_n) \downarrow$. $\quad\square$

Here is a characterization of $\mathfrak{P}_*$ that is identical in form to the characterizations of $\mathfrak{PR}_*$ and $\mathfrak{R}_*$ (Lemma I.8.20).

I.8.42 Corollary. $P(\vec{x}_n) \in \mathfrak{P}_*$ *iff, for some* $f \in \mathfrak{P}$,

$$P(\vec{x}_n) \leftrightarrow f(\vec{x}_n) = 0$$

[†] We are making this symbol up (it is not standard in the literature). We are motivated by comparing the contents of I.8.20 and of I.8.42 below.

Proof. Only-if part: Say $P(\vec{x}_n) \leftrightarrow g(\vec{x}_n) \downarrow$. Take $f = \lambda\vec{x}_n.0 \cdot g(\vec{x}_n)$.

If part: Let $f = \{a\}$. By I.8.37(1), $f(\vec{x}_n) = 0 \leftrightarrow (\exists z)(T^{(n)}(a, \vec{x}_n, z) \wedge (z)_0 = 0)$. We are done by strong projection. $\square$

The usual call-by-value semantics of $f(g(\vec{x}), \vec{y})$ require divergence if $g(\vec{x}) \uparrow$. That is, before we embark on calculating the value of $f(g(\vec{x}), \vec{y})$, we require the values of all the inputs. In particular, $0 \cdot g(\vec{x}_n) \uparrow$ iff $g(\vec{x}_n) \uparrow$.

Of course, "$0 \cdot g(\vec{x}_n)$" is convenient notation. If we set $\{b\} = g$, we can write instead

$$((\mu z)(T^{(n)}(b, \vec{x}_n, (z)_1) \wedge (z)_0 = 0))_0$$

We immediately obtain

I.8.43 Corollary. $\mathfrak{R}_* \subseteq \mathfrak{P}_*$.

Intuitively, for a predicate $R \in \mathfrak{R}_*$ we have an algorithm (that computes χ_R) that for any input $\vec{x}$ will halt and answer "yes" ($= 0$) or "no" ($= 1$) to the question "$\vec{x} \in R$?"

For a predicate $Q \in \mathfrak{P}_*$ we are only guaranteed the existence of a weaker algorithm (for $f \in \mathfrak{P}$ such that $\mathrm{dom}(f) = Q$). It will halt iff the answer to the question "$\vec{x} \in Q$?" is "yes" (and halting will amount to "yes"). If the answer is "no", it will never tell, because it will (as we say for non-halting) "loop for ever" (or diverge). Hence the name "semi-recursive" for such predicates.

I.8.44 Theorem. $R \in \mathfrak{R}_*$ *iff both R and $\neg R$ are in $\mathfrak{P}_*$.*

Proof. Only-if part. By I.8.43 and closure of $\mathfrak{R}_*$ under $\neg$.

If part. Let i and j be semi-recursive indices of R and $\neg R$ respectively, that is,

$$R(\vec{x}_n) \leftrightarrow (\exists z)T^{(n)}(i, \vec{x}_n, z)$$
$$\neg R(\vec{x}_n) \leftrightarrow (\exists z)T^{(n)}(j, \vec{x}_n, z)$$

Define

$$g = \lambda\vec{x}_n.(\mu z)(T^{(n)}(i, \vec{x}_n, z) \vee T^{(n)}(j, \vec{x}_n, z))$$

Trivially, $g \in \mathfrak{P}$. Hence, $g \in \mathfrak{R}$, since it is total (why?). We are done by noticing that $R(\vec{x}_n) \leftrightarrow T^{(n)}(i, \vec{x}_n, g(\vec{x}_n))$. $\square$

I.8.45 Definition (Unsolvable Problems; Halting Problem). A *problem* is a question "$\vec{x} \in R$?" for any predicate R. "The problem $\vec{x} \in R$ is *recursively* *unsolvable"*, or just *"unsolvable"*, means that $R \notin \mathfrak{R}_*$, that is, intuitively, there is no algorithmic solution to the problem.

The *halting problem* has central significance in recursion theory. It is the question whether program x will ever halt if it starts computing on input x. That is, setting $K = \{x : \{x\}(x) \downarrow\}$ we can then ask "$x \in K$?". This question is the halting problem.[†] We denote the complement of K by $\overline{K}$. We will refer to K as the *halting set*.

I.8.46 Theorem (Unsolvability of the Halting Problem). *The halting problem is unsolvable.*

Proof. It suffices to show that $\overline{K}$ is not semi-recursive. Suppose instead that i is a semi-recursive index of the set. Thus,

$$x \in \overline{K} \leftrightarrow (\exists z) T^{(1)}(i, x, z)$$

or, making the part $x \in \overline{K}$ – that is, $\{x\}(x) \uparrow$ – explicit,

$$\neg(\exists z) T^{(1)}(x, x, z) \leftrightarrow (\exists z) T^{(1)}(i, x, z) \tag{1}$$

Substituting i into x in (1), we get a contradiction. $\qquad\square$

I.8.47 Remark. Let us look at the above in the light of W_a-notation (p. 143).

Now, $\{x\}(x) \uparrow$ iff $x \notin W_x$; thus we want to show

$$\neg(\exists i)(W_i = \{x : x \notin W_x\}) \tag{2}$$

(2) says "$\{x : x \notin W_x\}$ is not a W_i". Well, if (2) is false, then, for some i,

$$x \in W_i \leftrightarrow x \notin W_x$$

and hence

$$i \in W_i \leftrightarrow i \notin W_i$$

– a contradiction. This is a classic application of Cantor *diagonalization*[‡] and is formally the same argument as in *Russell's paradox*, according to which $\{x : x \notin x\}$ is not a *set* – just omit the symbol "W" throughout.

The analogy is more than morphological: Our argument shows that $\{x : x \notin W_x\}$ is not an object of the *same type* as the rightmost object in the $\{\ \}$ brackets.

[†] "K" is a reasonably well-reserved symbol for the set $\{x : \{x\}(x) \downarrow\}$. Unfortunately, K is also used for the first projection of a pairing function, but the context easily decides which is which.

[‡] Cantor's theorem showed that if $(X_a)_{a \in I}$ is a family of sets, then $\{a : a \notin X_a\}$ is not an "X_i" – it is not in the family – for otherwise $i \in X_i$ iff $i \notin X_i$.

Russell's argument too shows that $\{x : x \notin x\}$ is not an object of the *same type* as the rightmost object in the $\{\ \}$ brackets. That is, unlike x, it is not a set. □

$K \in \mathfrak{P}_*$, of course, since $\{x\}(x) \downarrow \leftrightarrow (\exists z)T^{(1)}(x, x, z)$. We conclude that the inclusion $\mathfrak{R}_* \subseteq \mathfrak{P}_*$ is proper, i.e., $\mathfrak{R}_* \subset \mathfrak{P}_*$.

I.8.48 Theorem (Closure Properties of $\mathfrak{P}_*$). $\mathfrak{P}_*$ *is closed under* $\vee$, $\wedge$, $(\exists y)_{<z}$, $(\exists y)$, $(\forall y)_{<z}$. *It is* not *closed under either* $\neg$ *or* $(\forall y)$.

Proof. We will rely on the normal form theorem for semi-recursive relations and the strong projection theorem.

Given semi-recursive relations $P(\vec{x}_n)$, $Q(\vec{y}_m)$, and $R(y, \vec{u}_k)$ of semi-recursive indices p, q, r respectively.

$\vee$:

$$P(\vec{x}_n) \vee Q(\vec{y}_m) \leftrightarrow (\exists z)T^{(n)}(p, \vec{x}_n, z) \vee (\exists z)T^{(m)}(q, \vec{y}_m, z)$$
$$\leftrightarrow (\exists z)\big(T^{(n)}(p, \vec{x}_n, z) \vee T^{(m)}(q, \vec{y}_m, z)\big)$$

$\wedge$:

$$P(\vec{x}_n) \wedge Q(\vec{y}_m) \leftrightarrow (\exists z)T^{(n)}(p, \vec{x}_n, z) \wedge (\exists z)T^{(m)}(q, \vec{y}_m, z)$$
$$\leftrightarrow (\exists w)\big((\exists z)_{<w}T^{(n)}(p, \vec{x}_n, z) \wedge (\exists z)_{<w}T^{(m)}(q, \vec{y}_m, z)\big)$$

Breaking the pattern established by the proof for $\vee$, we may suggest a simpler proof for $\wedge$: $P(\vec{x}_n) \wedge Q(\vec{y}_m) \leftrightarrow ((\mu z)T^{(n)}(p, \vec{x}_n, z) + (\mu z)T^{(m)}(q, \vec{y}_m, z)) \downarrow$. Yet another proof, involving the decoding function $\lambda iz.(z)_i$ is

$$P(\vec{x}_n) \wedge Q(\vec{y}_m) \leftrightarrow (\exists z)T^{(n)}(p, \vec{x}_n, z) \wedge (\exists z)T^{(m)}(q, \vec{y}_m, z)$$
$$\leftrightarrow (\exists z)\big(T^{(n)}(p, \vec{x}_n, (z)_0) \wedge T^{(m)}(q, \vec{y}_m, (z)_1)\big)$$

There is a technical reason (to manifest itself in II.4.6) that we want to avoid "complicated" functions like $\lambda iz.(z)_i$ in the proof.

$(\exists y)_{<z}$:

$$(\exists y)_{<z}R(y, \vec{u}_k) \leftrightarrow (\exists y)_{<z}(\exists w)T^{(k+1)}(r, y, \vec{u}_k, w)$$
$$\leftrightarrow (\exists w)(\exists y)_{<z}T^{(k+1)}(r, y, \vec{u}_k, w)$$

$(\exists y)$:

$$(\exists y)R(y, \vec{u}_k) \leftrightarrow (\exists y)(\exists w)T^{(k+1)}(r, y, \vec{u}_k, w)$$
$$\leftrightarrow (\exists z)(\exists y)_{<z}(\exists w)_{<z}T^{(k+1)}(r, y, \vec{u}_k, w)$$

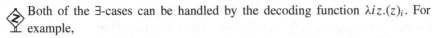 Both of the ∃-cases can be handled by the decoding function $\lambda iz.(z)_i$. For example,

$$(\exists y)R(y, \vec{u}_k) \leftrightarrow (\exists y)(\exists w)T^{(k+1)}(r, y, \vec{u}_k, w)$$
$$\leftrightarrow (\exists z)T^{(k+1)}(r, (z)_0, \vec{u}_k, (z)_1)$$

$(\forall y)_{<z}$:

$$(\forall y)_{<z}R(y, \vec{u}_k) \leftrightarrow (\forall y)_{<z}(\exists w)T^{(k+1)}(r, y, \vec{u}_k, w)$$
$$\leftrightarrow (\exists v)(\forall y)_{<z}(\exists w)_{<v}T^{(k+1)}(r, y, \vec{u}_k, w)$$

Think of v above as the successor $(+1)$ of the maximum of some set of w-values, $w_0, \ldots, w_{z-1}$, that "work" for $y = 0, \ldots, z-1$ respectively. The usual overkill proof of the above involves $(z)_i$ (or some such decoding scheme) as follows:

$$(\forall y)_{<z}R(y, \vec{u}_k) \leftrightarrow (\forall y)_{<z}(\exists w)T^{(k+1)}(r, y, \vec{u}_k, w)$$
$$\leftrightarrow (\exists w)(\forall y)_{<z}T^{(k+1)}(r, y, \vec{u}_k, (w)_y)$$

Regarding closure under $\neg$ and $\forall y$, K provides a counterexample to $\neg$, and $\neg T^{(1)}(x, x, y)$ provides a counterexample to $\forall y$. □

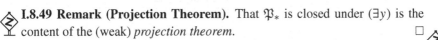 **I.8.49 Remark (Projection Theorem).** That $\mathfrak{P}_*$ is closed under $(\exists y)$ is the content of the (weak) *projection theorem*. □

I.8.50 Definition (Recursively Enumerable Predicates). A predicate $R(\vec{x}_n)$ is *recursively enumerable* (r.e.) iff $R = \emptyset$ or, for some $f \in \mathfrak{R}$ of one variable, $R = \{(\vec{x}_n) : (\exists m)f(m) = \langle \vec{x}_n \rangle\}$,† or, equivalently,

$$R(\vec{x}_n) \leftrightarrow (\exists m)f(m) = \langle \vec{x}_n \rangle \tag{1}$$

□

By (1) and strong projection (I.8.41), every r.e. relation is semi-recursive. The converse is also true.

I.8.51 Theorem. *Every semi-recursive R is r.e.*

Proof. Let a be a semi-recursive index of R. If $R = \emptyset$, then we are done. Suppose then $R(\vec{a}_n)$ for some $\vec{a}_n$. We define a function f by cases:

$$f(m) = \begin{cases} \langle (m)_0, \ldots, (m)_{n-1} \rangle & \text{if } T^{(n)}(a, (m)_0, \ldots, (m)_{n-1}, (m)_n) \\ \langle \vec{a}_n \rangle & \text{otherwise} \end{cases}$$

It is trivial that f is recursive and satisfies (1) above. Indeed, our f is in $\mathfrak{P}\mathfrak{R}$. □

---

† Cf. comment regarding rare use of round brackets for n-tuples, p. 136.

Suppose that i codes a program that acts on input variables x and y to compute a function $\lambda xy.f(x, y)$. It is certainly trivial to modify the program to compute $\lambda x.f(x, a)$ instead: In computer programming terms, we simply replace an instruction such as "read y" by one that says "$y := a$" (copy the value of a into y). From the original code, a new code (depending on i and a) ought to be trivially calculable.

This is the essence of Kleene's *iteration* or *S-m-n* theorem below.

I.8.52 Theorem (Kleene's Iteration or S-m-n Theorem). *There is a primitive recursive function $\lambda xy.\sigma(x, y)$ such that for all i, x, y,*

$$\{i\}(\langle x, y\rangle) = \{\sigma(i, y)\}(x)$$

Proof. Let a be a ϕ-index of $\lambda x.\langle x, 0\rangle$, and b a ϕ-index of $\lambda x.3x$. Next we find a primitive recursive $\lambda y.h(y)$ such that for all x and y

$$\{h(y)\}(x) = \langle x, y\rangle \qquad\qquad (*)$$

To achieve this observe that

$$\langle x, 0\rangle = \{a\}(x)$$

and

$$\langle x, y + 1\rangle = 3\langle x, y\rangle = \{b\}(\langle x, y\rangle)$$

Thus, it suffices to take

$$h(0) = a$$
$$h(y + 1) = \langle 1, 1, b, h(y)\rangle$$

Now that we have an h satisfying $(*)$, we note that

$$\sigma(i, y) \stackrel{\text{def}}{=} \langle 1, 1, i, h(y)\rangle$$

will do. $\qquad\qquad\square$

I.8.53 Corollary. *There is a primitive recursive function $\lambda iy.k(i, y)$ such that, for all i, x, y,*

$$\{i\}(x, y) = \{k(i, y)\}(x)$$

Proof. Let a_0 and a_1 be ϕ-indices of $\lambda z.(z)_0$ and $\lambda z.(z)_1$ respectively. Then $\{i\}(\langle(z)_0, (z)_1\rangle) = \{\langle 1, 1, i, a_0, a_1\rangle\}(z)$ for all z, i. Take $k(i, y) = \sigma(\langle 1, 1, i, a_0, a_1\rangle, y)$. $\qquad\square$

I.8.54 Corollary. *There is for each $m > 0$ and $n > 0$ a primitive recursive function $\lambda i \vec{y}_n . S_n^m(i, \vec{y}_n)$ such that, for all $i, \vec{x}_m, \vec{y}_n$,*

$$\{i\}(\vec{x}_m, \vec{y}_n) = \{S_n^m(i, \vec{y}_n)\}(\vec{x}_m)$$

Proof. Let a_r $(r = 0, \ldots, m-1)$ and b_r $(r = 0, \ldots, n-1)$ be ϕ-indices so that $\{a_r\} = \lambda xy.(x)_r$ $(r = 0, \ldots, m-1)$ and $\{b_r\} = \lambda xy.(y)_r$ $(r = 0, \ldots, n-1)$.

Set $c(i) = \langle 1, 2, i, a_0, \ldots, a_{m-1}, b_0, \ldots, b_{n-1} \rangle$, for all $i \in \mathbb{N}$, and let d be a ϕ-index of $\lambda \vec{x}_m . \langle \vec{x}_m \rangle$. Then,

$$
\begin{aligned}
\{i\}(\vec{x}_m, \vec{y}_n) &= \{c(i)\}(\langle \vec{x}_m \rangle, \langle \vec{y}_n \rangle) \\
&= \{k(c(i), \langle \vec{y}_n \rangle)\}(\langle \vec{x}_m \rangle) \qquad \text{by I.8.53} \\
&= \{\langle 1, m, k(c(i), \langle \vec{y}_n \rangle), d \rangle\}(\vec{x}_m)
\end{aligned}
$$

Take $\lambda i \vec{y}_n . S_n^m = \langle 1, m, k(c(i), \langle \vec{y}_n \rangle), d \rangle$. $\square$

 Since $\mathfrak{P}$-functions are closed under permutation of variables, there is no significance (other than notational convenience, and a random left vs. right choice) in presenting the S-m-n theorem in terms of a "neat" left-right partition of the variable list. Any variable sub-list can be parametrized.

I.8.55 Corollary (Kleene's Recursion Theorem). *If $\lambda z \vec{x} . f(z, \vec{x}_n) \in \mathfrak{P}$, then for some e,*

$$\{e\}(\vec{x}_n) = f(e, \vec{x}_n) \qquad \text{for all } \vec{x}_n$$

Proof. Let $\{a\} = \lambda z \vec{x}_n . f\left(S_1^n(z, z), \vec{x}_n\right)$. Then

$$
\begin{aligned}
f\left(S_1^n(a, a), \vec{x}_n\right) &= \{a\}(a, \vec{x}_n) \\
&= \{S_1^n(a, a)\}(\vec{x}_n) \qquad \text{by I.8.54}
\end{aligned}
$$

Take $e = S_1^n(a, a)$. $\square$

I.8.56 Definition. A *complete index set* is a set $A = \{x : \{x\} \in \mathfrak{Q}\}$ for some $\mathfrak{Q} \subseteq \mathfrak{P}$.

A is *trivial* iff $A = \emptyset$ or $A = \mathbb{N}$ (correspondingly, $\mathfrak{Q} = \emptyset$ or $\mathfrak{Q} = \mathfrak{P}$). Otherwise it is *non-trivial*. $\square$

I.8.57 Theorem (Rice). *A complete index set is recursive iff it is trivial.*

 Thus, algorithmically we can only decide trivial properties of programs.

Proof. (The idea of this proof is attributed in Rogers (1967) to G. C. Wolpin.)

If part: Immediate, since $\chi_\emptyset = \lambda x.1$ and $\chi_\mathbb{N} = \lambda x.0$.

Only-if part: By contradiction, suppose that $A = \{x : \{x\} \in \mathfrak{Q}\}$ is nontrivial, yet $A \in \mathfrak{R}_*$. So let $a \in A$ and $b \notin A$. Define f by

$$f(x) = \begin{cases} b & \text{if } x \in A \\ a & \text{if } x \notin A \end{cases}$$

Clearly,

$$x \in A \text{ iff } f(x) \notin A, \qquad \text{for all } x \tag{1}$$

By the recursion theorem, there is an e such that $\{f(e)\} = \{e\}$ (apply I.8.55 to $\lambda xy.\{f(x)\}(y)$).

Thus, $e \in A$ iff $f(e) \in A$, contradicting (1). $\square$

A few more applications of the recursion theorem will be found in the Exercises.

I.8.58 Example. Every function of $\mathfrak{P}$ has infinitely many indices (revisited). For suppose not, and let $A = \{x : \{x\} \in \mathfrak{Q}\}$ be finite, where $\mathfrak{Q} = \{f\}$. Then A is recursive (why?), contradicting Rice's theorem. $\square$

We have seen that progressing along $\mathfrak{P}\mathfrak{R}_*$, $\mathfrak{R}_*$, $\mathfrak{P}_*$ we obtain strictly more inclusive sets of relations, or, intuitively, progressively more "complex" predicates. For example, we can easily "solve" $\lambda xy.x < y$, we can only "half" solve $x \in K$, and we cannot even do that for $x \notin K$. The latter is beyond $\mathfrak{P}_*$. Still, all three are "arithmetic"[†] or "arithmetical" predicates in a sense that we make precise below. The interest immediately arises to classify arithmetic predicates according to increasing "complexity". This leads to the *arithmetic(al) hierarchy* of Kleene and Mostowski.

I.8.59 Definition. The set of all *arithmetic(al)[‡] predicates* is the least set that includes $\mathfrak{R}_*$ and is closed under $(\exists x)$ and $(\forall x)$. We will denote this set by Δ. $\square$

[†] Accent on "met".

[‡] We will adhere to the term "arithmetic", as in Smullyan (1992). The reader will note that these predicates are introduced in two different ways in Smullyan (1992), each different from the above. One is indicated by a capital "A" and the other by a lowercase "a". All three definitions are equivalent. We follow the standard definition given in works in recursion theory (Rogers (1967), Hinman (1978), Tourlakis (1984)).

We sort arithmetic relations into a *hierarchy* as follows (Kleene (1943), Mostowski (1947)).

I.8.60 Definition (The Arithmetic Hierarchy). We define by induction on $n \in \mathbb{N}$:

$$\Sigma_0 = \Pi_0 = \mathfrak{R}_*$$
$$\Sigma_{n+1} = \{(\exists x)P : P \in \Pi_n\}$$
$$\Pi_{n+1} = \{(\forall x)P : P \in \Sigma_n\}$$

The variable "x" above is generic.

We also define, for all n, $\Delta_n = \Sigma_n \cap \Pi_n$ and also set $\Delta_\infty = \bigcup_{n \geq 0}(\Sigma_n \cup \Pi_n)$. $\square$

I.8.61 Remark. Intuitively, the arithmetic hierarchy is composed of all relations of the form $(Q_1 x_1)(Q_2 x_2) \ldots (Q_n x_n)R$, where $R \in \mathfrak{R}_*$ and $Q_i \in \{\exists, \forall\}$ for $i = 1, \ldots, n$. If $n = 0$ there is no quantifier prefix. Since $\exists x \exists y$ and $\forall x \forall y$ can be "collapsed" into a single $\exists$ and single $\forall$ respectively,

Pause. Do you believe this?

one can think of the prefix as a sequence of alternating quantifiers. The relation is placed in a Π-set (respectively, Σ-set) iff the leftmost quantifier is a "$\forall$" (respectively, "$\exists$"). $\square$

I.8.62 Lemma. $R \in \Sigma_n$ *iff* $(\neg R) \in \Pi_n$. $R \in \Pi_n$ *iff* $(\neg R) \in \Sigma_n$.

Proof. We handle both equivalences simultaneously. For $n = 0$ this is so by closure properties of $\mathfrak{R}_*$.

Assuming the claim for n, we have

$$\begin{aligned}
R \in \Sigma_{n+1} \text{ iff } &R \leftrightarrow (\exists y)Q \text{ and } Q \in \Pi_n \\
\text{iff (I.H.)} &R \leftrightarrow \neg(\forall y)\neg Q \text{ and } (\neg Q) \in \Sigma_n \\
\text{iff } &\neg R \leftrightarrow (\forall y)\neg Q \text{ and } (\neg Q) \in \Sigma_n \\
\text{iff } &\neg R \in \Pi_{n+1}
\end{aligned}$$

and

$$\begin{aligned}
R \in \Pi_{n+1} \text{ iff } &R \leftrightarrow (\forall y)Q \text{ and } Q \in \Sigma_n \\
\text{iff (I.H.) } &R \leftrightarrow \neg(\exists y)\neg Q \text{ and } (\neg Q) \in \Pi_n \\
\text{iff } &\neg R \leftrightarrow (\exists y)\neg Q \text{ and } (\neg Q) \in \Pi_n \\
\text{iff } &\neg R \in \Sigma_{n+1}
\end{aligned}$$
$\square$

It is trivial that $\Delta_\infty \subseteq \Delta$. The following easy lemma yields the converse inclusion as a corollary.

I.8.63 Lemma. *We have the following closure properties*:

Δ_∞ *is closed under* (1)–(6) *from the following list.*
Δ_n $(n \geq 0)$ *is closed under* (1)–(4).
Σ_n $(n \geq 0)$ *is closed under* (1)–(3), *and, if* $n > 0$, *also* (5).
Π_n $(n \geq 0)$ *is closed under* (1)–(3), *and, if* $n > 0$, *also* (6).

(1) *Replacement of a variable by a recursive function*[†]
(2) $\vee, \wedge$
(3) $(\exists y)_{<z}, (\forall y)_{<z}$
(4) $\neg$
(5) $(\exists y)$
(6) $(\forall y)$

Proof. (1) follows from the corresponding closure of $\mathfrak{R}_*$. The rest follow at once from I.8.60 and the techniques of I.8.48 ((4) also uses I.8.62). $\square$

I.8.64 Corollary. $\Delta = \Delta_\infty$. *Hence, Definition* I.8.60 *classifies all arithmetic predicates according to definitional complexity.*

We next see that the complexity of arithmetic predicates increases in more than just "form" as n increases (i.e., as the form "$(Q_1 x_1)(Q_2 x_2) \ldots (Q_n x_n)R$" gets more complex with a longer alternating quantifier prefix, we do get new predicates defined).

I.8.65 Proposition. $\Sigma_n \cup \Pi_n \subseteq \Delta_{n+1}$ *for* $n \geq 0$.

Proof. Induction on n.
   Basis. For $n = 0$, $\Sigma_n \cup \Pi_n = \mathfrak{R}_*$. On the other hand, $\Sigma_1 = \mathfrak{P}_*$ (why?), while (by I.8.62) $\Pi_1 = \{Q : (\neg Q) \in \mathfrak{P}_*\}$. Thus $\Delta_1 = \Sigma_1 \cap \Pi_1 = \mathfrak{R}_*$ by I.8.44.

   We consider now the $n + 1$ case (under the obvious I.H.). Let $R(\vec{x}_r) \in \Sigma_{n+1}$. Then, by I.8.63(1), so is $\lambda z \vec{x}_r . R(\vec{x}_r)$, where z is not among the $\vec{x}_r$. Now, $R \leftrightarrow (\forall z)R$, but $(\forall z)R \in \Pi_{n+2}$.

[†] Since u_i^n, $\lambda x.0$, and $\lambda x.x + 1$ are recursive, this allows the full range of the Grzegorczyk substitutions (Exercise I.68), i.e., additionally to function substitution, also expansion of the variable list, permutation of the variable list, identification of variables, and substitution of constants into variables.

Next, let $R(\vec{x}_r) \leftrightarrow (\exists z)Q(z, \vec{x}_r)$, where $Q \in \Pi_n$. By the I.H., $Q \in \Delta_{n+1}$; hence $Q \in \Pi_{n+1}$. Thus, $R \in \Sigma_{n+2}$.

The argument is similar if we start the previous sentence with "Let $R(\vec{x}_r) \in \Pi_{n+1}$". $\qquad\square$

I.8.66 Corollary. $\Sigma_n \subseteq \Sigma_{n+1}$ *and* $\Pi_n \subseteq \Pi_{n+1}$, *for* $n \geq 0$.

I.8.67 Corollary. $\Delta_n \subseteq \Delta_{n+1}$ *for* $n \geq 0$.

We next sharpen the inclusions above to proper inclusions.

I.8.68 Definition (Kleene). For $n \geq 1$ we define $\lambda xy.E_n(x, y)$ by induction:

$$E_1(x, y) \leftrightarrow (\exists z)T^{(1)}(x, y, z)$$
$$E_{n+1}(x, y) \leftrightarrow (\exists z)\neg E_n(x, \langle z \rangle * y)$$

where "$\langle \rangle$" is our standard coding of p. 136. $\qquad\square$

Of course, $\langle z \rangle = 2^{z+1}$.

I.8.69 Lemma. $E_n \in \Sigma_n$ *and* $\neg E_n \in \Pi_n$, *for* $n \geq 1$.

Proof. A trivial induction (via I.8.62 and I.8.63). $\qquad\square$

I.8.70 Theorem (Enumeration or Indexing Theorem (Kleene)).

(1) $R(\vec{x}_r) \in \Sigma_{n+1}$ *iff* $R(\vec{x}_r) \leftrightarrow E_{n+1}(i, \langle \vec{x}_r \rangle)$ *for some* i.
(2) $R(\vec{x}_r) \in \Pi_{n+1}$ *iff* $R(\vec{x}_r) \leftrightarrow \neg E_{n+1}(i, \langle \vec{x}_r \rangle)$ *for some* i.

Proof. The if part is Lemma I.8.69 (with the help of I.8.63(1)). We prove the only-if part ((1) and (2) simultaneously) by induction on n.

Basis. $n = 0$. If $R(\vec{x}_r) \in \Sigma_1 = \mathfrak{P}_*$, then so is $R((u)_0, \dots, (u)_{r-1})$. Thus, for some i (semi-recursive index), $R((u)_0, \dots, (u)_{r-1}) \leftrightarrow (\exists z)T^{(1)}(i, u, z)$; hence $R(\vec{x}_r) \leftrightarrow (\exists z)T^{(1)}(i, \langle \vec{x}_r \rangle, z) \leftrightarrow E_1(i, \langle \vec{x}_r \rangle)$.

If $R(\vec{x}_r) \in \Pi_1$, then $(\neg R(\vec{x}_r)) \in \Sigma_1 = \mathfrak{P}_*$. Thus, for some e, $\neg R(\vec{x}_r) \leftrightarrow E_1(e, \langle \vec{x}_r \rangle)$; hence $R(\vec{x}_r) \leftrightarrow \neg E_1(e, \langle \vec{x}_r \rangle)$.

The induction step. (Based on the obvious I.H.) Let $R(\vec{x}_r) \in \Sigma_{n+2}$. Then $R(\vec{x}_r) \leftrightarrow (\exists z)Q(z, \vec{x}_r)$, where $Q \in \Pi_{n+1}$, hence (I.H.)

$$Q(z, \vec{x}_r) \leftrightarrow \neg E_{n+1}(e, \langle z, \vec{x}_r \rangle) \qquad \text{for some } e \qquad (i)$$

Since $\langle z, \vec{x}_r \rangle = \langle z \rangle * \langle \vec{x}_r \rangle$, (i) yields

$$R(\vec{x}_r) \leftrightarrow (\exists z)\neg E_{n+1}(e, \langle z \rangle * \langle \vec{x}_r \rangle)$$
$$\leftrightarrow E_{n+2}(e, \langle \vec{x}_r \rangle) \qquad \text{(by the definition of } E_{n+2}\text{)}$$

An entirely analogous argument takes care of "Let $R(\vec{x}_r) \in \Pi_{n+2}$". □

I.8.71 Corollary. *The same set of arithmetic relations,* Δ, *can be defined by setting* $\Sigma_0 = \Pi_0 = \mathfrak{PR}_*$. *Indeed, no sets in the hierarchy are affected by this change, except* $\Sigma_0 = \Pi_0$.

I.8.72 Theorem (Hierarchy Theorem (Kleene, Mostowski)).

(1) $\Sigma_{n+1} - \Pi_{n+1} \neq \emptyset$,
(2) $\Pi_{n+1} - \Sigma_{n+1} \neq \emptyset$.

Moreover, all the inclusions in I.8.65 $(n > 0)$, I.8.66 $(n \geq 0)$, *and* I.8.67 $(n > 0)$ *are proper.*

Proof. (1): $E_{n+1} \in \Sigma_{n+1} - \Pi_{n+1}$. Indeed (see also I.8.69), if $E_{n+1} \in \Pi_{n+1}$, then

$$E_{n+1}(x, \langle x \rangle) \leftrightarrow \neg E_{n+1}(i, \langle x \rangle) \qquad (1')$$

for some i (I.8.63(1) and, I.8.70). Letting $x = i$ in $(1')$, we get a contradiction.

(2): As in (1), but use $\neg E_{n+1}$ as the counterexample.

For I.8.65: Let $R(x, y) \leftrightarrow E_n(x, \langle x \rangle) \vee \neg E_n(y, \langle y \rangle)$. Since $E_n \in \Sigma_n$ and $(\neg E_n) \in \Pi_n$, I.8.63(1), I.8.65, and closure of Δ_{n+1} under $\vee$ yield $R \in \Delta_{n+1}$. Let x_0 and y_0 be such that $E_n(x_0, \langle x_0 \rangle)$ is false and $E_n(y_0, \langle y_0 \rangle)$ is true (if no such x_0, y_0 exist, then E_n is recursive (why?); not so for $n > 0$ (why?)).

Now, $R \notin \Sigma_n \cup \Pi_n$, for, otherwise, say it is in Σ_n. Then so is $R(x_0, y)$ (that is, $\neg E_n(y, \langle y \rangle)$), which is absurd. Similarly, if $R \in \Pi_n$, we are led to the absurd conclusion that $R(x, y_0) \in \Pi_n$.

For I.8.66: $K \in \Sigma_1 - \Sigma_0$. For $n > 0$, $\Sigma_n = \Sigma_{n+1}$ implies $\Pi_n = \Pi_{n+1}$ (why?); hence $\Sigma_n \cup \Pi_n = \Sigma_{n+1} \cup \Pi_{n+1} \supseteq \Sigma_{n+1} \cap \Pi_{n+1} = \Delta_{n+1}$; thus $\Sigma_n \cup \Pi_n = \Delta_{n+1}$ by I.8.65, contrary to what we have just established above. Similarly for the dual Π_n vs. Π_{n+1}.

For I.8.67: If $\Delta_{n+1} = \Delta_n$, then $\Delta_{n+1} = \Sigma_n \cap \Pi_n \subseteq \Sigma_n \cup \Pi_n$, which has been established as an absurdity for $n > 0$. □

I.9. Arithmetic, Definability, Undefinability, and Incompletableness

In this section we apply recursion-theoretic techniques to the proof theory of a certain axiomatic arithmetic in order to derive a major result (Gödel's first incompleteness theorem) regarding the inadequacy of the syntactic proof apparatus.

We have to overcome a small obstacle outright: Our recursion theory is a theory of number-theoretic functions and relations. We need some way to translate its results in the realm of strings (over some alphabet) so that our theory can handle recursive, r.e., primitive recursive, etc., functions and predicates that have inputs and outputs that are such strings. For example, a *recursively axiomatized* theory, we say, is one with a recursive set of axioms. But what do we mean by a recursive set of *strings*?

Well, we can code strings by numbers, and then use the numbers as proxies for the strings. This is the essence of *Gödel numbering*, invented by Gödel (1931) towards that end.

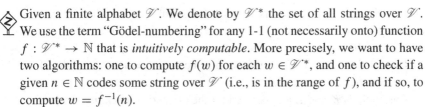

Given a finite alphabet $\mathscr{V}$. We denote by $\mathscr{V}^*$ the set of all strings over $\mathscr{V}$. We use the term "Gödel-numbering" for any 1-1 (not necessarily onto) function $f : \mathscr{V}^* \rightarrow \mathbb{N}$ that is *intuitively computable*. More precisely, we want to have two algorithms: one to compute $f(w)$ for each $w \in \mathscr{V}^*$, and one to check if a given $n \in \mathbb{N}$ codes some string over $\mathscr{V}$ (i.e., is in the range of f), and if so, to compute $w = f^{-1}(n)$.

We use special notation in the context of Gödel numberings. Rather than "$f(w)$" – the Gödel number of w – we write $\ulcorner w \urcorner$, allowing the context to tell which specific f we have in mind.

The above, of course, is not a precise *definition*, since the term "computable" (function) was never defined between "mixed" fields ($\mathscr{V}^*$ on the left and $\mathbb{N}$ on the right). This does not present a problem however, for in practice any specific Gödel numbering will be *trivially* seen (at the intuitive level) to satisfy the algorithmic coding-decoding requirement.

The heart of the matter in effecting a Gödel numbering of a given $\mathscr{V}^*$ is simple: We start with a "primitive" numbering by assigning distinct numbers from $\mathbb{N}$ to each symbol of $\mathscr{V}$. Then, *in principle*, we can extend this primitive numbering to the entire $\mathscr{V}^*$ by recursion on strings.

However, in the cases of interest to us, that is, terms, formulas, and sequences of such over some language L, we will prefer to do our recursion on terms, formulas, and sequences (rather than on generic strings).

For example, if the numbers $n_0, n_1, \ldots, n_m$ were already assigned to the formal function symbol g (m-ary) and to the terms $t_1, \ldots, t_m$ respectively, one would just need a way to obtain a number for $g t_1 \ldots t_m$ from the numbers we have just listed. This simply requires the presence of some *number*-coding scheme, "$\langle \rangle$", to compute $\langle n_0, n_1, \ldots, n_m \rangle$. Gödel used, essentially, the prime power coding scheme "$\langle \rangle$, lh, Seq, $(z)_i$" of p. 136.

We will not adopt prime power coding here, but instead we will use a different coding based on Gödel's β-*function*. Our approach is based on the exposition in Shoenfield (1967), the motivation being to do the number coding with as little number theory as possible,[†] so that when we have to revisit all this in the next chapter – formally this time – our task will be reasonably easy, and short.

We fix, for the balance of the chapter, a simple *pairing function*, that is, a total and 1-1 function $J : \mathbb{N}^2 \to \mathbb{N}$ that can be obtained via addition and multiplication. For example, let us set

$$J(x, y) = (x + y)^2 + x \qquad \text{for all } x, y \qquad (*)$$

J is trivially primitive recursive. Its 1-1-ness follows from the observation that $x + y < u + v$ implies $x + y + 1 \le u + v$ and hence $J(x, y) < (x + y + 1)^2 \le (u + v)^2 \le J(u, v)$.

Thus, if $J(x, y) = J(u, v)$, then it must be $x + y = u + v$, and therefore $x = u$. But then $y = v$.

The symbols K and L are reserved for the *projection functions* of a pairing function J.[‡]

We see at once that K, L are in $\mathfrak{PR}$; indeed,

$$K(z) = (\mu x)_{\le z} (\exists y)_{\le z} (J(x, y) = z)$$
$$L(z) = (\mu y)_{\le z} (\exists x)_{\le z} (J(x, y) = z)$$

Let us then address the task of finding a way to "algorithmically" code a number sequence $a_0, \ldots, a_n$ by a single number, so that each of the a_i can be algorithmically recovered from the code as long as we know its position i.

We start by setting

$$c = \max\{1 + J(i, a_i) : i \le n\} \qquad (**)$$

[†] Gödel's original approach needed enough number theory to include the prime factorization theorem. An alternative approach due to Quine, utilized in Smullyan (1961, 1992), requires the theorem that every number can be uniquely written as a string of base-b digits out of some fixed alphabet $\{0, 1, \ldots, b-1\}$. On the other hand, employing the variant of Gödel's β-function found in Shoenfield (1967), one does not even need the Chinese remainder theorem.

[‡] It is unfortunate, but a standard convention, that this "K" clashes with the usage of "K" as the name of the halting set (I.8.45). Fortunately, it is very unlikely that the context will allow the confusion of the two notations.

Next, let p be the *least common multiple* of the numbers $1, 2, \ldots, c + 1$, for short, $lcm\{1, 2, \ldots, c + 1\}$, that is,

$$p = (\mu z)\left(z > 0 \wedge (\forall i)_{\leq c}(i + 1)|z\right) \qquad (\text{***})$$

We recall the following definition from elementary number theory:

I.9.1 Definition. The *greatest common divisor*, or gcd, of two natural numbers a and b such that $a + b \neq 0$ is the largest d such that $d|a$ and $d|b$.[†] We write $d = \gcd(a, b) = \gcd(b, a)$. ☐

I.9.2 Lemma. *For each a and b in $\mathbb{N}$ ($a + b \neq 0$) there are integers u and v such that $\gcd(a, b) = au + bv$.*

Proof. The set $A = \{ax + by > 0 : x \in \mathbb{Z} \wedge y \in \mathbb{Z}\}$ is nonempty. For example, if $a \neq 0$, then $a1 + b0 > 0$ is in A. Let g be the least member of A. So, for some u and v in $\mathbb{Z}$,

$$g = au + bv > 0 \qquad (1)$$

We argue that $g|a$ (and similarly, $g|b$). If not, let

$$0 < r < g \qquad (2)$$

such that $a = gq + r$ ($q = \lfloor a/g \rfloor$). Thus, using (1), $r = a - gq = a(1 - uq) - bvq$, a member of A. (2) contradicts the minimality of g.

Thus, g is a common divisor of a and b. On the other hand, by (1), every common divisor d of a and b divides g; hence $d \leq g$. Thus, $g = \gcd(a, b)$. ☐

I.9.3 Definition. Two natural numbers a and b ($a + b \neq 0$) are *relatively prime* iff $\gcd(a, b) = 1$. ☐

The above is the standard definition of relative primality. However, an equivalent definition can save us from a lot of grief when we redo all this formally in the next chapter. This alternate definition is furnished by the statement of the following lemma, which we prove here (informally) only as a preview for things to come in the next chapter.

I.9.4 Lemma. *The natural numbers a and b ($a + b \neq 0$) are relatively prime iff for all $x > 0$, $a|bx$ implies $a|x$.*

[†] If $a = b = 0$, then such a largest d does not exist, since $(\forall x)(x > 0 \rightarrow x|0)$. Hence the restriction to $a + b \neq 0$.

Proof. Only-if part. By I.9.2, $1 = au + bv$ for some integers u and v. Thus, $x = axu + bxv$; hence $a|x$.

If part. We prove the *contrapositive.*[†] Let $1 < d = \gcd(a, b)$. Write $a = dk$ and $b = dm$. Now, $a|bk$, but $a > k$; hence $a \nmid k$.   □

We can now prove:

I.9.5 Lemma. *With $c > 0$ arbitrary, and p chosen as in $(***)$ on p. 156 above,*

$$\gcd\big(1 + (j + 1)p, 1 + (i + 1)p\big) = 1 \qquad \text{for all} \quad 0 \le i < j \le c + 1$$

Proof. We will proceed by contradiction. Let $d > 1$ divide both $1 + (j + 1)p$ and $1 + (i + 1)p$. Then $d > c + 1$ (otherwise $d|p$ and hence $d \nmid 1 + (j + 1)p$).

Now d divides any linear combination of $1 + (j + 1)p$ and $1 + (i + 1)p$, in particular $j - i$ (which is equal to $(1 + (i + 1)p)(j + 1) - (1 + (j + 1)p)(i + 1)$). Since $0 < j - i \le c + 1$, this is impossible.   □

I.9.6 Lemma. *If $d_0, \ldots, d_r, b$, with b greater than 1, are such that $\gcd(b, d_i) = 1$ for all $0 \le i \le r$, then $b \nmid lcm\{d_i : i = 0, \ldots, r\}$.*

Proof. Set $z = lcm\{d_i : i = 0, \ldots, r\}$, and suppose that $b|z$; thus $z = bk$ for some k. By I.9.4 (since $d_i|z$), we have $d_i|k$ for all $0 \le i \le r$. From $b > 1$ follows $k < z$, contradicting the minimality of z.   □

Armed with Lemmata I.9.5 and I.9.6, we can now code any nonempty sub*set* (the emphasis indicating disrespect for order) of numbers $i_j, 0 \le i_j \le c$ $(c > 0$ being an arbitrary integer), by the least common multiple q of the numbers $1 + (i_j + 1)p$, p being the *lcm* of $\{1, 2, \ldots, c + 1\}$. Indeed, *a number of the form $1 + (z + 1)p$, for $0 \le z \le c$, divides q precisely when z is one of the i_j.*

Thus, we can code the *sequence* $a_0, a_1, \ldots, a_n$ by coding the *set* of numbers $J(i, a_i)$ (position information i is packed along with value a_i), as suggested immediately above: We let

$$q = lcm\left\{1 + \big(1 + J(i, a_i)\big)p : i = 0, \ldots, n\right\}$$

With n, c, p, q as above (see $(**)$ and $(***)$), we recover a_i as

$$(\mu y)\big(1 + \big(1 + J(i, y)\big)p \mid q\big) \tag{1}$$

[†] The contrapositive of $\mathscr{A} \to \mathscr{B}$ is the tautologically equivalent $\neg \mathscr{B} \to \neg \mathscr{A}$.

or, setting, $a = J(c, q)$ and forcing the search to exit gracefully even when q is nonsense (that is, not a code),

$$(\mu y)\Big(y = a \vee 1 + \big(1 + J(i, y)\big)p \mid L(a)\Big) \tag{2}$$

Of course, p in (2) above is given in terms of c by (∗∗∗), that is, it equals the least common multiple of the numbers $\{1, 2, \dots, K(a) + 1\}$.

In what follows we will use the notation $\beta(a, i)$ for the expression (2) above. Thus,

$$\lambda ai.\beta(a, i) \in \mathfrak{PR}$$
$$\beta(a, i) \leq a \qquad \text{for all } a, i \tag{3}$$

The "$\leq$" in (3) is "$=$" iff the search in (1) above fails. Otherwise $y \leq J(i, y) < 1 + J(i, y) < L(a) \leq a + 1$; hence $y \leq J(i, y) < a$.

For any sequence $a_0, \dots, a_n$ there is an a such that $\beta(a, i) = a_i$ for $i \leq n$.

This is the version of Gödel's β-function in Shoenfield (1967). Since we will carefully prove all the properties of this β-function – and of its associated sequence coding scheme – later on, in the context of axiomatic arithmetic (Section II.2), we will avoid "obvious" details here.

I.9.7 Remark. (1) Why *lcm*? Both in (∗∗∗) and in the definition of q above, we could have used product instead. Thus, p could have been given as c-factorial ($c!$), and q as the product[†]

$$\prod \Big\{ 1 + \big(1 + J(i, a_i)\big)p : i = 0, \dots, n \Big\}$$

The seemingly more complex approach via *lcm* is actually formally simpler. The *lcm* is *explicitly* definable, while the product of a variable number of factors necessitates a (primitive) recursive definition. Alas, β will be employed in the next chapter precisely in order to show that recursive definitions are *allowed* in Peano arithmetic.

(2) Gödel's original β was less "elementary" than the above: Starting with a modified "c",

$$c' = \max\{1 + a_i : i \leq n\} \tag{∗∗′}$$

one then defines

$$p' = n!c' \tag{∗∗∗′}$$

As in Lemma I.9.5, $\gcd(1 + (1 + j)p', 1 + (1 + i)p') = 1$ for all $0 \leq j < i \leq n$.

[†] As a matter of fact, for a set of relatively prime numbers, their *lcm* is the same as their product.

By the *Chinese remainder theorem* (see, e.g., LeVeque (1956)), there is a $q' \in \mathbb{N}$ such that $q' \equiv a_i \bmod 1 + (1 + i)p'$ for $i = 0, \ldots, n$. Thus, q' codes the *sequence* $a_0, \ldots, a_n$. Since each a_i satisfies $a_i < 1 + (1 + i)p'$ by the choice of p', a_i is recovered from q' by $rem(q', 1 + (1 + i)p')$. One then sets $\beta'(q', p', i) = rem(q', 1 + (1 + i)p')$. This "$\beta'$" is Gödel's original "$\beta$".   □ ⚔

This ability to code sequences *without using (primitive) recursion* can be used to sharpen Corollary I.8.71, so that we can start the arithmetic hierarchy with even simpler relations $\Sigma_0 = \Pi_0$ than primitive recursive.

We will profit by a brief digression here from our discussion of Gödel numberings to show that any $R \in \mathfrak{PR}_*$ is a $(\exists x)$-projection of a very "simple" relation, one out of the set defined below. This result will be utilized in our study of (that is, in the metatheory of) formal arithmetic.

I.9.8 Definition (Constructive Arithmetic Predicates). (Smullyan (1961, 1992), Bennett (1962)). The *constructive arithmetic predicates* are the smallest set of predicates that includes $\lambda xyz.z = x + y$, $\lambda xyz.z = x \cdot y$, and is closed under $\neg$, $\vee$, $(\exists y)_{<z}$, and explicit transformations.

Explicit transformations (Smullyan (1961), Bennett (1962)) are exactly the following: Substitution of any constant into a variable, expansion of the variables list, permutation of variables, identification of variables.

We will denote the set of constructive arithmetic predicates by $\mathfrak{CA}$.[†]   □

Trivially,

I.9.9 Lemma. $\mathfrak{CA} \subseteq \mathfrak{PR}_*$.

and

I.9.10 Lemma. $\mathfrak{CA}$ *is closed under* $(\exists y)_{\leq z}$. *Conversely, the definition above could have been given in terms of* $(\exists y)_{\leq z}$.

Proof. $(\exists y)_{\leq z} R(y, \vec{w}) \leftrightarrow R(z, \vec{w}) \vee (\exists y)_{<z} R(y, \vec{w})$. Conversely, $(\exists y)_{<z} R(y, \vec{w}) \leftrightarrow (\exists y)_{\leq z} (\neg y = z \wedge R(y, \vec{w}))$. Of course, $y = z$ is an explicit transform of $x \cdot y = z$.   □

I.9.11 Lemma. $\mathfrak{C}\mathfrak{A}$ *is closed under substitution of the term* $x+1$ *into variables.*

Proof. We do induction on the definition of $\mathfrak{C}\mathfrak{A}$.

  Basis.

(1) $z = x + y + 1$: $\quad z = x + y + 1 \leftrightarrow (\exists w)_{\leq z}(z = x + w \wedge w = y + 1)$. Of course, $w = y + 1$ is an explicit transform of $w = y + u$.

(2) $z + 1 = x + y$: $\quad z + 1 = x + y \leftrightarrow (\exists w)_{\leq x}(x = w + 1 \wedge z = w + y) \vee (\exists w)_{\leq y}(y = w + 1 \wedge z = x + w)$.

(3) $z = x(y + 1)$: $\quad z = x(y+1) \leftrightarrow (\exists w)_{\leq z}(z = xw \wedge w = y + 1)$.

(4) $z + 1 = xy$: $\quad z + 1 = xy \leftrightarrow (\exists u)_{\leq x}(\exists w)_{\leq y}(x = u + 1 \wedge y = w + 1 \wedge z = uw + u + w)$. Of course, $z = uw + u + w \leftrightarrow (\exists x)_{\leq z}(\exists y)_{\leq z}(z = x + y \wedge x = uw \wedge y = u + w)$.

The property of $\mathfrak{C}\mathfrak{A}$ that we are proving trivially propagates with the Boolean operations. Moreover $(\exists y)_{< z+1} R(y, \vec{x}) \leftrightarrow (\exists y)_{\leq z} R(y, \vec{x})$. $\qquad\square$

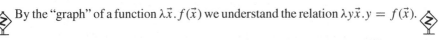 $(\exists y)_{\leq z}$ is a "derived operation" (I.9.10); thus it is not checked in the proof above. Needless to say, were it the principal (primary) bounded quantification *definitionally*, then we would have employed the argument $(\exists y)_{\leq z+1} R(y, \vec{x}) \leftrightarrow R(z + 1, \vec{x}) \vee (\exists y)_{\leq z} R(y, \vec{x})$ instead, and the I.H., to conclude the above proof.

I.9.12 Corollary. $\mathfrak{C}\mathfrak{A}$ *is closed under substitution into variables by functions* f *satisfying*

(*i*) $\lambda y \vec{x}_n . y = f(\vec{x}_n)$ *is in* $\mathfrak{C}\mathfrak{A}$, *and*

(*ii*) *for some* i, $f(\vec{x}_n) \leq x_i + 1$ *for all* $\vec{x}_n$.

Proof. Let $R(y, \vec{z})$ be in $\mathfrak{C}\mathfrak{A}$, and f be as described above; in particular let $f(\vec{x}_n) \leq x_i + 1$ for all $\vec{x}_n$. Then

$$R(f(\vec{x}_n), \vec{z}) \leftrightarrow (\exists y)_{\leq x_i + 1}(y = f(\vec{x}_n) \wedge R(y, \vec{z})) \qquad \square$$

I.9.13 Lemma. *The graphs of* J, K, *and* L (p. 156) *are in* $\mathfrak{C}\mathfrak{A}$.

By the "graph" of a function $\lambda \vec{x} . f(\vec{x})$ we understand the relation $\lambda y \vec{x} . y = f(\vec{x})$.

Proof. The case for J is trivial. K and L present some mild interest. For example,

$$y = K(x) \leftrightarrow \left(y = x + 1 \wedge \neg(\exists z)_{\leq x} J(y, z) = x \right) \vee (\exists z)_{\leq x} J(y, z) = x \qquad \square$$

I.9.14 Remark. We can substitute $K(x)$ and $L(x)$ for variables in $\mathfrak{CA}$ relations, by I.9.12, since $K(x) \leq x + 1$ for all x and $L(x) \leq x + 1$ for all x. $J(x, y)$ is "too big" for I.9.12 to apply to it, but that too can be substituted, as results of Bennett show (1962, retold in Tourlakis (1984)). However, we will not need this fact in this volume.   □

I.9.15 Proposition. $\lambda aiy.\beta(a, i) = y$ is in $\mathfrak{CA}$.

Proof. Recall that $\beta(a, i) = (\mu y)(y = a \vee 1 + (1 + J(i, y))p(a) \mid L(a))$, where

$$p(a) = (\mu z)(z > 0 \wedge (\forall i)_{\leq K(a)}(i + 1) \mid z)$$

We set for convenience $P(a, i, y) \leftrightarrow 1 + (1 + J(i, y))p(a) \mid L(a)$, and show that P is in $\mathfrak{CA}$. Now $P(a, i, y) \leftrightarrow (\exists u)_{\leq a}(L(a) = u(1 + (1 + J(i, y))p(a)))$, and in view of I.9.12 (cf. I.9.14) and closure under $(\exists y)_{\leq w}$, we need only show that

$$z = u\big(1 + (1 + J(i, y))p(a)\big)$$

is in $\mathfrak{CA}$. The above is equivalent to

$$(\exists v, x)_{\leq z}\Big(v = p(a) \wedge z = ux \wedge x = 1 + (1 + J(i, y))v\Big)$$

where we have used the shorthand "$(\exists v, x)_{\leq z}$" for "$(\exists v)_{\leq z}(\exists x)_{\leq z}$".

That $x = 1 + (1 + J(i, y))v$ is in $\mathfrak{CA}$ follows from

$$x = 1 + (1 + J(i, y))v \leftrightarrow (\exists u, w, r, s)_{\leq x}(w = J(i, y) \wedge u$$
$$= w + 1 \wedge r = uv \wedge x = r + 1)$$

We concentrate now on $v = p(a)$. We start with the predicate

$$H(z, w) \leftrightarrow z > 0 \wedge (\forall i)_{\leq w}(i + 1) \mid z$$

This is in $\mathfrak{CA}$ since $(i + 1) \mid z \leftrightarrow (\exists u)_{\leq z}z = u(i + 1)$. Now

$$v = p(a) \leftrightarrow H(v, K(a)) \wedge (\forall z)_{<v}\neg H(z, K(a))$$

and the case for $v = p(a)$ rests by I.9.14. This concludes the proof that $P(a, i, y)$ is in $\mathfrak{CA}$. Finally, setting

$$Q(a, i, y) \leftrightarrow y = a \vee P(a, i, y)$$

we have $Q(a, i, y)$ in $\mathfrak{C}\mathfrak{A}$. We are done by observing that

$$\beta(a, i) = y \leftrightarrow Q(a, i, y) \wedge (\forall z)_{<y} \neg Q(a, i, y) \qquad \square$$

The crucial step in order to achieve the earlier suggested projection representation of arbitrary primitive recursion relations (p. 160) is to express the arbitrary primitive recursive "computation"

$$(\exists c_0, \dots, c_x)\Big(z = c_x \wedge c_0 = h(\vec{y}) \wedge (\forall i)_{<x} c_{i+1} = g(i, \vec{y}, c_i)\Big) \qquad (1)$$

in the form $(\exists u)R$, where $R \in \mathfrak{C}\mathfrak{A}$. (1), of course, "computes" (or verifies) that $z = f(x, \vec{y})$, where, for all $x, \vec{y}$,

$$f(0, \vec{y}) = h(\vec{y})$$
$$f(x + 1, \vec{y}) = g(x, \vec{y}, f(x, \vec{y}))$$

I.9.16 Theorem. *If $R(\vec{x}) \in \mathfrak{PR}_*$, then, for some $Q(y, \vec{x}) \in \mathfrak{C}\mathfrak{A}$,*

$$R(\vec{x}) \leftrightarrow (\exists y)Q(y, \vec{x})$$

Proof. Since $R(\vec{x}) \in \mathfrak{PR}_*$ iff for some $f \in \mathfrak{PR}$ one has $R(\vec{x}) \leftrightarrow f(\vec{x}) = 0$, it suffices to show that *for all $f \in \mathfrak{PR}$ there is a $Q(z, y, \vec{x}) \in \mathfrak{C}\mathfrak{A}$ such that* $y = f(\vec{x}) \leftrightarrow (\exists z)Q(z, y, \vec{x})$.

We do induction on $\mathfrak{PR}$. The graphs of the initial functions are in $\mathfrak{C}\mathfrak{A}$ without the help of any $(\exists y)$-projection.

Let $f(\vec{x}) = g(h_1(\vec{x}), \dots, h_m(\vec{x}))$, and assume that (I.H.) $z = g(\vec{y}_m) \leftrightarrow (\exists u)G(u, z, \vec{y}_m)$ and $y = h_i(\vec{x}) \leftrightarrow (\exists z)H_i(z, y, \vec{x})$ for $i = 1, \dots, m$, where G and the H_i are in $\mathfrak{C}\mathfrak{A}$.

Let us write "$(\exists \vec{z}_r)$" and "$(\exists \vec{z}_r)_{\leq w}$" as short for "$(\exists z_1)(\exists z_2) \dots (\exists z_r)$" and "$(\exists z_1)_{\leq w}(\exists z_2)_{\leq w} \dots (\exists z_r)_{\leq w}$" respectively. Then

$$y = f(\vec{x}) \leftrightarrow (\exists \vec{u}_m)\Big(y = g(\vec{u}_m) \wedge u_1 = h_1(\vec{x}) \wedge \cdots \wedge u_m = h_m(\vec{x})\Big)$$

$$\leftrightarrow (\exists \vec{u}_m)\Big((\exists z)G(z, y, \vec{u}_m) \wedge (\exists z_1)H_1(z_1, u_1, \vec{x}) \wedge$$

$$\cdots \wedge (\exists z_m)H_m(z_m, u_m, \vec{x})\Big)$$

$$\leftrightarrow (\exists w)(\exists \vec{u}_m)_{\leq w}\Big((\exists z)_{\leq w}G(z, y, \vec{u}_m) \wedge (\exists z_1)_{\leq w}H_1(z_1, u_1, \vec{x}) \wedge$$

$$\cdots \wedge (\exists z_m)_{\leq w}H_m(z_m, u_m, \vec{x})\Big)$$

We finally turn to the graph $z = f(x, \vec{y})$, where f is defined by primitive recursion from h and g above. This graph is verified by the computation (1).

Assume that (this is the I.H.) $z = h(\vec{y}) \leftrightarrow (\exists w)H(w, z, \vec{y})$ and $u = g(x, \vec{y}, z) \leftrightarrow (\exists w)G(w, u, x, \vec{y}, z)$, where H and G are in $\mathfrak{CA}$. Then

$$z = f(x, \vec{y}) \leftrightarrow (\exists c)\Big(\beta(c, 0) = h(\vec{y}) \wedge \beta(c, x) = z \wedge$$
$$(\forall i)_{<x}\beta(c, i + 1) = g(i, \vec{y}, \beta(c, i))\Big)$$
$$\leftrightarrow \quad \text{(using the I.H.)}$$
$$(\exists c)\Big((\exists w)H(w, \beta(c, 0), \vec{y}) \wedge \beta(c, x) = z \wedge$$
$$(\forall i)_{<x}(\exists w)G(w, \beta(c, i + 1), i, \vec{y}, \beta(c, i))\Big)$$
$$\leftrightarrow \quad \text{(see the proof of I.8.48)}$$
$$(\exists u)(\exists c)_{<u}\Big((\exists w)_{<u}H(w, \beta(c, 0), \vec{y}) \wedge \beta(c, x) = z \wedge$$
$$(\forall i)_{<x}(\exists w)_{<u}G(w, \beta(c, i + 1), i, \vec{y}, \beta(c, i))\Big)$$

Of course things like $G(w, \beta(c, i + 1), i, \vec{y}, \beta(c, i))$ can be easily seen to be in $\mathfrak{CA}$, since $y = \beta(c, i)$ is by I.9.15, and $\beta(c, i) \leq c < u$. $\square$

I.9.17 Corollary. Δ *is the closure of* $z = x + y$ *and* $z = xy$ *under* $\vee, \neg, (\exists y)_{<z}$, $(\exists y)$, *and explicit transformations.*

Indeed, since $(\exists y)$ subsumes $(\exists y)_{<z}$,

I.9.18 Corollary. Δ *is the closure of* $z = x + y$ *and* $z = xy$ *under* $\vee, \neg, (\exists y)$, *and explicit transformations.*

The name "arithmetic" relations is now completely justified.

The following corollary is sufficiently important (even useful) to merit theorem status.

I.9.19 Theorem (A Constructive Arithmetic Kleene Predicate). *There is, for each* $n > 0$, *a constructive arithmetic predicate* $T_{CA}^{(n)}(i, \vec{x}_n, z)$ *such that*

$$\{i\}(\vec{x}_n) \downarrow \leftrightarrow (\exists z)T_{CA}^{(n)}(i, \vec{x}_n, z)$$

Moreover, there is a primitive recursive function U *of one argument, such that*

$$\{i\}(\vec{x}_n) = U\Big((\mu z)T_{CA}^{(n)}(i, \vec{x}_n, z)\Big)$$

Proof. By I.9.16, let, for every $n > 0$, $C^{(n)} \in \mathfrak{CA}$ be such that

$$T^{(n)}(i, \vec{x}_n, z) \leftrightarrow (\exists u)C^{(n)}(u, i, \vec{x}_n, z)$$

Set $T_{CA}^{(n)}(i, \vec{x}_n, z) \leftrightarrow (\exists u)_{\leq z}(\exists w)_{\leq z}(z = J(u, w) \wedge C^{(n)}(u, i, \vec{x}_n, w))$, in other words, $T_{CA}^{(n)}(i, \vec{x}_n, z) \leftrightarrow C^{(n)}(K(z), i, \vec{x}_n, L(z))$.

For U, set, for all z, $U(z) = (L(z))_0$. $\square$

After our brief digression to obtain I.9.16 and its corollaries, we now return to codings and Gödel numberings.

We have seen that any (fixed length) sequence $a_0, \ldots, a_n$ can be coded by a primitive recursive function of the a_i, namely, q of p. 158. Indeed, we can code *variable length sequences* $b_0, \ldots, b_{n-1}$ by appending the length information, n, at the beginning of the sequence.

I.9.20 Definition. (Following Shoenfield (1967).[†]) We now revise the symbol $\langle a_0, \ldots, a_{n-1} \rangle$ to mean, *for the balance of this chapter,* the smallest a such that

$$\beta(a, 0) = n \quad \text{and} \quad \beta(a, i + 1) = a_i + 1 \quad \text{for } i < n \qquad \square$$

We re-introduce also $Seq(z)$, $(z)_i$, $*$, and lh, but to mean something other than their homonyms on p. 136 (this time basing the definitions on the β-coding).

We let $lh(z) = \beta(z, 0)$ and $(z)_i = \beta(z, i + 1) \dot{-} 1$. Thus $\beta(z, i + 1) > 0$ implies $\beta(z, i + 1) = (z)_i + 1$.

$Seq(z)$ will say that z is some "beta-coded" sequence. We have analogous expectations on the "structure" of the number z (as in the case of the "Seq" of p. 136) as dictated by I.9.20. Thus a z is a "Seq" iff

(1) the sequence terms have been incremented by 1 before "packing" them into z, and
(2) any *smaller* x ($x < z$, that is) codes a *different sequence.*[‡]

That is,

$$Seq(z) \leftrightarrow (\forall i)_{<lh(z)} \beta(z, i + 1) > 0 \wedge (\forall x)_{<z}$$
$$(lh(x) \neq lh(z) \vee (\exists i)_{<lh(z)} (z)_i \neq (x)_i)$$

We also (re)define (see also II.2.16 and II.2.21)

$$a * b = (\mu z)(Seq(z) \wedge lh(z) = lh(a) + lh(b) \wedge$$
$$(\forall i)_{<lh(a)}(i > 0 \rightarrow (z)_i = (a)_i) \wedge$$
$$(\forall i)_{<lh(b)}(i > 0 \rightarrow (z)_{i+lh(a)} = (b)_i))$$

It is clear that $\langle a_0, \ldots, a_{n-1} \rangle * \langle b_0, \ldots, b_{m-1} \rangle = \langle a_0, \ldots, a_{n-1}, b_0, \ldots, b_{m-1} \rangle$. Note also that $Seq(z)$ implies $(z)_i < z$ for $i < lh(z)$.

[†] See however II.2.14, p. 242.
[‡] z is the *smallest* number that codes whatever it codes.

As was already suggested on p. 155, if $\ulcorner\ \urcorner : \mathscr{V} \to \mathbb{N}$ is any reasonable 1-1 total mapping from a finite symbol alphabet $\mathscr{V}$ to the natural numbers then we can use $\langle\ldots\rangle$ to extend it to a Gödel numbering $\ulcorner\ \urcorner : \mathscr{V}^+ \to \mathbb{N}$.[†]

Rather than attempting to implement that suggestion in the most general setting, we will shortly show exactly what we have in mind in the case of languages appropriate for formal arithmetic.

We now fix an alphabet over which we can define *a hierarchy of languages* where formal arithmetic can be spoken. For the balance of the chapter, the *absolutely essential nonlogical symbols*[‡] will be indicated in boldface type:

$$\mathbf{0}, \mathbf{S}, \mathbf{+}, \times, < \tag{NL}$$

The entire alphabet, *in the following fixed order*, is

$$\mathbf{0}, \square, \bigcirc, \triangle, =, \neg, \vee, \exists, (,), \# \tag{1}$$

The commas are just metalogical separators in (1) above. *All the rest are formal symbols*. The symbol "#" is "glue", and its purpose is to facilitate building an unlimited (enumerable) supply of predicate and function symbols, as we will shortly describe.

The brackets have their normal use, but will be also assisting the symbols "$\square$", "$\triangle$", and "$\bigcirc$" to build the object variables, function symbols, and predicate symbols of any language in the hierarchy.

The details are as follows:

Variables. The *argot* symbol "v_0" is short for "$(\square)$", "v_1" is short for "$(\square\square)$", and, in general,

$$\text{"}v_j\text{" is short for "}(\overbrace{\square \ldots \square}^{j+1})\text{"}$$

Function symbols. The *argot* symbol "f_j^n" – the jth function symbol ($j \geq 0$) with arity $n > 0$ – is a short name for the string

$$\text{"}(\overbrace{\triangle \ldots \triangle}^{j+1} \# \overbrace{\triangle \ldots \triangle}^{n})\text{"}$$

[†] $\mathscr{V}^+$ denotes the set of all nonempty strings of $\mathscr{V}$.

[‡] That is, all languages where arithmetic is spoken and practised formally will include these symbols.

Predicate symbols. The *argot* symbol "P_j^n" – the jth predicate symbol
($j \geq 0$) with arity $n > 0$ – is a short name for the string

$$\overset{j+1}{\overbrace{\qquad}} \quad \overset{n}{\overbrace{\qquad}}$$
$$\text{``}(\bigcirc \ldots \bigcirc \# \bigcirc \ldots \bigcirc)\text{''}$$

Some symbols in the list (*NL*) *are abbreviations* (*argot*) that will be used
throughout, in conformity with standard mathematical practice:

S	names	$(\triangle \# \triangle)$,	that is, f_0^1
$+$	names	$(\triangle \# \triangle \triangle)$,	that is, f_0^2
$\times$	names	$(\triangle \triangle \# \triangle \triangle)$,	that is, f_1^2
$<$	names	$(\bigcirc \# \bigcirc \bigcirc)$,	that is, P_0^2

I.9.21 Definition. A language $L_{\mathfrak{A}}$ of arithmetic is a first order language over
the alphabet (1), with the stated understanding of the ontology of v_i's, f_i^n's,
and P_i^n's.

$L_{\mathfrak{A}}$ *must* contain the nonlogical symbols (*NL*).

We will normally write (*argot*) $t < s$, $t + s$ and $t \times s$ (t, s are terms) rather
than $<ts$, $+ts$, and $\times ts$ respectively. □

The subscript $\mathfrak{A}$ (for "$\mathfrak{A}$rithmetic") in $L_{\mathfrak{A}}$ reflects the particular standard
structure (appropriate for the language) "$\mathfrak{A}$" that we have in mind. In what
follows we will study extensively the language for the *standard structure*[†] $\mathfrak{N} =$
$(\mathbb{N}; 0; S, +, \times; <)$.

Our notation for the language under discussion will normally indicate the
structure we have in mind, e.g., by writing $L_{\mathfrak{N}}$.

$L_{\mathfrak{N}}$ *has no nonlogical symbols beyond the ones listed under* (*NL*).

Still, occasionally we will just write $L_{\mathfrak{A}}$ regardless, and let the context reveal
what we are thinking.

The "natural" structures for extensions of $L_{\mathfrak{N}}$ will be expansions of $\mathfrak{N}$ (see
Definition I.5.3, p. 54). Such expansions will normally be denoted by $\mathfrak{N}' =$
$(\mathbb{N}; 0; S, +, \times, \ldots; <, \ldots)$, or just $\mathfrak{N}'$. We will never need to specify the "$\ldots$"
part in $\mathfrak{N}'$.

The extended language, correspondingly, may be denoted by $L_{\mathfrak{N}'}$, if the
correspondence needs to be emphasized, else we will fall back on the generic
$L_{\mathfrak{A}}$.

[†] Also sometimes called the "standard model", although this latter term implies the presence of
some nonlogical axioms for $\mathfrak{N}$ to be a model of.

Note that all languages $L_{\mathfrak{A}}$ will have just one constant symbol, $\mathbf{0}$.

Each bold nonlogical symbol of the language is interpreted as its lightface "real" counterpart (as usual, $S = \lambda x.x + 1$), that is, $\mathbf{0}^{\mathfrak{N}} = 0$, $<^{\mathfrak{N}} = <$, $S^{\mathfrak{N}} = S$, $+^{\mathfrak{N}} = +$, etc.

Of course, in any expansion $\mathfrak{N}'$, all we do is give meaning to *new* nonlogical symbols leaving everything else alone. In particular, in any expansion it is still the case that $\mathbf{0}^{\mathfrak{N}'} = 0$, $<^{\mathfrak{N}'} = <$, $S^{\mathfrak{N}'} = S$, $+^{\mathfrak{N}'} = +$, etc.

Gödel numbering of $L_{\mathfrak{A}}$: At this point we fix our Gödel numbering for all possible languages $L_{\mathfrak{A}}$. Referring to (1) of p. 166, we assign

$$\ulcorner \mathbf{0} \urcorner = \langle 0, 1 \rangle$$
$$\ulcorner \square \urcorner = \langle 0, 2 \rangle$$
$$\ulcorner \triangle \urcorner = \langle 0, 3 \rangle$$
$$\ulcorner \bigcirc \urcorner = \langle 0, 4 \rangle$$
$$\ulcorner = \urcorner = \langle 0, 5 \rangle$$
$$\ulcorner \neg \urcorner = \langle 0, 6 \rangle$$
$$\ulcorner \vee \urcorner = \langle 0, 7 \rangle$$
$$\ulcorner \exists \urcorner = \langle 0, 8 \rangle$$
$$\ulcorner (\urcorner = \langle 0, 9 \rangle$$
$$\ulcorner) \urcorner = \langle 0, 10 \rangle$$
$$\ulcorner \# \urcorner = \langle 0, 11 \rangle$$

We then set, for all j, n in $\mathbb{N}$,

$$\ulcorner (\overbrace{\square \ldots \square}^{j+1}) \urcorner = \langle 1, \ulcorner (\urcorner, \overbrace{\ulcorner \square \urcorner, \ldots, \ulcorner \square \urcorner}^{j+1}, \ulcorner) \urcorner \rangle$$

$$\ulcorner (\overbrace{\triangle \ldots \triangle}^{j+1} \# \overbrace{\triangle \ldots \triangle}^{n+1}) \urcorner = \langle 2, \ulcorner (\urcorner, \overbrace{\ulcorner \triangle \urcorner, \ldots, \ulcorner \triangle \urcorner}^{j+1}, \ulcorner \# \urcorner, \overbrace{\ulcorner \triangle \urcorner, \ldots, \ulcorner \triangle \urcorner}^{n+1}, \ulcorner) \urcorner \rangle$$

and

$$\ulcorner (\overbrace{\bigcirc \ldots \bigcirc}^{j+1} \# \overbrace{\bigcirc \ldots \bigcirc}^{n+1}) \urcorner$$
$$= \langle 3, \ulcorner (\urcorner, \overbrace{\ulcorner \bigcirc \urcorner, \ldots, \ulcorner \bigcirc \urcorner}^{j+1}, \ulcorner \# \urcorner, \overbrace{\ulcorner \bigcirc \urcorner, \ldots, \ulcorner \bigcirc \urcorner}^{n+1}, \ulcorner) \urcorner \rangle$$

and by recursion on terms and formulas

1. If $\boldsymbol{g}$ and $\boldsymbol{Q}$ are function and predicate symbols respectively of arity n, and $t_1, \ldots, t_n$ are terms, then

$$\ulcorner \boldsymbol{g} t_1 \ldots t_n \urcorner = \langle \ulcorner \boldsymbol{g} \urcorner, \ulcorner t_1 \urcorner, \ldots, \ulcorner t_n \urcorner \rangle$$

and

$$\ulcorner Qt_1 \ldots t_n \urcorner = \langle \ulcorner Q \urcorner, \ulcorner t_1 \urcorner, \ldots, \ulcorner t_n \urcorner \rangle$$

Moreover, for terms t and s, $\ulcorner t = s \urcorner = \langle \ulcorner = \urcorner, \ulcorner t \urcorner, \ulcorner s \urcorner \rangle$.

2. If $\mathscr{A}$ and $\mathscr{B}$ are formulas, then

$$\ulcorner (\neg \mathscr{A}) \urcorner = \langle \ulcorner \neg \urcorner, \ulcorner \mathscr{A} \urcorner \rangle$$

$$\ulcorner ((\exists x) \mathscr{A}) \urcorner = \langle \ulcorner \exists \urcorner, \ulcorner x \urcorner, \ulcorner \mathscr{A} \urcorner \rangle$$

and

$$\ulcorner (\mathscr{A} \vee \mathscr{B}) \urcorner = \langle \ulcorner \vee \urcorner, \ulcorner \mathscr{A} \urcorner, \ulcorner \mathscr{B} \urcorner \rangle$$

Pause. What happened to the brackets in case 2 above?

It is clear that, *intuitively*, this numbering is algorithmic both ways. Given a *basic symbol* ((1) of p. 166), term or formula, we can apply the above rules to come up with its Gödel number.

Conversely, given a number n we can test first of all if it satisfies *Seq* (for all numbers "$\ulcorner \urcorner$" necessarily satisfy *Seq*). If so, and if it also satisfies $lh(n) = 2$, $(n)_0 = 0$ and $1 \leq (n)_1 \leq 11$, then n is the Gödel number of a basic symbol (1) of the alphabet, and we can tell which one it is.

If now $(n)_0 = 1$, we further check to see if we got a code for a variable, and, if so, of which "v_i". This will entail ascertaining that $lh(n) \geq 4$, that $(n)_1$ "is" [†] a "(", $(n)_{lh(n)-1}$ is a ")", and all the other $(n)_i$ are $\square$'s. If all this testing passes, then we have a code for v_i with $i = lh(n) - 4$.

We can similarly check if n codes a function or predicate symbol, and if so, which one.

As one more example,[‡] let $Seq(n)$, $lh(n) = 3$ and $(n)_0 = \ulcorner \exists \urcorner$. Then we need to ascertain that $(n)_1$ is some v_i and – benefiting from an I.H. that we can decode numbers $< n$ – that $(n)_2$ is some formula $\mathscr{A}$.

Pause. This I.H. hinges on the crucial "if $Seq(n)$, then $(n)_i < n$".

We have thus decoded n into $((\exists v_i) \mathscr{A})$.

We note the following table, which shows that codes of distinct constructs do not clash.

[†] More accurately, "codes".

[‡] We will formalize this discussion in the next chapter, so there is little point to pursue it in detail here.

Code (n) attribute	*Potential* expression
$(n)_0 = 0$	Basic symbol
$(n)_0 = 1$	Variable
$(n)_0 = 2$	Function symbol
$(n)_0 = 3$	Predicate symbol
$(n)_0 > 3 \wedge ((n)_0)_0 = 2$	Term
$(n)_0 > 3 \wedge ((n)_0)_0 = 3$	Atomic formula
$(n)_0 > 3 \wedge ((n)_0)_0 = 0$	
$\quad ((n)_0)_1 = 5$	Atomic formula $(=)$
$\quad ((n)_0)_1 = 6$	Negation
$\quad ((n)_0)_1 = 7$	Disjunction
$\quad ((n)_0)_1 = 8$	Existential quantification

Note that $(n)_0 > 3$ for those items claimed above stems from $(z)_i < z$ for codes z, and the presence of, say, "(" – $\ulcorner (\urcorner > 9$ – in the syntax of functions and predicates, while the symbols $=, \neg, \vee, \exists$ have each a code greater than 5.

Having arithmetized $L_{\mathfrak{A}}$, we can now apply the tools of analysis of number sets to sets of expressions (strings) over $L_{\mathfrak{A}}$. Thus,

I.9.22 Definition. We call a set A of expressions over $L_{\mathfrak{A}}$ *constructive arithmetic, primitive recursive, recursive, semi-recursive,* or *definable over $L_{\mathfrak{A}}$* (in some structure) iff $\{\ulcorner x \urcorner : x \in A\}$ is constructive arithmetic, primitive recursive, recursive, semi-recursive, or definable over $L_{\mathfrak{A}}$, respectively. $\square$

We have seen that one way to generate theories is to start with some set of nonlogical axioms Γ and then build **Thm$_\Gamma$**. Another way is to start with a structure $\mathfrak{M}$ and build $\mathscr{T}(\mathfrak{M}) = \{\mathscr{A} : \models_{\mathfrak{M}} \mathscr{A}\}$.[†]

We will be interested here, among other theories, in the *complete arithmetic*, that is, $\mathscr{T}(\mathfrak{N})$. One of our results will be Tarski's undefinability of arithmetical truth, that is, intuitively, that the set $\mathscr{T}(\mathfrak{N})$ is *not* definable in the structure $\mathfrak{N}$.

Recall that (refer to Definition I.5.15, p. 60) $R \subseteq M^n$ is definable over L in $\mathfrak{M} = (M, \mathscr{I})$ iff, for some formula $\mathscr{R}(x_1, \ldots, x_n)$ of L,

$$R(i_1, \ldots, i_n) \quad \text{iff} \quad \models_{\mathfrak{M}} \mathscr{R}(\bar{i}_1, \ldots, \bar{i}_n) \quad \text{for all } i_j \in M$$

where $\bar{i}_j$ are imported constants (see I.5.4).

[†] Cf. I.6.1.

We say that $\mathscr{R}(x_1, \ldots, x_n)$ *is a regular formula for* R iff the variables $x_1, \ldots, x_n$ are the formal variables $v_0, \ldots, v_{n-1}$. Applying "dummy renaming" (I.4.13) judiciously, followed by the substitution metatheorem (I.4.12), we can convert every formula to a logically equivalent regular formula. For example, $v_{13} = v_{99}$ is not regular, but it says the same thing as the regular formula $v_0 = v_1$, i.e, $v_{13} = v_{99} \models v_0 = v_1$ and $v_0 = v_1 \models v_{13} = v_{99}$.

Henceforth we will tacitly assume that formulas representing relations are always chosen to be regular.

Turning our attention to $\mathfrak{N}'$, we have an interesting variant to the notion of definability over (any) $L_{\mathfrak{N}'}$, due to the richness of $L_{\mathfrak{N}'}$ in certain types of terms.

I.9.23 Definition (Numerals). A term such as "$\overbrace{S \ldots S}^{n} 0$" for $n \geq 1$ is called a *numeral*. We use the shorthand notation $\tilde{n}$ for this term. We also let (the case of $n = 0$) $\tilde{0}$ be an alias for the formal **0**. $\qquad\square$

$\tilde{n}$ must *not* be confused with the imported constants $\bar{n}$ of $L_{\mathfrak{A}}(\mathfrak{N})$ (or $L_{\mathfrak{A}}(\mathfrak{N}')$). The former are (*argot* names of) terms over $L_{\mathfrak{A}}$; the latter are (*argot* names of) *new* constants, *not* present in the original $L_{\mathfrak{A}}$ (which only has the constant **0**).

The usefulness of numerals stems from the following trivial lemma.

I.9.24 Lemma. *For all* $n \in \mathbb{N}$, $\tilde{n}^{\mathfrak{N}} = n$.

Proof. Induction on n. For $n = 0$, $\tilde{0}^{\mathfrak{N}} = \mathbf{0}^{\mathfrak{N}} = 0$.

Now,

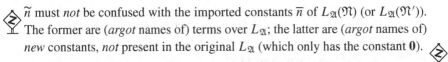

$$\widetilde{n+1}^{\,\mathfrak{N}} = \left(S \overbrace{S \ldots S}^{n} 0 \right)^{\mathfrak{N}}$$
$$= S^{\mathfrak{N}} \left(\tilde{n}^{\mathfrak{N}} \right)$$
$$= S(n) \qquad \text{since } S^{\mathfrak{N}} = S, \text{ using the I.H.}$$
$$= n + 1 \qquad\qquad\qquad\qquad\qquad\qquad\square$$

I.9.25 Corollary. *For all* $n \in \mathbb{N}$, $\models_{\mathfrak{N}} \tilde{n} = \bar{n}$.

Thus, for any formula $\mathscr{A}$ over $L_{\mathfrak{A}}$ and any n, $\models_{\mathfrak{A}} \mathscr{A}[\tilde{n}] \leftrightarrow \mathscr{A}[\bar{n}]$, since $\models \tilde{n} = \bar{n} \rightarrow (\mathscr{A}[\tilde{n}] \leftrightarrow \mathscr{A}[\bar{n}])$.

Therefore, a relation $R \subseteq \mathbb{N}^n$ is definable over $L_{\mathfrak{A}}$ in an appropriate expansion of $\mathfrak{N}$, say $\mathfrak{N}'$, iff there is a regular formula $\mathscr{R}$ in $L_{\mathfrak{A}}$ such that, for all m_j in $\mathbb{N}$,

$$R(m_1, \ldots, m_n) \quad \text{iff} \quad \models_{\mathfrak{A}} \mathscr{R}(\tilde{m}_1, \ldots, \tilde{m}_n) \qquad (D)$$

The above expresses definability over $L_\mathfrak{A}$ without using any "syntactic materials" (such as imported constants $\bar{n}$) that are *outside* the language $L_\mathfrak{A}$.

Lemmata I.9.24–I.9.25 and (D) above go through, of course, for any expansion $\mathfrak{N}'$ of $\mathfrak{N}$.

The following lemma will be used shortly:

I.9.26 Lemma. *The function* $\lambda n. \ulcorner \tilde{n} \urcorner$ *is primitive recursive.*

Proof.

$$\ulcorner \tilde{0} \urcorner = \langle 0, 1 \rangle$$
$$\ulcorner \widetilde{n+1} \urcorner = \langle \ulcorner S \urcorner, \ulcorner \tilde{n} \urcorner \rangle$$

where, of course, $\ulcorner S \urcorner = \langle 2, \ulcorner (\urcorner, \ulcorner \triangle \urcorner, \ulcorner \# \urcorner, \ulcorner \triangle \urcorner, \ulcorner) \urcorner \rangle$. $\square$

Next, through the concept of definability over the *minimum language* $L_\mathfrak{N}$, we assign (finite) *string representations* to each arithmetic formula. Through this device, and Gödel numbering of the string representations, we assign (indirectly) Gödel numbers to those formulas. Thus, we can speak of recursive, semi-recursive, arithmetic, etc., subsets of (arithmetic) relations.

I.9.27 Definition. The set $\mathbf{\Delta_0}(L_\mathfrak{A})$ of formulas over $L_\mathfrak{A}$ is the *smallest* set of formulas over the language that includes the atomic formulas and satisfies the closure conditions: If $\mathscr{A}$ and $\mathscr{B}$ are in $\mathbf{\Delta_0}(L_\mathfrak{A})$, then so are $\neg\mathscr{A}$, $\mathscr{A} \vee \mathscr{B}$, and $(\exists x)_{<t}\mathscr{A}$, where "$(\exists x)_{<t}\mathscr{A}$" is short for $(\exists x)(x < t \wedge \mathscr{A})$.

In $(\exists x)_{<t}\mathscr{A}$ we require that the variable x does not occur in the term t. In the case that $L_\mathfrak{A} = L_\mathfrak{N}$ we often simply write $\mathbf{\Delta_0}$ rather than $\mathbf{\Delta_0}(L_\mathfrak{N})$. $\square$

 We do hope (by virtue of choosing boldface type for $\mathbf{\Delta_0}$) that we will not confuse this set of formulas over $L_\mathfrak{N}$ with the Δ_0 relations of the arithmetic hierarchy.

I.9.28 Lemma. *Every relation in* $\mathfrak{C}\mathfrak{A}$ *is definable in* $\mathfrak{N}$ *over* $L_\mathfrak{N}$ *(in the sense of (D) above) by some formula in* $\mathbf{\Delta_0}$.

Proof. We do induction on $\mathfrak{C}\mathfrak{A}$. The basis contains two cases, $z = x + y$ and $z = xy$.

$v_0 = v_1 + v_2$ defines $z = x + y$, since for any a, b, c in $\mathbb{N}$,

$$(\tilde{a} = \tilde{b} + \tilde{c})^{\mathfrak{N}} = \mathbf{t} \leftrightarrow (\overline{a} = \overline{b} + \overline{c})^{\mathfrak{N}} = \mathbf{t}$$
$$\leftrightarrow \overline{a}^{\mathfrak{N}} = \overline{b}^{\mathfrak{N}} + {}^{\mathfrak{N}}\overline{c}^{\mathfrak{N}}$$
$$\leftrightarrow a = b + c$$

An analogous case can be made for $z = xy$.

We leave it to the reader to verify that if R and Q are defined by $\mathscr{R}$ and $\mathscr{Q}$ respectively, then $\neg R$ and $R \vee Q$ are defined by $\neg \mathscr{R}$ and $\mathscr{R} \vee \mathscr{Q}$ respectively.

Finally, we show that $(\exists x)_{<y} R(\vec{z}_r, x)$ is defined by $(\exists x)_{<y}\mathscr{R}$, that is,

$$(\exists v_{r+1})(v_{r+1} < v_0 \wedge \mathscr{R}(v_1, \ldots, v_r, v_{r+1}))$$

Fix $b_0, \ldots, b_r$ in $\mathbb{N}$. Trivially, $x < y$ defines $x < y$; thus (using the I.H. for R) for any fixed a in $\mathbb{N}$,

$$a < b_0 \wedge R(\vec{b}_r, a) \quad \text{iff} \quad \models_{\mathfrak{N}} \tilde{a} < \tilde{b}_0 \wedge \mathscr{R}(\tilde{b}_1, \ldots, \tilde{b}_r, \tilde{a}) \qquad (i)$$

Let $(\exists x)_{<b_0} R(\vec{b}_r, x)$. Then the left side of "iff" in (i) holds for some a; hence so does the right side. Therefore,

$$\models_{\mathfrak{N}} \tilde{a} < \tilde{b}_0 \wedge \mathscr{R}(\tilde{b}_1, \ldots, \tilde{b}_r, \tilde{a}) \qquad (ii)$$

Hence

$$\models_{\mathfrak{N}} (\exists v_{r+1})(v_{r+1} < \tilde{b}_0 \wedge \mathscr{R}(\tilde{b}_1, \ldots, \tilde{b}_r, v_{r+1})) \qquad (iii)$$

Conversely, if (iii) holds, then so does (ii) for some a, and hence also the right hand side of "iff" in (i). This yields the left hand side; thus $(\exists x)_{<b_0} R(\vec{b}_r, x)$. $\square$

I.9.29 Lemma. *Every relation that is definable in $\mathfrak{N}$ by some formula in Δ_0 over $L_{\mathfrak{N}}$ is in $\mathfrak{PR}_*$.*

Proof. Induction on Δ_0.

Basis. An atomic formula is $t = s$ or $t < s$. It is a trivial matter to verify that a relation defined by such formulas is obtained by $\lambda xy.x = y$ and $\lambda xy.x < y$ by a finite number of explicit transformations *and* substitutions of the functions $\lambda xy.x + y$ and $\lambda xy.xy$ for variables. We know that this does not lead beyond $\mathfrak{PR}_*$.

Induction step. Relations defined by combining their *defining formulas* from Δ_0, using (the formal) $\neg, \vee, (\exists x)_{<y}$ also stay in $\mathfrak{PR}_*$ by the closure of the latter set under (the informal) $\neg, \vee, (\exists x)_{<y}$. $\square$

The above lemma can be sharpened to replace $\mathfrak{PR}_*$ by $\mathcal{CA}$. This follows from results of Bennett (1962) (retold in Tourlakis (1984)), that $\mathcal{CA}$ predicates are closed under replacement of variables by so-called *rudimentary functions* (Smullyan (1961), Bennett (1962), Tourlakis (1984)).

I.9.30 Theorem. *A relation is arithmetic iff it is definable in $\mathfrak{N}$ over $L_{\mathfrak{N}}$.*

Proof. If part. We do induction on the complexity of the defining formula. Atomic formulas define relations in $\mathfrak{PR}_*$, as was seen in the basis step above (indeed, relations in $\mathcal{CA}$ by the previous remarks), and hence in Σ_0 (Definition I.8.60). Since arithmetic relations are closed under $\neg$, $\vee$, $(\exists x)$, the property (that relations which are first order definable in $\mathfrak{N}$ are arithmetic) propagates with (the formal) $\neg$, $\vee$, $(\exists x)$.

Only-if part. This follows because $z = x + y$ and $z = xy$ are definable (I.9.28), and definability is preserved by Boolean operations and $(\exists x)$. $\quad\square$

We now turn to Tarski's theorem.

I.9.31 Theorem (Tarski). *The set $\mathbf{T} = \{\ulcorner \mathscr{A} \urcorner : \mathscr{A} \in \mathscr{T}(\mathfrak{N})\}$ is not arithmetic; therefore it is not definable in $\mathfrak{N}$ (over $L_{\mathfrak{N}}$).*

Proof. Suppose that $\mathbf{T}$ *is* arithmetic, say,

$$(\lambda x. x \in \mathbf{T}) \in \Sigma_m \cup \Pi_m$$

Pick $R(x) \in \Delta_{m+1} - \Sigma_m \cup \Pi_m$ (cf. I.8.72), and let the regular formula $\mathscr{R}(v_0)$ – or, simply, $\mathscr{R}$ – define R over $L_{\mathfrak{N}}$.

Clearly, the formula

$$(\exists v_0)(v_0 = v_1 \wedge \mathscr{R}(v_0))$$

also defines $R(x)$, since

$$\models \mathscr{R}(v_1) \leftrightarrow (\exists v_0)(v_0 = v_1 \wedge \mathscr{R}(v_0))$$

Thus,

$$R(n) \leftrightarrow \models_{\mathfrak{N}} (\exists v_0)(v_0 = \tilde{n} \wedge \mathscr{R})$$
$$\leftrightarrow \left\langle \ulcorner \exists \urcorner, \ulcorner v_0 \urcorner, \langle \ulcorner \wedge \urcorner, \langle \ulcorner = \urcorner, \ulcorner v_0 \urcorner, \ulcorner \tilde{n} \urcorner \rangle, \ulcorner \mathscr{R} \urcorner \rangle \right\rangle \in \mathbf{T}$$

By I.9.26, $g = \lambda n.\langle \ulcorner \exists \urcorner, \ulcorner v_0 \urcorner, \langle \ulcorner \wedge \urcorner, \langle \ulcorner = \urcorner, \ulcorner v_0 \urcorner, \ulcorner \tilde{n} \urcorner \rangle, \ulcorner \mathscr{R} \urcorner \rangle \rangle$ is in $\mathfrak{R}$.[†]

Thus, $R(x)$ iff $g(x) \in \mathbf{T}$; hence $R(x) \in \Sigma_m \cup \Pi_m$, which is absurd. □

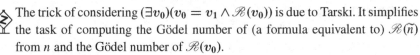 The trick of considering $(\exists v_0)(v_0 = v_1 \wedge \mathscr{R}(v_0))$ is due to Tarski. It simplifies the task of computing the Gödel number of (a formula equivalent to) $\mathscr{R}(\tilde{n})$ from n and the Gödel number of $\mathscr{R}(v_0)$.

Gödel's original approach to computing such numbers was more complicated, utilizing a *substitution function – sub*[‡] (see next chapter) – that was analogous to Kleene's later invention, the "S-m-n" functions of recursion theory.

We have cheated a bit in the above proof, in pretending that "$\wedge$" was a *basic symbol* of the alphabet. The reader can easily fix this by invoking De Morgan's laws.

We have just seen that the set of all formulas of arithmetic that are true in the standard structure $\mathfrak{N}$ is really "hard".[§] Certainly it is recursively unsolvable (not even semi-recursive, not even arithmetic, ...).

What can we say about *provable* formulas of arithmetic? But then, "provable" under what (nonlogical) axioms?

I.9.32 Definition (Formal Arithmetic(s)). We will use the ambiguous phrase "a Formal Arithmetic" for any first order theory over a language $L_{\mathfrak{A}}$ for arithmetic (Definition I.9.21), that contains *at least* the following nonlogical axioms (due to R. M. Robinson (1950)).

The specific Formal Arithmetic over $L_{\mathfrak{N}}$ that has *precisely* the following nonlogical axioms we will call **ROB**. The name "**ROB**" will also apply to the list of the axioms below:

ROB-1. (Regarding S)

    $S1.$ $\neg Sx = 0$ (for any variable x)

    $S2.$ $Sx = Sy \rightarrow x = y$ (for any variables x, y)

ROB-2. (Regarding $+$)

    $+1.$ $x + 0 = x$ (for any variable x)

    $+2.$ $x + Sy = S(x + y)$ (for any variables x, y)[¶]

[†] It is trivial to see that the $\lambda \vec{x}.\langle \vec{x} \rangle$ that was introduced on p.165 is in $\mathfrak{R}$. With slightly more effort (Exercise I.106) one can see that it is even in $\mathfrak{PR}$, and therefore so is g.

[‡] The function sub is primitive recursive, of two arguments, such that $\ulcorner \mathscr{R}(\tilde{n}) \urcorner = sub(g, n)$, where $g = \ulcorner \mathscr{R}(v_0) \urcorner$.

[§] These are the formulas we often call "really true".

[¶] If we had stayed with the formal notation of p. 167, we would not have needed brackets here. We would have simply written $+xSy = S + xy$.

ROB-3. (Regarding ×)

    ×**1.** $x \times 0 = 0$ (for any variable x)

    ×**2.** $x \times Sy = (x \times y) + x$ (for any variables x, y)

ROB-4. (Regarding <)

    <**1.** $\neg x < 0$ (for any variable x)

    <**2.** $x < Sy \leftrightarrow x < y \vee x = y$ (for any variables x, y)

    <**3.** $x < y \vee x = y \vee y < z$ (for any variables x, y).

Let us abstract (i.e., generalize) the situation, for a while, and assume that we have the following:

(1) An *arbitrary* first order language L over some finite alphabet $\mathscr{V}$ (not necessarily *the* one for formal arithmetic).

(2) A Gödel numbering $\ulcorner \, \urcorner$ of $\mathscr{V}^+$ that we extend to terms and formulas, exactly as we described starting on p. 168, where we extended the specific Gödel numbering on the alphabet of arithmetic (p. 166). We are continuing to base our coding on the $\langle \rangle$ of p. 165.

(3) A theory Γ over L with a recursive set of axioms, Γ, that is, one for which the set $A = \{\ulcorner \mathscr{F} \urcorner : \mathscr{F} \in \Gamma\}$ is recursive.

(4) The theory Γ has just two rules of inference,

$$\mathbf{I}_1 : \frac{\mathscr{A}}{\mathscr{B}}$$

$$\mathbf{I}_2 : \frac{\mathscr{A}, \mathscr{B}}{\mathscr{C}}$$

(5) There are *recursive relations* $I_1(x, y)$, $I_2(x, y, z)$, corresponding to $\mathbf{I}_1$ and $\mathbf{I}_2$, that mean

$$I_1(\ulcorner \mathscr{X} \urcorner, \ulcorner \mathscr{Y} \urcorner) \quad \text{iff} \quad \mathscr{X} \vdash \mathscr{Y} \text{ via } \mathbf{I}_1$$

and

$$I_2(\ulcorner \mathscr{X} \urcorner, \ulcorner \mathscr{Y} \urcorner, \ulcorner \mathscr{Z} \urcorner) \quad \text{iff} \quad \mathscr{X}, \mathscr{Y} \vdash \mathscr{Z} \text{ via } \mathbf{I}_2$$

That is, intuitively, we require the rules of inference to be "computable" (or algorithmic).

We call such a theory *recursively axiomatized*.

A proof over Γ is a sequence of formulas $\mathscr{A}_1, \mathscr{A}_2, \ldots, \mathscr{A}_r$, and it is assigned the Gödel number $\langle \ulcorner \mathscr{A}_1 \urcorner, \ulcorner \mathscr{A}_2 \urcorner, \ldots, \ulcorner \mathscr{A}_r \urcorner \rangle$.

The following is at the root of Gödel's *incompletableness* result.

I.9.33 Theorem. *If* Γ *is a recursively axiomatized theory over L, then* **Thm**$_\Gamma$
is semi-recursive.

 Of course, this means that *the set of Gödel numbers of theorems*, $\Theta = \{\ulcorner \mathscr{A} \urcorner :$
$\mathscr{A} \in$ **Thm**$_\Gamma\}$, is semi-recursive.

Proof. We set $A = \{\ulcorner \mathscr{A} \urcorner : \mathscr{A} \in \Gamma\}$ and $B = \{\ulcorner \mathscr{A} \urcorner : \mathscr{A} \in \Lambda\}$, where Λ is
our "standard" set of logical axioms. We defer to the reader the proof that B is
recursive.

 Actually, we suggest the following easier approach. Show that the set of Gödel
numbers of an *equivalent* set of logical axioms, Λ_2 (Exercises I.26–I.41) is
recursive (indeed, primitive recursive).

Let *Proof*(*u*) mean that *u* is the Gödel number of a Γ-proof. Then

$$Proof(u) \leftrightarrow Seq(u) \wedge (\forall i)_{<lh(u)}\Big((u)_i \in A \vee (u)_i \in B \vee$$
$$(\exists j)_{<i} I_1((u)_j, (u)_i) \vee$$
$$(\exists j)_{<i}(\exists k)_{<i} I_2((u)_j, (u)_k, (u)_i)\Big)$$

Thus, *Proof*(*u*) is in $\mathfrak{R}_*$.

Finally, if $\Theta = \{\ulcorner \mathscr{A} \urcorner : \Gamma \vdash \mathscr{A}\}$ (the set of Gödel numbers of Γ-theorems),
then Θ is semi-recursive, since

$$x \in \Theta \leftrightarrow (\exists u)\big(Proof(u) \wedge (\exists i)_{<lh(u)} x = (u)_i\big) \qquad \square$$

I.9.34 Corollary. *The set of theorems of any recursively axiomatized formal
arithmetic is semi-recursive. In particular, the set of theorems of* **ROB** *is semi-
recursive.*

I.9.35 Definition. Smullyan (1992). Let us call a formal arithmetic Γ over $L_\mathfrak{N}$
correct iff $\Gamma \subseteq \mathscr{T}(\mathfrak{N})$. $\qquad \square$

 Many people call such an arithmetic "sound". We opted for the nomenclature
"correct", proposed in Smullyan (1992), because we have defined "sound" in
such a manner that *all* first order theories (recursively axiomatizable or not) are
sound (that is due to the "easy half" of Gödel's completeness theorem).

Correctness along with soundness implies **Thm**$_\Gamma \subseteq \mathscr{T}(\mathfrak{N})$.

Thus, we can rephrase the above corollary as

I.9.36 Corollary (Gödel's First Incompleteness Theorem). [Semantic ver-
sion]. **ROB** *is* incompletable *as long as we extend it correctly and recursively*

over the language $L_{\mathfrak{N}}$. That is, every correct recursively axiomatized formal arithmetic Γ over $L_{\mathfrak{N}}$ is simply incomplete: There are infinitely many sentences $\mathscr{F}$ over $L_{\mathfrak{N}}$ that are undecidable by Γ.

Proof. Since Γ is correct, $\mathbf{Thm}_\Gamma \subseteq \mathscr{T}(\mathfrak{N})$. This is an inequality because $\mathbf{Thm}_\Gamma$ is semi-recursive, while $\mathscr{T}(\mathfrak{N})$ is not even arithmetic.

Every one of the infinitely many sentences $\mathscr{F} \in \mathscr{T}(\mathfrak{N}) - \mathbf{Thm}_\Gamma$,

Pause. Why "infinitely many"?

is unprovable. By correctness, $\neg \mathscr{F}$ is not provable either, since it is false. $\square$

I.9.37 Remark. (1) **ROB** has $\mathfrak{N}$ as a model. Thus it is correct and, of course, consistent.

(2) There are some "really true" sentences that we cannot prove in **ROB** (and, by correctness of **ROB**, we cannot prove their negations either). For example, $(\forall x)\neg x < x$ is not provable (see Exercise I.108).

Similarly, $(\forall x)(\forall y)x + y = y + x$ is not provable (see Exercise I.109).

Thus, it may not have come as a surprise that this formal arithmetic is incomplete. Gödel's genius came in showing that it is impossible to complete **ROB** by throwing axioms at it (a recursive set thereof). It is *incompletable*.

(3) It turns out that there is an "algorithm" to generate sentences $\mathscr{F} \in \mathscr{T}(\mathfrak{N}) -$ $\mathbf{Thm}_\Gamma$ for any recursive Γ that has been "given", say as a W_m (that is, $W_m = \{\ulcorner \mathscr{F} \urcorner : \mathscr{F} \in \Gamma\}$).

First of all (cf. proof of I.9.33), we revise *Proof* (u) to λum. *Proof* (u, m), given by

$$Proof\,(u, m) \leftrightarrow Seq(u) \wedge (\forall i)_{<lh(u)} \Big((\exists z)T^{(1)}(m, (u)_i, z) \vee (u)_i \in B \,\vee$$
$$(\exists j)_{<i}\, I_1((u)_j, (u)_i) \,\vee$$
$$(\exists j)_{<i}(\exists k)_{<i}\, I_2((u)_j, (u)_k, (u)_i) \Big)$$

We set $\Theta_m = \{\ulcorner \mathscr{A} \urcorner : \Gamma_{(m)} \vdash \mathscr{A}\}$, where the subscript (m) of Γ simply reminds us how the axioms are "given", namely: $W_m = A = \{\ulcorner \mathscr{F} \urcorner : \mathscr{F} \in \Gamma_{(m)}\}$.

Then $x \in \Theta_m \leftrightarrow (\exists u)(Proof\,(u, m) \wedge x = (u)_{lh(u) \dot{-} 1})$, a semi-recursive relation of x and m, is equivalent to "$\{a\}(u, m) \downarrow$" for some a, and therefore to "$\{S_1^1(a, m)\}(u) \downarrow$". Setting $f(m) = S_1^1(a, m)$ for all m, we have

There is a primitive recursive function f such that $\Theta_m = W_{f(m)}$ (*)

Pause. A by-product worth verbalizing is that *if the set of axioms is semi-recursive, then so is the set of theorems, and from a semi-recursive index of the former, a semi-recursive index of the latter is primitive recursively computable.*

Let now $\mathscr{H}(v_0)$ define $\lambda x.\neg(\exists z)T^{(1)}(x, x, z)$. Clearly, it is also the case that

$$\neg(\exists z)T^{(1)}(x, x, z) \quad \text{iff} \quad \models_{\mathfrak{N}} (\exists v_0)\big(v_0 = \tilde{x} \wedge \mathscr{H}(v_0)\big)$$

We next consider the set $\tilde{K} = \{\ulcorner(\exists v_0)(v_0 = \tilde{x} \wedge \mathscr{H}(v_0))\urcorner : x \in \overline{K}\}$.
  Set, for all x,

$$g(x) = \ulcorner(\exists v_0)\big(v_0 = \tilde{x} \wedge \mathscr{H}(v_0)\big)\urcorner$$

Then $g \in \mathfrak{PR}$, for the reason we saw in the proof of I.9.31, and

$$x \in \overline{K} \quad \text{iff} \quad g(x) \in \tilde{K}$$

We also consider the set $Q = \{\ulcorner(\exists v_0)(v_0 = \tilde{x} \wedge \mathscr{H}(v_0))\urcorner : x \in \mathbb{N}\}$. Since $x \in \mathbb{N}$ iff $g(x) \in Q$, and g is strictly increasing, Q is recursive (Exercise I.95).

Actually, all we will need here is that Q is semi-recursive, and this it readily is, since it is r.e. (Definition I.8.50) via g (strictly increasing or not). Trivially,

$$\tilde{K} = Q \cap \mathbf{T} \tag{$\ast\ast$}$$

where, as in I.9.31, we have set $\mathbf{T} = \{\ulcorner \mathscr{A}\urcorner : \mathscr{A} \in \mathscr{T}(\mathfrak{N})\}$.

We make a few observations:

(i) If $W_i \subseteq \overline{K}$, then $i \in \overline{K} - W_i$ (else, $i \in K \cap \overline{K}$). Sets such as $\overline{K}$ that come equipped with an algorithm for producing counterexamples to a claim $\overline{K} = W_i$ are called *productive* (Dekker 1995). That is, a set S is productive with *productive function* $h \in \mathfrak{R}$ iff for all i, $W_i \subseteq S$ implies $h(i) \in S - W_i$. In particular, $h = \lambda x.x$ works in the case of $\overline{K}$.

(ii) We saw that $\overline{K} = g^{-1}[\tilde{K}]$, where $g^{-1}[\dots]$ denotes inverse image. We show that $\tilde{K}$ is productive as well. We just need to find a productive function for $\tilde{K}$. Let $W_i \subseteq \tilde{K}$. Then $g^{-1}[W_i] \subseteq g^{-1}[\tilde{K}] = \overline{K}$. By the S-m-n theorem, there is an $h \in \mathfrak{PR}$ such that $g^{-1}[W_i] = W_{h(i)}$. Indeed,

$$\begin{aligned} x \in g^{-1}[W_i] &\leftrightarrow g(x) \in W_i \\ &\leftrightarrow \{i\}(g(x)) \downarrow \\ &\leftrightarrow \{e\}(x, i) \downarrow \qquad \text{for some } e \text{ (why?)} \\ &\leftrightarrow \{S_1^1(e, i)\}(x) \downarrow \end{aligned}$$

Take $h = \lambda x.S_1^1(e, x)$. Thus, $h(i) \in \overline{K} - W_{h(i)} = g^{-1}[\tilde{K} - W_i]$ (g is 1-1). Therefore $\lambda x.g(h(x))$ is the sought productive function.

(iii) We finally show that $\mathbf{T}$ is productive, using $(\ast\ast)$ above. First, we ask the reader (Exercise I.98) to show that there is a $k \in \mathfrak{PR}$ such that

$$W_{k(i)} = W_i \cap Q \tag{1}$$

(recall that Q is semi-recursive). We can now find a productive function for **T**. Let $W_i \subseteq$ **T**. Thus, $W_{k(i)} \subseteq$ **T** $\cap Q = \widetilde{K}$, from which

$$g(h(k(i))) \in \mathbf{T} \cap Q - W_i \cap Q = \mathbf{T} \cap Q - W_i \subseteq \mathbf{T} - W_i$$

We return now to what this was all about: Suppose we have started with a correct recursively axiomatized extension of **ROB**, Γ (over $L_{\mathfrak{N}}$), where W_i is the set of Gödel numbers of the nonlogical axioms. By $(*)$, the set of Gödel numbers of the Γ-theorems is $W_{f(i)}$. By correctness, $W_{f(i)} \subseteq$ **T**.

Then $g(h(k(f(i))))$ is the Gödel number of a true unprovable sentence. A ϕ-index of $\lambda i.g(h(k(f(i))))$ is an algorithm that produces this sentence for any set of axioms (coded by) i. □

We saw in I.9.37(2) above that **ROB** cannot prove some startlingly simple (and useful) formulas. On the other hand, it turns out that it has sufficient power to "define syntactically" all recursive relations. We now turn to study this phenomenon, which will lead to a syntactic version of Gödel's first incompleteness theorem.

I.9.38 Definition. Let Γ be a formal arithmetic over some language $L_{\mathfrak{A}}(\supseteq L_{\mathfrak{N}})$. We say that a relation $R \subseteq \mathbb{N}^r$ is *formally definable in* Γ, or Γ-*definable*, iff there is a formula $\mathcal{R}(v_0, \dots, v_{r-1})$ over $L_{\mathfrak{A}}$ such that, for all $\vec{a}_r$ in $\mathbb{N}$,

$$\text{if} \quad R(\vec{a}_r), \quad \text{then} \quad \Gamma \vdash \mathcal{R}(\widetilde{a}_1, \dots, \widetilde{a}_r)$$

and

$$\text{if} \quad \neg R(\vec{a}_r), \quad \text{then} \quad \Gamma \vdash \neg \mathcal{R}(\widetilde{a}_1, \dots, \widetilde{a}_r)$$

Of course, the left "$\neg$" is informal (metamathematical); the right one is formal.

We say that $\mathcal{R}$ *formally defines* R, but often, just that it *defines* R. The context will determine if we mean formally or in the semantic sense. □

The terminology that names the above concept varies quite a bit in the literature. Instead of "formally definable" some say just "definable" and let the context fix the meaning. Out of a fear that the context might not always successfully do so, we added the obvious qualifier "formally", since this type of definability is about provability, while the other one (I.5.15) is about truth. Terms such as "representable" (or "Γ-representable") are also used elsewhere instead of "formally definable".

It is clear that if R is **ROB**-definable and Γ extends **ROB** (over the same or over an extended language), then R is also Γ-definable.

I.9.39 Lemma. $x = y$ *is* **ROB**-*definable.*

Proof. We show that $v_0 = v_1$ defines $x = y$.

Let $a = b$ (be true). Thus $\tilde{a}$ and $\tilde{b}$ are identical strings over $L_{\mathfrak{A}}$. Therefore $\vdash \tilde{a} = \tilde{b}$ by substitution in the logical axiom $x = x$.

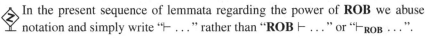 In the present sequence of lemmata regarding the power of **ROB** we abuse notation and simply write "$\vdash \ldots$" rather than "**ROB** $\vdash \ldots$" or "$\vdash_{\mathbf{ROB}} \ldots$".

Let next $a \neq b$. We use (metamathematical) induction on b to show $\vdash \neg\tilde{a} = \tilde{b}$.

Basis. If $b = 0$, then $a = c + 1$ for some c, under the assumption. We want $\vdash \neg\widetilde{c+1} = \tilde{0}$, i.e., $\vdash \neg S\tilde{c} = \tilde{0}$, which we have, by axiom **S1** (I.9.32) and substitution.

Induction step. We now go to the case $a \neq b + 1$. If $a = 0$, then we are back to what we have already argued (using $\vdash x = y \rightarrow y = x$). Let then $a = c + 1$. Thus, $c \neq b$, and hence (I.H.), $\vdash \neg\tilde{c} = \tilde{b}$. By **S2** (and $\models_{\text{Taut}}$) $\vdash \neg S\tilde{c} = S\tilde{b}$, i.e., $\vdash \neg\widetilde{c+1} = \widetilde{b+1}$. □

I.9.40 Corollary. *Let t be a term with the free variables $v_0, \ldots, v_{n-1}$ and f a total function of n arguments such that, for all $\vec{a}_n, b$, if $f(\vec{a}_n) = b$, then*

$$\vdash t(\tilde{a}_1, \ldots, \tilde{a}_n) = \tilde{b}$$

Then the formula $t(v_0, \ldots, v_{n-1}) = v_n$ defines the graph of f.

Proof. It suffices to show that if $f(\vec{a}_n) \neq b$, then

$$\vdash \neg t(\tilde{a}_1, \ldots, \tilde{a}_n) = \tilde{b} \tag{1}$$

Well, the assumption means that for some $c \neq b$, $f(\vec{a}_n) = c$ (f is total), and hence $\vdash t(\tilde{a}_1, \ldots, \tilde{a}_n) = \tilde{c}$.

If we add $t(\tilde{a}_1, \ldots, \tilde{a}_n) = \tilde{b}$ to our assumptions, then $\vdash \tilde{c} = \tilde{b}$ (by $\vdash x = y \rightarrow y = z \rightarrow x = z$ and substitution), which contradicts $\vdash \neg\tilde{c} = \tilde{b}$, yielded by $c \neq b$. Via proof by contradiction we have established (1). □

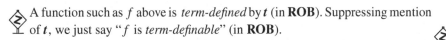 A function such as f above is *term-defined* by t (in **ROB**). Suppressing mention of t, we just say "f is *term-definable*" (in **ROB**).

I.9.41 Lemma. $x + y = z$ *is* **ROB**-*definable.*

Proof. The formula $v_0 + v_1 = v_2$ fills the bill. Indeed, by Corollary I.9.40 we only need to prove that

$$a + b = c \quad \text{implies} \quad \vdash \tilde{a} + \tilde{b} = \tilde{c} \tag{2}$$

We do induction on b.

Basis. Let $b = 0$. Then, $a = c$; thus

$$\vdash \tilde{a} = \tilde{c} \tag{3}$$

By axiom $+\mathbf{1}$ (I.9.32) and substitution, $\vdash \tilde{a} + \tilde{0} = \tilde{a}$. Transitivity of formal equality and (3) yield $\vdash \tilde{a} + \tilde{0} = \tilde{c}$.

Induction step. Let $a + (b + 1) = c$. Then $c = d + 1$ for some d; hence $a + b = d$. By I.H.,

$$\vdash \tilde{a} + \tilde{b} = \tilde{d}$$

Hence

$$\vdash S(\tilde{a} + \tilde{b}) = S\tilde{d} \tag{4}$$

by the Leibniz axiom (substitution and modus ponens).

By axiom $+\mathbf{2}$, we also have (via substitution)

$$\vdash \tilde{a} + S\tilde{b} = S(\tilde{a} + \tilde{b})$$

so that transitivity of equality and (4) result in

$$\vdash \tilde{a} + S\tilde{b} = S\tilde{d}$$

that is, $\vdash \widetilde{a} + \widetilde{b + 1} = \widetilde{d + 1}$. $\qquad\qquad\qquad\qquad\qquad\qquad\quad \square$

I.9.42 Lemma. $x \times y = z$ *is* **ROB**-*definable.*

Proof. Exercise I.110. $\qquad\qquad\qquad\qquad\qquad\qquad\qquad\qquad\qquad\qquad\quad \square$

I.9.43 Lemma. $x < y$ *is* **ROB**-*definable.*

Proof. By induction on b we prove simultaneously

$$a < b \quad \text{implies} \quad \vdash \tilde{a} < \tilde{b} \tag{i}$$

and

$$a \not< b \quad \text{implies} \quad \vdash \neg \tilde{a} < \tilde{b} \tag{ii}$$

Basis. For $b = 0$, (i) is vacuously satisfied, while (ii) follows from axiom $<\mathbf{1}$ and substitution.

Induction step. Let $a < b + 1$. Thus, $a < b$ or $a = b$. One case yields (I.H.) $\vdash \tilde{a} < \tilde{b}$, and the other $\vdash \tilde{a} = \tilde{b}$ (I.9.39).

By tautological implication, $\vdash \tilde{a} < \tilde{b} \vee \tilde{a} = \tilde{b}$ in either case. Hence $\vdash \tilde{a} <$ $S\tilde{b}$, by substitution, $\models_{\text{Taut}}$, and axiom $< \mathbf{2}$. That is, $\vdash \tilde{a} < \widetilde{b+1}$.

Let next $a \not< b + 1$. Thus, $a \not< b$ and $a \neq b$. Thus we have both $\vdash \neg\tilde{a} < \tilde{b}$ (by I.H.) and $\vdash \neg\tilde{a} = \tilde{b}$ (by I.9.39); hence $(\models_{\text{Taut}})$

$$\vdash \neg(\tilde{a} < \tilde{b} \vee \tilde{a} = \tilde{b}) \qquad\qquad (iii)$$

Via the equivalence theorem (I.4.25), $\models_{\text{Taut}}$, and axiom $<\mathbf{2}$, (iii) yields $\vdash$ $\neg\tilde{a} < S\tilde{b}$, that is, $\vdash \neg\tilde{a} < \widetilde{b+1}$. $\qquad\qquad \square$

I.9.44 Lemma. *For any formula $\mathcal{A}$ and number n,* **ROB** *can prove*

$$\mathcal{A}[\tilde{0}], \dots, \mathcal{A}[\widetilde{n-1}] \vdash x < \tilde{n} \to \mathcal{A}$$

where $n = 0$ means that we have nothing nonlogical to the left of "$\vdash$" (beyond the axioms of **ROB***).*

Proof. Induction on n.

Basis. If $n = 0$, then we need $\vdash x < \tilde{0} \to \mathcal{A}$, which is a tautological implication of axiom $<\mathbf{1}$.

Induction step. We want

$$\mathcal{A}[\tilde{0}], \dots, \mathcal{A}[\tilde{n}] \vdash x < S\tilde{n} \to \mathcal{A}$$

By axiom $<\mathbf{2}$ and the equivalence theorem (I.4.25) we need to just prove

$$\mathcal{A}[\tilde{0}], \dots, \mathcal{A}[\tilde{n}] \vdash (x < \tilde{n} \vee x = \tilde{n}) \to \mathcal{A}$$

This follows at once by proof by cases (I.4.26) since $\mathcal{A}[\tilde{0}], \dots, \mathcal{A}[\widetilde{n-1}] \vdash$ $x < \tilde{n} \to \mathcal{A}$ (by I.H.) and $\mathcal{A}[\tilde{n}] \vdash x = \tilde{n} \to \mathcal{A}$ by tautological implication from the Leibniz axiom and modus ponens. $\qquad\qquad \square$

By the deduction theorem we also get

$$\mathbf{ROB} \vdash \mathcal{A}[\tilde{0}] \to \cdots \to \mathcal{A}[\widetilde{n-1}] \to x < \tilde{n} \to \mathcal{A}$$

I.9.45 Lemma. **ROB***-definable relations are closed under Boolean operations.*

Proof. If R and Q are defined by $\mathcal{R}$ and $\mathcal{Q}$ respectively, then it is a trivial exercise (Exercise I.111) to show that $\neg R$ and $R \vee Q$ are defined by $\neg\mathcal{R}$ and $\mathcal{R} \vee \mathcal{Q}$ respectively. $\qquad\qquad \square$

I.9.46 Lemma. **ROB***-definable relations are closed under explicit transformations.*

Proof. (Mostly a task for the reader, Exercise I.112).

Substitution by a constant. Let $R(z, \vec{x}_m)$ be defined by $\mathscr{R}(z, \vec{x}_m)$ (where we have used syntactic variables $z, \vec{x}_m$ for $v_0, \ldots, v_m$).

Then, for any $a \in \mathbb{N}$, $R(a, \vec{x}_m)$ is defined by $\mathscr{R}(\tilde{a}, \vec{x}_m)$. Indeed, for any $\vec{b}_m$, if $R(a, \vec{b}_m)$, then $\vdash \mathscr{R}(\tilde{a}, \tilde{b}_1, \ldots, \tilde{b}_m)$, and if $\neg R(a, \vec{b}_m)$, then $\vdash \neg\mathscr{R}(\tilde{a}, \tilde{b}_1, \ldots, \tilde{b}_m)$.

Introduction of a new variable. Let $Q(\vec{x}_n)$ be defined by $\mathscr{Q}(\vec{x}_n)$ and let $S(z, \vec{x}_n) \leftrightarrow Q(\vec{x}_n)$ for all $z, \vec{x}_n$, where z is a new variable. That is, $S(z, \vec{x}_n) \leftrightarrow Q(\vec{x}_n) \wedge z = z$; hence it is definable (by $\mathscr{Q}(\vec{x}_n) \wedge z = z$, incidentally) due to I.9.39, and I.9.45.   □

I.9.47 Lemma. **ROB**-*definable relations are closed under* $(\exists x)_{<y}$.

Proof. Let $Q(x, \vec{z}_n)$ be defined by $\mathscr{Q}(x, \vec{z}_n)$. We show that $(\exists x)_{<y}\mathscr{Q}(x, \vec{z}_n)$ – that is,

$$(\exists x)\big(x < y \wedge \mathscr{Q}(x, \vec{z}_n)\big)$$

– defines $(\exists x)_{<y} Q(x, \vec{z}_n)$.

Let $(\exists x)_{<a} Q(x, \vec{b}_n)$. Then, for some $c < a$, we have $Q(c, \vec{b}_n)$. By I.9.43, I.9.45, the assumption, and $\models_{\textbf{Taut}}$,

$$\vdash \tilde{c} < \tilde{a} \wedge \mathscr{Q}(\tilde{c}, \tilde{b}_1 \ldots, \tilde{b}_n)$$

Hence (by the substitution axiom and modus ponens)

$$\vdash (\exists x)\big(x < \tilde{a} \wedge \mathscr{Q}(x, \tilde{b}_1 \ldots, \tilde{b}_n)\big)$$

Next, let $\neg(\exists x)_{<a} Q(x, \vec{b}_n)$, that is,

$$Q(0, \vec{b}_n), \ldots, Q(a - 1, \vec{b}_n) \text{ are all false}$$

By assumption on Q,

$$\vdash \neg\mathscr{Q}(\tilde{i}, \tilde{b}_1, \ldots, \tilde{b}_n), \qquad \text{for } i = 0, 1, \ldots, a - 1$$

Hence (I.9.44 followed by generalization)

$$\vdash (\forall x)\big(x < \tilde{a} \rightarrow \neg\mathscr{Q}(x, \tilde{b}_1 \ldots, \tilde{b}_n)\big)$$

In short, removing the abbreviation "$\forall$", and using the logical axiom (tautology)

$$\big(x < \tilde{a} \rightarrow \neg\mathscr{Q}(x, \tilde{b}_1 \ldots, \tilde{b}_n)\big) \leftrightarrow \neg\big(x < \tilde{a} \wedge \mathscr{Q}(x, \tilde{b}_1 \ldots, \tilde{b}_n)\big)$$

and the equivalence theorem, we have

$$\vdash \neg(\exists x)\big(x < \tilde{a} \wedge \mathscr{Q}(x, \tilde{b}_1 \ldots, \tilde{b}_n)\big)   □$$

We have established

I.9.48 Proposition. *Every relation in* $\mathfrak{C}\mathfrak{A}$ *is* **ROB***-definable.*

I.9.49 Lemma (Separation Lemma). *Let R and Q be any two* disjoint *semi-recursive sets. Then there is formula* $\mathscr{F}(x)$ *such that, for all* $m \in \mathbb{N}$,

$$m \in R \quad implies \quad \textbf{ROB} \vdash \mathscr{F}(\tilde{m})$$

and

$$m \in Q \quad implies \quad \textbf{ROB} \vdash \neg\mathscr{F}(\tilde{m})$$

Proof. By I.9.19, there are relations $A(x, y)$ and $B(x, y)$ in $\mathfrak{C}\mathfrak{A}$ such that

$$m \in R \leftrightarrow (\exists x)A(x, m) \tag{1}$$

and

$$m \in Q \leftrightarrow (\exists x)B(x, m) \tag{2}$$

for all m. By I.9.48, let $\mathscr{A}$ and $\mathscr{B}$ define A and B respectively.

We define $\mathscr{F}(y)$ to stand for

$$(\exists x)\Big(\mathscr{A}(x, y) \wedge (\forall z)(z < x \rightarrow \neg\mathscr{B}(z, y))\Big) \tag{3}$$

and proceed to show that it satisfies the conclusion of the lemma.

Let $m \in R$. Then $m \notin Q$; hence (by (1) and (2)), $A(n, m)$ holds for some n, while $B(i, m)$, for all $i \geq 0$, are false. Thus,

$$\vdash \mathscr{A}(\tilde{n}, \tilde{m}) \tag{4}$$

and $\vdash \neg\mathscr{B}(\tilde{i}, \tilde{m})$, for $i \geq 0$.[†] By I.9.44, $\vdash z < \tilde{n} \rightarrow \neg\mathscr{B}(z, \tilde{m})$; hence by generalization followed by $\models_{\textbf{Taut}}$ (using (4)),

$$\vdash \mathscr{A}(\tilde{n}, \tilde{m}) \wedge (\forall z)\Big(z < \tilde{n} \rightarrow \neg\mathscr{B}(z, \tilde{m})\Big)$$

The substitution axiom and $\models_{\textbf{Taut}}$ yield

$$\vdash (\exists x)\Big(\mathscr{A}(x, \tilde{m}) \wedge (\forall z)(z < x \rightarrow \neg\mathscr{B}(z, \tilde{m}))\Big)$$

i.e., $\vdash \mathscr{F}(\tilde{m})$.

Let next $m \in Q$. We want to show that $\vdash \neg\mathscr{F}(\tilde{m})$, which – referring to (3) and using $\models_{\textbf{Taut}}$, using the Leibniz rule, and inserting/removing the defined symbol "$\forall$" – translates to[‡]

$$(\forall x)\Big(\neg\mathscr{A}(x, y) \vee (\exists z)(z < x \wedge \mathscr{B}(z, y))\Big) \tag{5}$$

[†] Throughout, "$\vdash$" is short for "$\vdash_{\textbf{ROB}}$", of course.

[‡] "Translates to" means "is provably equivalent to, without the assistance of any nonlogical axioms".

The assumption yields $m \notin R$; hence (by (1) and (2)), $B(n, m)$ holds for some n, while $A(i, m)$, for all $i \geq 0$, are false. Thus,

$$\vdash \mathscr{B}(\tilde{n}, \tilde{m}) \tag{6}$$

and $\vdash \neg \mathscr{A}(\tilde{i}, \tilde{m})$ for $i \geq 0$. The latter yields (by I.9.44) $\vdash x < S\tilde{n} \to \neg \mathscr{A}(x, \tilde{m})$, or, using the Leibniz rule and axiom $<\mathbf{2}$,

$$\vdash (x < \tilde{n} \vee x = \tilde{n}) \to \neg \mathscr{A}(x, \tilde{m}) \tag{7}$$

(6) and $\models_{\text{Taut}}$ yield $\vdash \tilde{n} < x \to \tilde{n} < x \wedge \mathscr{B}(\tilde{n}, \tilde{m})$; therefore

$$\vdash \tilde{n} < x \to (\exists z)(z < x \wedge \mathscr{B}(z, \tilde{m})) \tag{8}$$

by the substitution axiom and $\models_{\text{Taut}}$.

Proof by cases (I.4.26) and (7) and (8) yield

$$\vdash (x < \tilde{n} \vee x = \tilde{n} \vee \tilde{n} < x) \to \neg \mathscr{A}(x, \tilde{m}) \vee (\exists z)(z < x \wedge \mathscr{B}(z, \tilde{m}))$$

Hence (axiom $<\mathbf{3}$)

$$\vdash \neg \mathscr{A}(x, \tilde{m}) \vee (\exists z)(z < x \wedge \mathscr{B}(z, \tilde{m}))$$

The generalization of the above is (5). $\square$

The above lemma trivially holds for predicates of any number of arguments. That is, we may state and prove, with only notational changes ($\vec{y}$ for y, $\vec{m}$ for m, $\tilde{\vec{m}}$ for $\tilde{m}$), the same result for disjoint semi-recursive subsets of $\mathbb{N}^n$, for any $n > 0$.

I.9.50 Corollary. *Every recursive predicate is definable in* **ROB**.

Proof. If $R \subseteq \mathbb{N}^n$ is recursive, then it and $\mathbb{N}^n - A$ are semi-recursive. $\square$

We can say a bit more. First, a definition:

I.9.51 Definition. A predicate $R(\vec{x}_n)$ is *strongly formally definable in* Γ, or just *strongly definable in* Γ, iff, for some formula $\mathscr{R}(\vec{x}_n)$, both of the following equivalences hold (for all $\vec{b}_n \in \mathbb{N}$):

$$R(\vec{b}_n) \quad \text{iff} \quad \Gamma \vdash \mathscr{R}(\tilde{b}_1, \dots, \tilde{b}_n)$$

and

$$\neg R(\vec{b}_n) \quad \text{iff} \quad \Gamma \vdash \neg \mathscr{R}(\tilde{b}_1, \dots, \tilde{b}_n)$$

We say that $\mathscr{R}$ *strongly defines* R (in Γ). $\square$

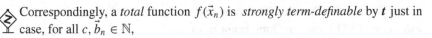 Correspondingly, a *total* function $f(\vec{x}_n)$ is *strongly term-definable* by t just in case, for all $c, \vec{b}_n \in \mathbb{N}$,

$$f(\vec{b}_n) = c \quad \text{iff} \quad \Gamma \vdash t(\tilde{b}_1, \ldots, \tilde{b}_n) = \tilde{c}$$

I.9.52 Corollary. *Every recursive predicate is strongly definable in* **ROB**.

Proof. Let the recursive $R \subseteq \mathbb{N}^n$ be defined (in **ROB**) by $\mathscr{R}(\vec{x}_n)$. We already have the $\rightarrow$-directions for the positive and negative cases (by "weak" definability). We need the $\leftarrow$-directions.

Let then $\vdash \mathscr{R}(\widetilde{m}_1, \ldots, \widetilde{m}_n)$. If $\neg R(\vec{m}_n)$, then also $\vdash \neg\mathscr{R}(\widetilde{m}_1, \ldots, \widetilde{m}_n)$, contradicting the consistency of **ROB**.[†] Thus, $R(\vec{m}_n)$. A similar argument works for the negative case. □

We have at once

I.9.53 Corollary. *Every recursive predicate is strongly definable in any consistent extension of* **ROB**. □

I.9.54 Definition. Two disjoint semi-recursive subsets R and Q of $\mathbb{N}$ are *recursively inseparable* iff there is *no* recursive set S such that $R \subseteq S$ and $S \subseteq \overline{Q}$ (where $\overline{Q} = \mathbb{N} - Q$). □

I.9.55 Lemma. $R = \{x : \{x\}(x) = 0\}$ *and* $Q = \{x : \{x\}(x) = 1\}$ *are recursively inseparable.*

Proof. Clearly R and Q are disjoint semi-recursive sets (why?).

Let S be recursive, and suppose that it separates R and Q, i.e.,

$$R \subseteq S \subseteq \overline{Q} \tag{1}$$

Let $\{i\} = \chi_{\overline{S}}$.

Assume that $i \in S$. Then $\{i\}(i) = 1$, hence $i \in Q$ (definition of Q). But also $i \in \overline{Q}$, by (1).

Thus, $i \in \overline{S}$, after all. Well, if so, then $\{i\}(i) = 0$, hence $i \in R$ (definition of R). By (1), $i \in S$ as well.

Thus $i \notin S \cup \overline{S} = \mathbb{N}$, which is absurd. □

[†] Having adopted a Platonist's metatheory, the fact that $\mathfrak{N}$ is a model of **ROB** establishes consistency. A constructive proof of the consistency of **ROB** is beyond our scope. For an outline see Shoenfield (1967).

I.9.56 Theorem (Church's Theorem). *(Church (1936)).* For *any consistent extension of* **ROB**, *the set* **Thm$_\Gamma$** *is not recursive.* $\square$

Proof. Let $\Theta = \{\ulcorner \mathscr{A} \urcorner : \mathscr{A} \in \mathbf{Thm}_\Gamma\}$, where Γ consistently extends **ROB**.

Applying I.9.49 to the R and Q of I.9.55, we obtain a formula $\mathscr{F}$ of a single free variable v_0 such that

$$m \in R \quad \text{implies} \quad \Gamma \vdash \mathscr{F}(\tilde{m}) \tag{1}$$

and

$$m \in Q \quad \text{implies} \quad \Gamma \vdash \neg \mathscr{F}(\tilde{m}) \tag{2}$$

Let $S = \{m : \Gamma \vdash \mathscr{F}(\tilde{m})\}$. By (1), $R \subseteq S$. By (2) and consistency, $Q \cap S = \emptyset$, i.e., $S \subseteq \overline{Q}$.

By I.9.55,

$$S \notin \mathfrak{R}_* \tag{3}$$

By definition of S,

$$m \in S \quad \text{iff} \quad \Gamma \vdash \mathscr{F}(\tilde{m}) \quad \text{iff} \quad \Gamma \vdash (\exists v_0)(v_0 = \tilde{m} \wedge \mathscr{F}) \tag{4}$$

using the Tarski trick once more (p. 175).

We already know that there is a primitive recursive g such that $g(m) = \ulcorner (\exists v_0)(v_0 = \tilde{m} \wedge \mathscr{F}) \urcorner$ for all m.

Suppose now that $\Theta \in \mathfrak{R}_*$. Then so is S, since (by (4))

$$m \in S \quad \text{iff} \quad g(m) \in \Theta \tag{5}$$

This contradicts (3). $\square$

The above theorem, published some three years after Gödel published his incompleteness results, shows that the decision problem for any consistent theory – not just for one that is recursively axiomatized – that contains arithmetic (**ROB**) is recursively unsolvable. In particular, we rediscover that $\mathscr{T}(\mathfrak{N})$, which extends **ROB** consistently, is not recursive. Of course, we already know much more than this about $\mathscr{T}(\mathfrak{N})$.

Church's result shattered Hilbert's belief that the *Entscheidungsproblem* (decision problem) of any axiomatic theory ought to be solvable by "mechanical means".

On the other hand, Gödel's incompleteness theorems had already shown the untenability of Hilbert's hope to address the consistency of axiomatic theories by finitary means.[†]

[†] Finitary means can be formalized within Peano arithmetic using, if necessary, arithmetization. However, the consistency of Peano arithmetic is not provable from within.

The following is a strengthened (by Rosser (1936)) *syntactic* version of Gödel's first incompleteness theorem. It makes no reference to correctness; instead it relies on the concept of consistency.

Gödel's original syntactic version was proved under stronger consistency assumptions, that the formal arithmetic under consideration was *ω-consistent.*[†]

I.9.57 Theorem (Gödel-Rosser First Incompleteness Theorem) [Syntactic Version]. *Any consistent recursive formal arithmetic* Γ *is simply incomplete.*

Proof. We work with the same R, Q, S as in I.9.55–I.9.56. By I.9.34, Θ is semi-recursive, and hence, so is S by (5) in the previous proof.

How about $\overline{S}$?

At this point we add the assumption (that we expect to contradict) that Γ *is* simply complete. Then (referring to the proof of I.9.56),

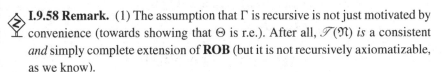

$$m \in \overline{S} \quad \text{iff} \quad \Gamma \nvdash \mathscr{F}(\widetilde{m}) \quad \overset{\text{completeness}}{\underset{\text{consistency}}{\rightleftarrows}} \quad \Gamma \vdash \neg \mathscr{F}(\widetilde{m})$$

As in the proof of I.9.56, setting, for all m, $f(m) = \ulcorner(\exists v_0)(v_0 = \widetilde{m} \wedge \neg \mathscr{F})\urcorner$, we get

$$m \in \overline{S} \quad \text{iff} \quad f(m) \in \Theta$$

Since f is in $\mathfrak{PR}$, $\overline{S}$ is semi-recursive; hence S is recursive, contradicting I.9.55. □

I.9.58 Remark. (1) The assumption that Γ is recursive is not just motivated by convenience (towards showing that Θ is r.e.). After all, $\mathscr{T}(\mathfrak{N})$ *is* a consistent *and* simply complete extension of **ROB** (but it is not recursively axiomatizable, as we know).

(2) We can retell the above proof slightly differently:

Let $S' = \{m : \Gamma \vdash \neg \mathscr{F}(\widetilde{m})\}$. Then S' is semi-recursive. $S \cap S' = \emptyset$ by consistency. Thus, $S \cup S' \neq \mathbb{N}$ (otherwise $\overline{S'} = S$, making S recursive).

Let $n \in \mathbb{N} - S \cup S'$. Then $\nvdash \mathscr{F}(\widetilde{n})$ and $\nvdash \neg \mathscr{F}(\widetilde{n})$.

It is clear that the set $\mathbb{N} - S \cup S'$ is infinite (why?); hence *we have infinitely many undecidable sentences.*

(3) For each $n \in \mathbb{N} - S \cup S'$, exactly one of $\mathscr{F}(\widetilde{n})$ and $\neg \mathscr{F}(\widetilde{n})$ is true (in $\mathfrak{N}$). Which one?

[†] A formal arithmetic is *ω-inconsistent* iff for some formula $\mathscr{F}$ of a single free variable x, $(\exists x)\mathscr{F}$ and all of $\neg \mathscr{F}(\widetilde{m})$ – for all $m \geq 0$ – are provable. Otherwise, it is *ω-consistent*. Of course, an *ω*-consistent theory, as it fails to prove something, is consistent. For Gödel's original formulation of the first theorem see Exercise I.116.

We have defined $\mathscr{F}(y)$ ((3) on p. 185 originally; see also the proof of I.9.56) to stand for

$$(\exists x)\big(\mathscr{A}(x,y) \wedge (\forall z)(z < x \rightarrow \neg\mathscr{B}(z,y))\big)$$

where, with R and Q as in I.9.55–I.9.56,

$$k \in R \quad \text{iff} \quad \models_{\mathfrak{N}} (\exists x)\mathscr{A}(x,\widetilde{k})$$

and[†]

$$k \in Q \quad \text{iff} \quad \models_{\mathfrak{N}} (\exists x)\mathscr{B}(x,\widetilde{k})$$

We also note that

$$\models \neg\mathscr{F}(y) \leftrightarrow (\forall x)\big(\neg\mathscr{A}(x,y) \vee (\exists z)(z < x \wedge \mathscr{B}(z,y))\big) \qquad (i)$$

Now, if $n \in \mathbb{N} - S \cup S'$, then $n \notin R$; hence $\models_{\mathfrak{N}} \neg(\exists x)\mathscr{A}(x,\widetilde{n})$, that is,

$$\models_{\mathfrak{N}} (\forall x)\neg\mathscr{A}(x,\widetilde{n}) \qquad (ii)$$

Noting that (I.4.24, p. 50)

$$\models (\forall x)\,\mathscr{C} \rightarrow (\forall x)(\mathscr{C} \vee \mathscr{D})$$

for any $\mathscr{C}$ and $\mathscr{D}$, (ii) yields

$$\models_{\mathfrak{N}} \neg\mathscr{F}(\widetilde{n}) \qquad (iii)$$

(4) *ω-Incompleteness of arithmetic.* Now, (iii) implies

$$\models_{\mathfrak{N}} \mathscr{G}(\widetilde{k},\widetilde{n}) \qquad \text{for all } k \geq 0$$

by (i) and substitution, where $\mathscr{G}(x,y)$ abbreviates

$$\neg\mathscr{A}(x,y) \vee (\exists z)(z < x \wedge \mathscr{B}(z,y))$$

Thus,

$$\Gamma \vdash \mathscr{G}(\widetilde{k},\widetilde{n}) \qquad \text{for all } k \geq 0$$

since $A(x,y)$ and $B(x,y)$ (p. 185) are in $\mathfrak{CA}$.

Thus we have the phenomenon of *ω-incompleteness* in *any* consistent (and recursive) extension Γ of **ROB**. That is, there is a formula $\mathscr{H}(x)$ such that while $\Gamma \vdash \mathscr{H}(\widetilde{k})$ for all $k \geq 0$, yet $\Gamma \nvdash (\forall x)\mathscr{H}(x)$. An example of such an $\mathscr{H}$ is $\mathscr{G}(x,\widetilde{n})$. Therefore, **Thm**$_\Gamma$ for such Γ is *not* closed under the "ω-rule"

$$\mathscr{H}(\widetilde{0}), \mathscr{H}(\widetilde{1}), \mathscr{H}(\widetilde{2}), \ldots \vdash (\forall x)\mathscr{H}(x)$$

[†] This and the previous equivalence are just restatements of (1) and (2) on p. 185.

(5) *A consistent but ω-inconsistent arithmetic*. Next, consider the theory $\Gamma + \mathscr{F}(\tilde{n})$, for some fixed $n \in \mathbb{N} - S \cup S'$. Since $\Gamma \nvdash \neg\mathscr{F}(\tilde{n})$, this theory is consistent. However, it is ω-inconsistent:

Indeed, by (*ii*),

$$\models_{\mathfrak{N}} \neg\mathscr{A}(\tilde{k}, \tilde{n}) \qquad \text{for all } k \geq 0$$

Hence, for reasons already articulated,

$$\Gamma \vdash \neg\mathscr{A}(\tilde{k}, \tilde{n}) \qquad \text{for all } k \geq 0$$

and, trivially,

$$\Gamma + \mathscr{F}(\tilde{n}) \vdash \neg\mathscr{A}(\tilde{k}, \tilde{n}) \quad \text{for all } k \geq 0 \tag{iv}$$

But also $\Gamma + \mathscr{F}(\tilde{n}) \vdash \mathscr{F}(\tilde{n})$, so that, along with (*iv*), we have derived the ω-inconsistency of $\Gamma + \mathscr{F}(\tilde{n})$; for

$$\Gamma + \mathscr{F}(\tilde{n}) \vdash (\exists x)\mathscr{A}(x, \tilde{n})$$

– the latter due to $\vdash (\exists x)(\mathscr{C} \wedge \mathscr{D}) \rightarrow (\exists x)\mathscr{C}$ and the definition of $\mathscr{F}$.

(6) *A consistent but incorrect arithmetic*. The consistent formal arithmetic $\Gamma + \mathscr{F}(\tilde{n})$ is *not* correct, since $\not\models_{\mathfrak{N}} \mathscr{F}(\tilde{n})$.

(7) From what we have seen here (cf. also Exercise I.113) we can obtain an alternative foundation of computability via **ROB**: We can *define* a recursive predicate to be one that is strongly definable in **ROB** (I.9.52).

We can also define a semi-recursive predicate $P(\vec{x}_n)$ to be one that is *positively strongly definable* in **ROB**, i.e., for some $\mathscr{P}$ and all $\vec{a}_n$,

$$P(\vec{a}_n) \quad \text{iff} \quad \textbf{ROB} \vdash \mathscr{P}(\tilde{a}_1, \dots, \tilde{a}_n)$$

We can then say that a partial recursive function is one whose graph is positively strongly definable, while a recursive function is a total partial recursive function.

With this understanding, *uncomputability* coincides with *undecidability* (of **ROB**) and hence with its incomplete(able)ness (unprovability): There are sets that are positively strongly definable but not strongly definable (e.g., the set K; see also Exercise I.116). Thus, our claim at the beginning of this volume, – that not only is there an intimate connection between uncomputability and unprovability, but also you cannot have one without having the other – is now justified. □

I.10. Exercises

I.1. Prove that the closure obtained by $\mathscr{I} = \{3\}$ and the two relations $z = x + y$ and $z = x - y$ is the set $\{3k : k \in \mathbb{Z}\}$.

I.2. The pair that effects the definition of **Term** (I.1.5, p. 14) is unambiguous.

I.3. The pair that effects the definition of **Wff** (I.1.8, p. 15) is unambiguous.

I.4. With reference to I.2.13 (p. 25), prove that if all the g_Q and h are defined everywhere on their input sets (i.e., they are *total*) – these are $\mathscr{I}$ for h and $A \times Y^r$ for g_Q and $(r+1)$-ary Q – then f is defined everywhere on $\mathrm{Cl}(\mathscr{I}, \mathscr{R})$.

I.5. Let us define inductively the set of formulas Ar by:

(1) 1, 2, and 3 are in Ar.

(2) If a and b stand for formulas of Ar, then so do $a + b$ and $a \times b$. (N.B. It is the intention here *not* to utilize brackets.)

(3) Define a *function* (intuitively speaking), $Eval(x)$ by $Eval(x) = x$ if x is 1, 2, or 3, and (inductive step) $Eval(a + b) = Eval(a) + Eval(b)$, $Eval(a \times b) = Eval(a) \times Eval(b)$ for all a, b in Ar, defined via (1) and (2).

Show that the definition (1)–(2) above allows more than one possible parse, which makes (3) ill-defined; indeed, show that for some $a \in Ar$, $Eval(a)$ has more than one possible value, so it is not a function after all.

I.6. Prove that for every formula $\mathscr{A}$ in **Prop** (I.3.2, p. 28) the following is true: Every nonempty proper prefix (I.1.4, p. 13) of the string A has an excess of left brackets.

I.7. Prove that any non-prime $\mathscr{A}$ in **Prop** has uniquely determined immediate predecessors.

I.8. For any formula $\mathscr{A}$ and any two valuations v and v', $\bar{v}(\mathscr{A}) = \bar{v}'(\mathscr{A})$ if v and v' agree on all the propositional variables that occur in $\mathscr{A}$.

I.9. Prove that $\mathscr{A}[x \leftarrow t]$ is a formula (whenever it is defined) if t is a term.

I.10. Prove that Definition I.3.12 does not depend on our choice of *new variables* $\vec{z}_r$.

I.11. Prove that $\vdash (\forall x)(\forall y)\mathscr{A} \leftrightarrow (\forall y)(\forall x)\mathscr{A}$.

I.12. Prove I.4.23.

I.13. Prove I.4.24.

I.14. (1) Show that $x < y \vdash y < x$ ($<$ is some binary predicate symbol; the choice of symbol here is meant to provoke).

(2) Show informally that $\nvdash x < y \rightarrow y < x$ (*Hint*: Use the assumption that our logic "does not lie" (soundness theorem).)

(3) Does this invalidate the deduction theorem? Explain.

I.15. Prove I.4.25

I.16. Prove that $\vdash x = y \rightarrow y = x$.

I.17. Prove that $\vdash x = y \land y = z \rightarrow x = z$.

I.18. Suppose that $\Gamma \vdash t_i = s_i$ for $i = 1, \ldots, m$, where the t_i, s_i are arbitrary terms. Let $\mathscr{F}$ be a formula, and $\mathscr{F}'$ be obtained from it by replacing any number of occurrences of t_i in $\mathscr{F}$ (*not necessarily all*) by s_i. Prove that $\Gamma \vdash \mathscr{F} \leftrightarrow \mathscr{F}'$.

I.19. Suppose that $\Gamma \vdash t_i = s_i$ for $i = 1, \ldots, m$, where the t_i, s_i are arbitrary terms. Let r be a term, and r' be obtained from it by replacing any number of occurrences of t_i in r (*not necessarily all*) by s_i.
Prove that $\Gamma \vdash r = r'$.

I.20. Settle the Pause following I.4.21.

I.21. Prove I.4.27.

I.22. Suppose that x is not free in $\mathscr{A}$. Prove that $\vdash \mathscr{A} \rightarrow (\forall x)\mathscr{A}$ and $\vdash (\exists x)\mathscr{A} \rightarrow \mathscr{A}$.

I.23. Prove the distributive laws :
$\vdash (\forall x)(\mathscr{A} \wedge \mathscr{B}) \leftrightarrow (\forall x)\mathscr{A} \wedge (\forall x)\mathscr{B}$ and $\vdash (\exists x)(\mathscr{A} \vee \mathscr{B}) \leftrightarrow (\exists x)\mathscr{A} \vee (\exists x)\mathscr{B}$.

I.24. Prove $\vdash (\exists x)(\forall y)\mathscr{A} \rightarrow (\forall y)(\exists x)\mathscr{A}$ with two methods: first using the auxiliary constant method, next exploiting monotonicity.

I.25. Prove $\vdash (\exists x)(\mathscr{A} \rightarrow (\forall x)\mathscr{A})$.

In what follows let us denote by Λ_1 the *pure* logic of Section I.3 (I.3.13 and I.3.15).

Let us now introduce a *new* pure logic, which we will call Λ_2. This is exactly the same as our Λ_1 (which we have called just Λ until now), except that we have a different axiom group **Ax1**. Instead of adopting *all* tautologies, we only adopt the following four *logical axiom schemata of* group **Ax1** in Λ_2:[†]

(1) $\mathscr{A} \vee \mathscr{A} \rightarrow \mathscr{A}$
(2) $\mathscr{A} \rightarrow \mathscr{A} \vee \mathscr{B}$
(3) $\mathscr{A} \vee \mathscr{B} \rightarrow \mathscr{B} \vee \mathscr{A}$
(4) $(\mathscr{A} \rightarrow \mathscr{B}) \rightarrow (\mathscr{C} \vee \mathscr{A} \rightarrow \mathscr{C} \vee \mathscr{B})$

Λ_2 is due to Hilbert (actually, he also included *associativity* in the axioms, but, as Gentzen has proved, this was deducible from the system as here given; therefore it was not an *independent* axiom – see Exercise I.35). In the exercises below we write $\vdash_i$ for $\vdash_{\Lambda_i}$, $i = 1, 2$.

I.26. Show that for all $\mathscr{F}$ and every set of formulas Γ, if $\Gamma \vdash_2 \mathscr{F}$ holds, then so does $\Gamma \vdash_1 \mathscr{F}$.

Our aim is to see that the logics Λ_1 and Λ_2 are equivalent, i.e., have exactly the same theorems. In view of the trivial Exercise I.26 above, what remains to be shown is that every tautology is a theorem of Λ_2. One particular way to prove this is through the following sequence of Λ_2-facts.

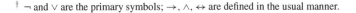

[†] $\neg$ and $\vee$ are the primary symbols; $\rightarrow$, $\wedge$, $\leftrightarrow$ are defined in the usual manner.

I.27. Show the transitivity of $\to$ in Λ_2:

$$\mathscr{A} \to \mathscr{B}, \mathscr{B} \to \mathscr{C} \vdash_2 \mathscr{A} \to \mathscr{C} \qquad \text{for all } \mathscr{A}, \mathscr{B}, \text{ and } \mathscr{C}.$$

I.28. Show that $\vdash_2 \mathscr{A} \to \mathscr{A}$ (i.e., $\vdash_2 \neg \mathscr{A} \vee \mathscr{A}$), for any $\mathscr{A}$.

I.29. For all $\mathscr{A}, \mathscr{B}$ show that $\vdash_2 \mathscr{A} \to \mathscr{B} \vee \mathscr{A}$.

I.30. Show that for all $\mathscr{A}$ and $\mathscr{B}$, $\mathscr{A} \vdash_2 \mathscr{B} \to \mathscr{A}$.

I.31. Show that for all $\mathscr{A}$, $\vdash_2 \neg\neg \mathscr{A} \to \mathscr{A}$ and $\vdash_2 \mathscr{A} \to \neg\neg \mathscr{A}$.

I.32. For all $\mathscr{A}$ and $\mathscr{B}$, show that $\vdash_2 (\mathscr{A} \to \mathscr{B}) \to (\neg \mathscr{B} \to \neg \mathscr{A})$. Conclude that $\mathscr{A} \to \mathscr{B} \vdash_2 \neg \mathscr{B} \to \neg \mathscr{A}$.

   (*Hint.* $\vdash_2 \mathscr{A} \to \neg\neg \mathscr{A}$.)

I.33. Show that $\mathscr{A} \to \mathscr{B} \vdash_2 (\mathscr{B} \to \mathscr{C}) \to (\mathscr{A} \to \mathscr{C})$ for all $\mathscr{A}, \mathscr{B}, \mathscr{C}$.

I.34. (Proof by cases in Λ_2.) Show for all $\mathscr{A}, \mathscr{B}, \mathscr{C}, \mathscr{D}$,

$$\mathscr{A} \to \mathscr{B}, \mathscr{C} \to \mathscr{D} \vdash_2 \mathscr{A} \vee \mathscr{C} \to \mathscr{B} \vee \mathscr{D}$$

I.35. Show for all $\mathscr{A}, \mathscr{B}, \mathscr{C}$ that

   (1) $\vdash_2 \mathscr{A} \vee (\mathscr{B} \vee \mathscr{C}) \to (\mathscr{A} \vee \mathscr{B}) \vee \mathscr{C}$ and

   (2) $\vdash_2 (\mathscr{A} \vee \mathscr{B}) \vee \mathscr{C} \to \mathscr{A} \vee (\mathscr{B} \vee \mathscr{C})$.

I.36. Deduction theorem in "propositional" Λ_2. Prove that if $\Gamma, \mathscr{A} \vdash_2 \mathscr{B}$ using only modus ponens, then also $\Gamma \vdash_2 \mathscr{A} \to \mathscr{B}$ using only modus ponens, for any formulas $\mathscr{A}, \mathscr{B}$ and set of formulas Γ.

   (*Hint.* Induction on the length of proof of $\mathscr{B}$ from $\Gamma \cup \{\mathscr{A}\}$, using the results above.)

I.37. Proof by contradiction in "propositional" Λ_2. Prove that if $\Gamma, \neg \mathscr{A}$ derives a contradiction in Λ_2 using only modus ponens,[†] then $\Gamma \vdash_2 \mathscr{A}$ using only modus ponens, for any formulas $\mathscr{A}$ and set of formulas Γ. Also prove the converse.

We can now prove the *completeness theorem* (Post's theorem) for the "propositional segment" of Λ_2, that is, the logic Λ_3 – so-called *propositional logic* (or *propositional calculus*) – obtained from Λ_2 by keeping only the "propositional axioms" (1)–(4) and modus ponens, dropping the remaining axioms and the $\exists$-introduction rule. (Note. It is trivial that if $\Gamma \vdash_3 \mathscr{A}$, then $\Gamma \vdash_2 \mathscr{A}$.) Namely, we will prove that, for any $\mathscr{A}$ and Γ, if $\Gamma \models_{\text{Taut}} \mathscr{A}$, then $\Gamma \vdash_3 \mathscr{A}$.

   First, a definition:

I.10.1 Definition (Complete Sets of Formulas). A set Γ is *complete* iff for every $\mathscr{A}$, at least one of $\mathscr{A}$ or $\neg \mathscr{A}$ is a member of Γ. □

[†] That is, it proves some $\mathscr{B}$ but also proves $\neg \mathscr{B}$.

I.38. Let $\Gamma \nvdash_3 \mathscr{A}$. Prove that there is a *complete* $\Delta \supseteq \Gamma$ such that also $\Delta \nvdash_3 \mathscr{A}$. This is a *completion* of Γ.

(*Hint.* Let $\mathscr{F}_0, \mathscr{F}_1, \mathscr{F}_3, \ldots$ be an enumeration of all formulas. (There *is* such an enumeration. Define Δ_n by induction on n:

$$\Delta_0 = \Gamma$$

$$\Delta_{n+1} = \begin{cases} \Delta_n \cup \{\mathscr{F}_n\} & \text{if } \Delta_n \cup \{\mathscr{F}_n\} \nvdash_3 \mathscr{A} \\ \text{otherwise} \\ \Delta_n \cup \{\neg\mathscr{F}_n\} & \text{if } \Delta_n \cup \{\neg\mathscr{F}_n\} \nvdash_3 \mathscr{A} \end{cases}$$

To make sense of the above definition, show the impossibility of having both $\Delta_n \cup \{\mathscr{F}_n\} \vdash_3 \mathscr{A}$ and $\Delta_n \bigcup \{\neg\mathscr{F}_n\} \vdash_3 \mathscr{A}$. Then show that $\Delta = \bigcup_{n \geq 0} \Delta_n$ is as needed.)

I.39. (Post.) If $\Gamma \models \mathscr{A}$, then $\Gamma \vdash_3 \mathscr{A}$.

(*Hint.* Prove the contrapositive. If $\Gamma \nvdash_3 \mathscr{A}$, let Δ be a completion (Exercise I.38) of Γ such that $\Delta \nvdash_3 \mathscr{A}$. Now, for every *prime formula* (cf. I.3.1, p. 28) $\mathscr{P}$, *exactly one* of $\mathscr{P}$ or $\neg\mathscr{P}$ (why exactly one?) is in Δ. Define a *valuation* (cf. I.3.4, p. 29) v on all prime formulas by

$$v(\mathscr{P}) = \begin{cases} 0 & \text{if } \mathscr{P} \in \Delta \\ 1 & \text{otherwise} \end{cases}$$

Of course, "0" codes, intuitively, "**true**", while "1" codes "**false**". To conclude, prove by induction on the formulas of **Prop** (cf. I.3.2, p. 28) that the extension $\bar{v}$ of v satisfies, for all formulas $\mathscr{B}$, $\bar{v}(\mathscr{B}) = 0$ iff $\mathscr{B} \in \Delta$. Argue that $\mathscr{A} \notin \Delta$.)

I.40. If $\Gamma \models_{\mathbf{Taut}} \mathscr{A}$, then $\Gamma \vdash_2 \mathscr{A}$.

I.41. For any formula $\mathscr{F}$ and set of formulas Γ, one has $\Gamma \vdash_1 \mathscr{F}$ iff $\Gamma \vdash_2 \mathscr{F}$.

I.42. (Compactness of propositional logic.) We say that Γ is *finitely satisfiable* (*in the propositional sense*) iff every finite subset of Γ is satisfiable (cf. I.3.6, p. 30). Prove that Γ is satisfiable iff it is finitely satisfiable.

(*Hint.* Only the if part is nontrivial. It uses Exercise I.39. Further hint: If Γ is unsatisfiable, then $\Gamma \models_{\mathbf{Taut}} \mathscr{A} \wedge \neg\mathscr{A}$ for some formula $\mathscr{A}$.)

I.43. Prove semantically, without using soundness, that $\mathscr{A} \models (\forall x)\mathscr{A}$.

I.44. Give a semantic proof of the deduction theorem.

I.45. Show semantically that for all $\mathscr{A}$ and $\mathscr{B}$, $\mathscr{A} \to \mathscr{B} \models (\forall x)\mathscr{A} \to (\forall x)\mathscr{B}$.

I.46. Show semantically that for all $\mathscr{A}$ and $\mathscr{B}$, $\mathscr{A} \to \mathscr{B} \models (\exists x)\mathscr{A} \to (\exists x)\mathscr{B}$.

I.47. Prove the claim in Remark I.6.7.

I.48. Prove that the composition of two embeddings $\phi : \mathfrak{M} \to \mathfrak{K}$ and $\psi : \mathfrak{K} \to \mathfrak{L}$ is an embedding.

I.49. Find two structures that are elementarily equivalent, but not isomorphic.

I.50. Let $\phi : \mathfrak{M} \to \mathfrak{M}$ be an isomorphism. We say also (since we have the same structure on both sides of "$\to$") that ϕ is an *automorphism*. Prove that if $S \subseteq |\mathfrak{M}|^n$ is first order definable (cf. I.5.15) in $\mathfrak{M}$, then, for all $\vec{i}_n$ in $|\mathfrak{M}|$,

$$\langle \vec{i}_n \rangle \in S \quad \text{iff} \quad \langle \phi(i_1), \dots, \phi(i_n) \rangle \in S$$

I.51. Prove that $\mathbb{N}$ is not first order definable in the structure $\mathfrak{R} = (\mathbb{R}, <)$ ($\mathbb{R}$ is the set of reals).

(*Hint.* Use Exercise I.50 above.)

I.52. Prove that addition is not definable in $(\mathbb{N}, \times)$. More precisely, show that the set $\{\langle x, y, z \rangle \in \mathbb{N}^3 : z = x + y\}$ is not definable.

(*Hint.* Define a function $\phi : \mathbb{N} \to \mathbb{N}$ by $\phi(x) = x$ if x is not divisible by 2 or 3. Otherwise $\phi(x) = y$, where y has the same prime number decomposition as x, except that the primes 2 and 3 are interchanged. For example, $\phi(6) = 6$, $\phi(9) = 4$, $\phi(12) = 18$. Prove that ϕ is an automorphism on $(\mathbb{N}, \times)$, and then invoke Exercise I.50 above.)

I.53. Prove the only-if part of the Loś-Tarski theorem (I.6.26).

I.54. Prove Theorem I.6.29.

I.55. Prove the if part of Theorem I.6.30.

I.56. Prove by a direct construction, without using the upward Löwenheim-Skolem theorem, that there are nonstandard models for arithmetic.

(*Hint.* Work with the theory $\text{Th}(\mathfrak{N}) \cup \{\tilde{n} < C : n \in \mathbb{N}\}$ where C is a new constant added to the language of arithmetic $L_\mathfrak{N} = \{0, S, +, \times, <\}$. Use compactness and the consistency theorem.)

I.57. Prove that if Γ has arbitrarily large finite models, then it has an infinite model.

I.58. Prove by a direct construction, without using the upward Löwenheim-Skolem theorem, that there is a nonstandard model for all true first order sentences about the reals.

I.59. Prove Proposition I.6.45.

I.60. Prove the pinching lemma (I.6.53).

I.61. Conclude the proof of I.6.55.

I.62. Let N, a unary predicate of the extended language L of the reals, have as its interpretation $N^\mathfrak{R} = \mathbb{N}$, the set on natural numbers. Use it to prove that there are *infinite natural numbers* in *$\mathfrak{R}$.

(*Hint.* Use the true (in $\mathfrak{R}$) sentence $(\forall x)(\exists y)(N(y) \land x < y)$.)

I.63. Prove that if h is an infinite natural number, then so is $h - 1$. A side effect of this is an infinite descending chain of (infinite) natural numbers in *$\mathbb{R}$. Hence there are nonempty sets of natural numbers in *$\mathbb{R}$ with no minimum element. Does this contradict the transfer principle (p. 99)? Why?

I.64. Prove the *extreme value theorem*: Every real function of one real variable that is continuous on the real interval $[a, b]$ attains its maximum. That is, $(\exists x \in [a, b])(\forall y \in [a, b]) f(y) \leq f(x)$.

(*Hint*. Subdivide $[a, b]$ into $n > 0$ subintervals of equal length, $[a_i, a_{i+1}]$, where $a_i = a + (b - a)i/n$ for $i = 0, \ldots, n$. Formulate as a first order sentence over L the true (why true?) statement that "for all choices of $n > 0$ (in $\mathbb{N}$) there is an $i \in \mathbb{N}$ such that $f(a + (b - a)i/n)$ is maximum among the values $f(a+(b-a)k/n), k = 0, \ldots, n$". Transfer to $^*\mathfrak{R}$. This is still true. Now take n to be an infinite natural number K. Let I be a (hyperreal) natural number that makes $f(a + (b - a)I/K)$ maximum among the $f(a+(b-a)i/K), 0 \leq i \leq K$. See if $f(\mathrm{st}(a+(b-a)I/K))$ is what you want.)

I.65. Use the technique of the previous problem to prove the *intermediate value theorem*: Every real function of one real variable that is continuous on the real interval $[a, b]$ attains every value between its minimum and its maximum.

I.66. Prove the existence of *infinite primes* in $^*\mathbb{R}$.

I.67. Let $\mathfrak{T}$ be a pure theory over some language L. Form the theory $\mathfrak{T}'$ over $L \cup \{\tau\}$ by adding the schemata

$$(\forall x)(\mathscr{A} \leftrightarrow \mathscr{B}) \rightarrow (\tau x).\mathscr{A} = (\tau x).\mathscr{B}$$

and

$$(\exists x).\mathscr{A}[x] \rightarrow \mathscr{A}[(\tau x).\mathscr{A}]$$

where τ is a new symbol used to build terms: If $\mathscr{A}$ is a wff, then $(\tau x).\mathscr{A}$ is a term. Prove that $\mathfrak{T}'$ extends $\mathfrak{T}$ conservatively (τ may be interpreted as that of p. 123).

I.68. Prove that in the presence of the *initial functions* of $\mathfrak{PR}$ (I.8.10, p. 128) the following Grzegorczyk (1953) substitution operations can be simulated by composition (I.8.6):

(a) Substitution of a function into a variable

(b) Substitution of a constant into a variable

(c) Identification of any two variables

(d) Permutation of any two variables

(e) Introduction of new ("dummy") variables (i.e., forming $\lambda\vec{x}\vec{y}.g(\vec{y})$).

I.69. Prove that if f is total and $\lambda\vec{x}y.y = f(\vec{x})$ is in $\mathfrak{R}_*$, then $f \in \mathfrak{R}$. Is the assumption that f is total necessary?

I.70. Show that both $\mathfrak{PR}$ and $\mathfrak{R}$ are closed under bounded summation, $\sum_{y<z} f(y, \vec{x})$ and bounded product, $\prod_{y<z} f(y, \vec{x})$.

I.71. Prove that if f is total and $\lambda \vec{x} y . y = f(\vec{x})$ is in $\mathfrak{P}_*$ (semi-recursive), then $f \in \mathfrak{R}$.

I.72. Prove I.8.27 using the if-then-else function rather than addition and multiplication.

I.73. Prove that $p_n \leq 2^{2^n}$ for $n \geq 0$.
(*Hint.* Course-of-values induction on n. Work with $p_0 p_1 \ldots p_n + 1$.)

I.74. Prove, without using the fact that $\lambda n . p_n \in \mathfrak{PR}$, that $\pi \in \mathfrak{PR}$, where the π-function is given by $\pi(n) =$ (the number of primes $\leq n$).

I.75. Prove, using Exercise I.74 above, but without using the fact that $\lambda n . p_n \in \mathfrak{PR}$, that the predicate $\lambda y n . y = p_n$ is in $\mathfrak{PR}_*$. Conclude, using Exercise I.73, that $\lambda n . p_n \in \mathfrak{PR}$.

I.76. The *Ackermann function*[†] is given by double recursion by

$$A(0, x) = x + 2$$
$$A(n + 1, 0) = 2$$
$$A(n + 1, x + 1) = A(n, A(n + 1, x))$$

Prove that $\lambda n x . A(n, x) \in \mathfrak{R}$.
(*Hint.* Define

$$F(z, n, x) = \begin{cases} x + 2 & \text{if } n = 0 \\ 2 & \text{if } n > 0 \wedge x = 0 \\ \{z\}(n \dot- 1, \{z\}(n, x \dot- 1)) & \text{if } n > 0 \wedge x > 0 \end{cases}$$

Now apply the recursion theorem.)

I.77. Prove that there exists a partial recursive h that satisfies

$$h(y, x) = \begin{cases} y & \text{if } x = y + 1 \\ h(y + 1, x) & \text{otherwise} \end{cases}$$

Which function is $\lambda x . h(0, x)$?
(*Hint.* Use the recursion theorem, imitating your solution to Exercise I.76 above.)

I.78. Prove that there exists a partial recursive k that satisfies

$$k(y, x) = \begin{cases} 0 & \text{if } x = y + 1 \\ k(y, x) + 1 & \text{otherwise} \end{cases}$$

Which function is $\lambda x . k(0, x)$?

[†] There are many versions, their origin part of computability folklore. The one here is not the original one.

I.79. Given $\lambda y\vec{x}.f(y,\vec{x}) \in \mathfrak{P}$. Prove that there exists a partial recursive g that satisfies

$$g(y,\vec{x}) = \begin{cases} y & \text{if } f(y,\vec{x}) = 0 \\ g(y+1,x) & \text{otherwise} \end{cases}$$

How can you express $\lambda x.g(0,x)$ in terms of f?

I.80. Prove that $K = \{x : \{x\}(x) \downarrow\}$ is not a *complete index set*, that is, there is no $\mathfrak{D} \subseteq \mathfrak{P}$ such that $K = \{x : \{x\} \in \mathfrak{D}\}$.
(*Hint.* Show that there is an $e \in \mathbb{N}$ such that $\{e\}(e) = 0$, while $\{e\}(x) \uparrow$ if $x \neq e$.)

I.81. Is $\{x : \{x\}(x) = 0\}$ a complete index set? Recursive? Semi-recursive? (Your answer should not leave any dangling "why"s.)

I.82. Let $f \in \mathfrak{R}$. Is $\{x : \{f(x)\}(x) \downarrow\}$ a complete index set? Why?

I.83. Consider a complete index set $A = \{x : \{x\} \in \mathfrak{D}\}$ such that there are two functions $\{a\}$ and $\{b\}$, the first in $\mathfrak{D}$, the second *not* in $\mathfrak{D}$, and $\{a\} \subseteq \{b\}$. Prove that there is a 1-1 recursive function f such that

$$x \in \overline{K} \leftrightarrow f(x) \in A$$

Conclude that A is not semi-recursive.
(*Hint.* To find f use the S-m-n theorem to show that you can have

$$\{f(x)\}(y) = \begin{cases} \{a\}(y) & \text{if computation } \{a\}(y) \text{ takes } \leq \text{ steps than } \{x\}(x) \\ \{b\}(y) & \text{otherwise} \end{cases}$$

The wordy condition above can be made rigorous by taking "steps" to be intuitively measured by the size of z in $T(i,x,z)$. "$\leq$" is understood to be fulfilled if both $\{a\}(y)$ and $\{x\}(x)$ are undefined.)

I.84. An A as the above is productive.

I.85. Prove that there is an $f \in \mathfrak{P}$ such that $W_x \neq \emptyset$ iff $f(x) \downarrow$, and $W_x \neq \emptyset$ iff $f(x) \in W_x$.

I.86. *Selection theorem.* Prove that for every $n > 0$, there is a function $\lambda a\vec{y}_n.$ $Sel^{(n)}(a,\vec{y}_n)$ such that

$$(\exists x)(\{a\}(x,\vec{y}_n) \downarrow) \leftrightarrow Sel^{(n)}(a,\vec{y}_n) \downarrow$$

and

$$(\exists x)(\{a\}(x,\vec{y}_n) \downarrow) \leftrightarrow \{a\}(Sel^{(n)}(a,\vec{y}_n),\vec{y}_n) \downarrow$$

(*Hint.* Expand on the proof idea of Exercise I.85.)

I.87. Prove that $f \in \mathfrak{P}$ iff its graph $\lambda y\vec{x}.y = f(\vec{x})$ is in $\mathfrak{P}_*$.
(*Hint.* For the if part, one, but not the only, way is to apply the selection theorem.)

I.88. *Definition by positive cases.* Let $R_i(\vec{x}_n)$, $i = 1, \ldots, k$, be mutually exclusive relations in $\mathfrak{P}_*$, and $\lambda \vec{x}_n . f_i(\vec{x}_n)$, $i = 1, \ldots, k$, functions in $\mathfrak{P}$. Then f defined below is in $\mathfrak{P}$:

$$f(\vec{x}_n) = \begin{cases} f_1(\vec{x}_n) & \text{if } R_1(\vec{x}_n) \\ \vdots & \vdots \\ f_k(\vec{x}_n) & \text{if } R_k(\vec{x}_n) \\ \uparrow & \text{otherwise} \end{cases}$$

(*Hint.* The if-then-else function will not work here. (Why? I thought $\mathfrak{P}$ was closed under if-then-else.) Either use the selection theorem directly, or use Exercise I.87.)

I.89. Prove that every r.e. relation $R(\vec{x})$ can be enumerated by a partial recursive function.

(*Hint.* Modify the "otherwise" part in the proof of I.8.51.)

I.90. Prove that there is an $h \in \mathfrak{PR}$ such that $W_x = \operatorname{ran}(\{h(x)\})$.

(*Hint.* Use Exercise I.89, but include x in the active arguments. Then use the S-m-n theorem.)

I.91. Sharpen Exercise I.90 above as follows: Ensure that h is such that $W_x \neq \emptyset$ implies $\{h(x)\} \in \mathfrak{R}$.

(*Hint.* Use Exercise I.85 for the "otherwise".)

I.92. Prove that for some $\sigma \in \mathfrak{PR}$, $\operatorname{ran}(\{x\}) = \operatorname{dom}(\{\sigma(x)\})$.

(*Hint.* Show that $y \in \operatorname{ran}(\{x\})$ is semi-recursive, and then use S-m-n.)

I.93. Prove that there is a $\tau \in \mathfrak{PR}$ such that $\operatorname{ran}(\{x\}) = \operatorname{ran}(\{\tau(x)\})$ and, moreover, $\operatorname{ran}(\{x\}) \neq \emptyset$ implies that $\{\tau(x)\} \in \mathfrak{R}$.

I.94. Prove that a set is recursive and nonempty iff it is the range of a nondecreasing recursive function.

(*Hint.* To check for $a \in \operatorname{ran}(f)$, f non-decreasing, find the smallest i such that $a \leq f(i)$, etc.)

I.95. Prove that a set is recursive and infinite iff it is the range of a strictly increasing recursive function.

I.96. Prove that every infinite semi-recursive set has an infinite recursive subset.

(*Hint.* Effectively define a strictly increasing subsequence.)

I.97. Prove that there is an $m \in \mathfrak{PR}$ such that W_x infinite implies that $W_{m(x)} \subseteq W_x$ and $W_{m(x)}$ is an infinite recursive set.

(*Hint.* Use Exercise I.91 and I.96.)

I.98. Prove that there is an h in $\mathfrak{PR}$ such that $W_x \cap W_y = W_{h(x,y)}$ (all x, y).

I.99. Prove that there is an $k \in \mathfrak{PR}$ such that $W_x \cup W_y = W_{k(x,y)}$ (all x, y).

I.100. Prove that there is an $g \in \mathfrak{PR}$ such that $W_x \times W_y = W_{g(x,y)}$ (all x, y).

I.101. Prove that $\mathfrak{P}$ is not closed under min by finding a function $f \in \mathfrak{P}$ such that

$$\lambda x. \begin{cases} \min\{y : f(x, y) = 0\} & \text{if min exists} \\ \uparrow & \text{otherwise} \end{cases}$$

is total and 0-1-valued, but not recursive.
(*Hint.* Try $f(x, y)$ that yields 0 if $(y = 0 \wedge \{x\}(x) \downarrow) \vee y = 1$ and is undefined for all other inputs.)

I.102. Prove that $\lambda x.\{x\}(x)$ has no (total) recursive extension.

I.103. Express the projections K and L of $J(x, y) = (x + y)^2 + x$ in *closed form* – that is, without using $(\mu y)_{<z}$ or bounded quantification.
(*Hint.* Solve for x and y the Diophantine equation $z = (x + y)^2 + x$. The term $\lfloor \sqrt{z} \rfloor$ is involved in the solution.)

I.104. Prove that the pairing function $J(x, y) = (x + y)(x + y + 1)/2 + x$ is onto (of course, the division by 2 is exact), and find its projections K and L in closed form.
(*Hint.* For the onto part you may convince yourself that J enumerates-spairs as follows (starting from the 0th pair, $(0, 0)$: It enumerates by ascending *group number*, where the group number of the pair (a, b) is $a + b$. In each group it enumerates by ascending first component; thus (a, b) is ath in the group of number $a + b$.)

I.105. Find a *polynomial* onto pairing function via the following enumeration: Enumerate by *group number*. Here the group number of (a, b) is $\max(a, b)$, that is, $a \overset{\cdot}{-} b + b$. In group i the enumeration is

$$(0, i), (1, i), (2, i), \ldots, (i - 1, i), (i, i), (i, i - 1), (i, i - 2),$$
$$(i, i - 3), \ldots, (i, 1), (i, 0)$$

Find also the projections K and L in closed form.
By the way, what makes this J "polynomial" is that (like the one in Exercise I.103 above) it does not involve division. It only involves $+, \times, \overset{\cdot}{-}$, and substitutions.

I.106. Prove that $\lambda \vec{x}_n.\langle \vec{x}_n \rangle$, where $\langle \ldots \rangle$ is that defined on p. 165, is in $\mathfrak{PR}$.

I.107. Prove that if t is a closed term of $L_{\mathfrak{N}}$, then

$$\mathbf{ROB} \vdash t = \widetilde{t^{\mathfrak{N}}}$$

(*Hint.* Induction on terms.)

I.108. Prove, by constructing an appropriate model, that $(\exists x)x < x$ is consistent with **ROB**, and therefore $\mathbf{ROB} \nvdash \neg x < x$.
(*Hint.* For example, you can build a model on the set $\mathbb{N} \cup \{\infty\}$, where "$\infty$" is a new symbol (not in $\mathbb{N}$).)

202   *I. Basic Logic*

I.109. Prove, by constructing an appropriate model, that $(\exists x)(\exists y)x + y \neq y + x$ is consistent with **ROB**, and therefore **ROB** $\nvdash x + y = y + x$.

I.110. Prove that $x \times y = z$ is **ROB**-definable.

I.111. Prove I.9.45.

I.112. Complete the proof of I.9.46.

I.113. Prove that if $A \subseteq \mathbb{N}$ is *positively* strongly formally definable in **ROB** – that is, for some $\mathscr{A}$ and all n, $n \in A$ iff **ROB** $\vdash \mathscr{A}(\tilde{n})$ – then it is semi-recursive. What can you say if we drop "positively"? Is the converse true?

I.114. Is either of the two sets in I.9.55 recursive? Why?

I.115. Prove that if Γ is a complete extension of **ROB** and $\{\ulcorner \mathscr{A} \urcorner : \mathscr{A} \in \Gamma\}$ is recursive, then Γ has a decidable decision problem, i.e., $\{\ulcorner \mathscr{A} \urcorner : \Gamma \vdash \mathscr{A}\}$ is recursive.

I.116. *Gödel's first incompleteness theorem – original version.* Prove that if Γ is a ω-consistent extension of **ROB** and $\{\ulcorner \mathscr{A} \urcorner : \mathscr{A} \in \Gamma\}$ is recursive, then Γ is incomplete.

(*Hint.* This is a suggestion for "proof by hindsight", using recursion-theoretic techniques (not Gödel's original proof). So, prove, under the stated assumptions, that every semi-recursive $A \subseteq \mathbb{N}$ is *positively strongly definable* (cf. Exercise I.113) in Γ (ω-consistency helps one direction). Thus, the halting set K is so definable. What does this do to **Thm**$_\Gamma$? Etc.)

We explore here some related formal definability concepts for functions.

We say that a total $\lambda \vec{x}_n . f(\vec{x}_n)$ is *formally functionally definable* (in some extension of **ROB**, Γ) iff for some $\mathscr{F}(y, \vec{x}_n)$ the following holds for all b, $\vec{a}_n$ in $\mathbb{N}$:

$$b = f(\vec{a}_n) \quad \text{implies} \quad \Gamma \vdash \mathscr{F}(y, \tilde{a}_1, \ldots, \tilde{a}_n) \leftrightarrow y = \tilde{b} \qquad (1)$$

I.117. Prove that if Γ is a consistent extension of **ROB** (for example, **ROB** itself or a conservative extension), then, in the definition above, the informal "implies" can be strengthened to "iff".

I.118. Prove that a total $\lambda \vec{x}_n . f(\vec{x}_n)$ is formally functionally definable (in some extension of **ROB**, Γ) iff for some $\mathscr{F}(y, \vec{x}_n)$ the following hold for all b, $\vec{a}_n$ in $\mathbb{N}$:
(i) The *graph* of $f - \lambda y \vec{x}_n . y = f(\vec{x}_n)$ – is formally defined in Γ by $\mathscr{F}$ in the sense of I.9.38, and
(ii) $b = f(\vec{a}_n)$ implies $\Gamma \vdash \mathscr{F}(y, \tilde{a}_1, \ldots, \tilde{a}_n) \to y = \tilde{b}$.

I.119. Actually, the above was just a warm-up and a lemma. Prove that a total f is formally functionally definable in **ROB** (or extension thereof) iff its graph is just definable (I.9.38).

(*Hint.* For the hard direction, let (using a single argument for notational convenience) $\mathscr{F}(x, y)$ define $y = f(x)$ in the sense of I.9.38. Prove that

$\mathscr{G}(x, y)$, a short name for

$$\mathscr{F}(x, y) \wedge (\forall z < y)\neg\mathscr{F}(x, z)$$

also defines $y = f(x)$ and moreover satisfies

$$\text{if} \quad f(a) = b \quad \text{then} \quad \vdash \mathscr{G}(\tilde{a}, y) \rightarrow y = \tilde{b}$$

To this end assume $f(a) = b$ and prove, first, that $\vdash \mathscr{G}(\tilde{a}, y) \rightarrow y < S\tilde{b}$, and second (using I.9.44) that $\vdash y < S\tilde{b} \rightarrow \mathscr{G}(\tilde{a}, y) \rightarrow y = \tilde{b}$.)

I.120. Let the total f of one variable be *weakly definable* in **ROB** (or extension thereof), that is, its graph is formally definable in the sense of I.9.38. Let $\mathscr{A}(x)$ be any formula. Then prove that, for some well-chosen formula $\mathscr{B}(x)$,

$$(\forall a \in \mathbb{N}) \vdash \mathscr{B}(\tilde{a}) \leftrightarrow \mathscr{A}(\widetilde{f(a)})$$

(*Hint.* By Exercise I.119, there is a $\mathscr{F}(y, x)$ that *functionally* defines f ((1) above). Take for $\mathscr{B}$ the obvious: $(\exists y)(\mathscr{F}(y, x) \wedge \mathscr{A}(y))$.)

I.121. Let $A \subseteq \mathbb{N}$ be positively strongly definable in **ROB**, and the total $\lambda x. f(x)$ be functionally definable in **ROB**. Prove that $f^{-1}[A]$, the inverse image of A under f, is also positively strongly definable. Give two proofs: one using the connection between **ROB** and recursion theoretic concepts, the second ignorant of recursion theory.
(*Hint.* Use Exercise I.120.)

We have used the formula $(\exists v_0)(v_0 = \tilde{n} \wedge \mathscr{F}(v_0))$ – which is logically equivalent to $\mathscr{F}(\tilde{n})$ by the one point rule (I.7.2) – on a number of occasions (e.g., p. 175), notably to "easily" obtain a Gödel number of (a formula equivalent to) $\mathscr{F}(\tilde{n})$ from a Gödel number of $\mathscr{F}(v_0)$ and $\tilde{n}$. This Gödel number is (pretending that $\wedge$ is a primitive symbol so that we do not obscure the notation)

$$\langle \ulcorner\exists\urcorner, \ulcorner v_0\urcorner, \langle \ulcorner \wedge \urcorner, \langle \ulcorner = \urcorner, \ulcorner v_0\urcorner, \ulcorner\tilde{n}\urcorner\rangle, \ulcorner\mathscr{F}(v_0)\urcorner\rangle\rangle$$

The above, with n as the only variable, is recursive (indeed, primitive recursive). So trivially, is, the function s of *two* (informal) variables n and x over $\mathbb{N}$:

$$s = \lambda n x\langle \ulcorner\exists\urcorner, \ulcorner v_0\urcorner, \langle \ulcorner \wedge \urcorner, \langle \ulcorner = \urcorner, \ulcorner v_0\urcorner, \ulcorner\tilde{n}\urcorner\rangle, x\rangle\rangle$$

Clearly,

$$s(n, \ulcorner\mathscr{F}(v_0)\urcorner) = \ulcorner(\exists v_0)(v_0 = \tilde{n} \wedge \mathscr{F}(v_0))\urcorner$$

A *fixed point*, or *fixpoint*, of a formula $\mathscr{A}(v_0)$ in an extension of **ROB**, Γ, is a sentence $\mathscr{F}$ such that

$$\Gamma \vdash \mathscr{F} \leftrightarrow \mathscr{A}(\ulcorner\widetilde{\mathscr{F}}\urcorner)$$

I.122. Let **ROB** $\leq \Gamma$ and $\mathscr{A}(v_0)$ be any formula in the language of Γ. Prove that $\mathscr{A}$ has a fixpoint $\mathscr{F}$ in Γ.

(*Hint.* The function s above is recursive; thus so is $D = \lambda x.s(x, x)$. Therefore, D is functionally definable in Γ. Now use Exercise I.120 with $f = D$. See if you can use a "good" $a \in \mathbb{N}$.)

I.123. *Tarski's "undefinability of truth" again* (cf. I.9.31). Prove that $\mathbf{T} = \{\ulcorner \mathscr{A} \urcorner : \mathscr{A} \in \mathscr{T}(\mathfrak{N})\}$ is *not* (semantically) definable in $\mathfrak{N}$, basing the argument on the existence of fixpoints in **ROB** and on the latter's correctness.

(*Hint.* Suppose that some $L_{\mathfrak{N}}$ formula, say $\mathscr{A}(v_0)$, defines **T**. Use the previous exercise to find a sentence that says "I am false".)

I.124. Let us use the term *strongly μ-recursive* functions for the smallest set of functions, $\mathfrak{R}'$, that includes the initial functions of $\mathfrak{P}\mathfrak{R}$ *and* the functions $\lambda xy.x + y$, $\lambda xy.xy$, and $\lambda xy.x \overset{.}{-} y$, and is closed under composition and (μy) applied on *total, regular functions* $\lambda \vec{x} y.g(\vec{x}, y)$, that is, total functions satisfying

$$(\forall \vec{x})(\exists y)g(\vec{x}, y) = 0$$

Prove that $\mathfrak{R}' = \mathfrak{R}$.

(*Hint.* Use coding and decoding, e.g., via the β-function, to implement primitive recursion.)

I.125. Prove (again) that all recursive functions are functionally definable in **ROB**. Do so via Exercise I.124, by induction on $\mathfrak{R}'$, without using the separation lemma (I.9.49).

I.126. (Craig.) Prove that a recursively enumerable set of sentences $\mathscr{T}$ over a finitely generated language (e.g., like that of arithmetic) admits a recursive set of axioms, i.e., for some recursive Γ, $\mathscr{T} = \mathbf{Thm}_\Gamma$.

(*Hint.* Note that for any $\mathscr{A} \in \mathscr{T}$, any two sentences in the sequence

$$\mathscr{A}, \mathscr{A} \wedge \mathscr{A}, \mathscr{A} \wedge \mathscr{A} \wedge \mathscr{A}, \dots$$

are logically equivalent. Now see if Exercises I.94 and I.95 can be of any help.)

I.127. Let $\mathfrak{T}$ be the *pure* theory over the language that contains precisely the following nonlogical symbols: One constant, one unary function, two binary functions, and one binary predicate. Prove that $\mathfrak{T}$ has an undecidable decision problem, that is, $\{\ulcorner \mathscr{A} \urcorner : \mathscr{A} \in \mathfrak{T}\}$ is not recursive.

II

The Second Incompleteness Theorem

Our aim in the previous section was to present Gödel's first incompleteness theorem in the context of recursion theory. Much as this "modern" approach is valuable for showing the links between unprovability and uncomputability, it has obscured the simplicity of Gödel's ingenious idea (as it was carried out in his original paper (1931)).

What he had accomplished in that paper, through arithmetization of formulas and proofs, was to build a *sentence* of arithmetic, $\mathscr{F}$, that said "I am not a theorem". One can easily prove, metamathematically, that such an $\mathscr{F}$ is undecidable, *if arithmetic is ω-consistent.*

To see this at the intuitive level, let us replace ω-consistency by correctness. Then surely $\mathscr{F}$ is not provable, for if it is, then it *is* a theorem, and hence false (contradicting correctness).

On the other hand, we have just concluded that $\mathscr{F}$ is true! Hence, $\neg\mathscr{F}$ is false, and therefore not provable either (by correctness).

This simple application of the "liar's paradox"[†] is at the heart of the first incompleteness theorem.

Imagine now that the arithmetization is actually carried out *within* (some) formal arithmetic, and that with some effort we have managed to embed *into formal arithmetic* the metamathematical argument that leads to the assertion "if arithmetic is consistent, then $\nvdash \mathscr{F}$"[‡]. The quoted statement is formalized by "Con $\to \mathscr{F}$", where "Con" is some (formal) sentence that says that arithmetic is consistent. This is so, intuitively, since $\mathscr{F}$ "says": "$\mathscr{F}$ is not a theorem".

[†] Attributed in its original form to Epimenides of Crete, who proclaimed: "All Cretans are liars". Is this true? Gödel's version is based on the variation: "I am lying". Where does such a proclamation lead?

[‡] Where, of course, "$\vdash$" is using the nonlogical axioms of our formal arithmetic.

It follows that $\nvdash$ Con (why?).

This is Gödel's second incompleteness theorem, that if a recursively axiomatized theory is a consistent extension of (Peano) arithmetic, then it cannot prove the sentence that asserts its own consistency.

In order to prove this theorem we will need to develop enough formal arithmetic to be able to carry out elementary arguments in it. We will also need to complete the details of our earlier arithmetization, *within formal arithmetic* this time.

II.1. Peano Arithmetic

We start by extending **ROB** with the addition of the *induction axiom schema*, to obtain *Peano arithmetic*. Within this arithmetic we will perform all our formal reasoning and constructions (arithmetization).

As a by-product of the required extra care that we will exercise in this section, regarding arithmetization details, we will be able to see through some technicalities that we have suppressed in I.9.33–I.9.34 and I.9.37. For example, we had accepted there, without explicitly stating so, that our standard rules of inference, modus ponens and ($\exists$)-introduction, are recursive, and therefore fit the general description of the unspecified rules $\mathbf{I}_1$ and $\mathbf{I}_2$ on p. 176.

We will soon see below why this is so. Similarly, we have said that Λ is recursive. *At the intuitive level* this is trivial, since some logical axioms are recognized by their form (e.g., the axiom $x = x$), while the tautologies can be recognized by constructing a truth table. While this ability to recognize tautologies can be demonstrated rigorously, that is far too tedious an undertaking. Instead, we opt for a more direct avenue. In the exercises (Exercises I.26–I.41) we led the reader to adopt a finite set of axiom schemata in lieu of the infinitely many schemata whose instances are tautologies. This makes it much easier to see that this amended Λ is recursive (even primitive recursive).

II.1.1 Definition (Peano Arithmetic). We extend **ROB** over the same language, $L_{\mathfrak{N}}$, by adding the *induction axiom* (in reality, an *induction schema*), **Ind**, below:

$$\mathscr{A}[x \leftarrow 0] \wedge (\forall x)(\mathscr{A} \rightarrow \mathscr{A}[x \leftarrow Sx]) \rightarrow \mathscr{A} \qquad \textbf{(Ind)}$$

We often write more simply[†]

$$\mathscr{A}[0] \wedge (\forall x)(\mathscr{A}[x] \rightarrow \mathscr{A}[Sx]) \rightarrow \mathscr{A}[x]$$

[†] Cf. p. 33.

or just

$$\mathscr{A}[0] \wedge (\forall x)(\mathscr{A} \rightarrow \mathscr{A}[Sx]) \rightarrow \mathscr{A}$$

The theory **ROB** + **Ind** is called *Peano arithmetic*, for short **PA**. □

II.1.2 Remark (A Note on Nonlogical Schemata and Defined Symbols).
Metatheorems I.7.1 and I.7.3 ensure that the addition of defined predicates, functions, and constants to any language and theory results in a conservative extension of the theory, that is, any theorem of the new theory over the old (original) language is also provable in the old theory. Moreover, we saw how any formula $\mathscr{A}$ of the new language can be naturally transformed to a formula $\mathscr{A}^*$ of the old language (eliminating defined symbols), so that

$$\mathscr{A} \leftrightarrow \mathscr{A}^* \tag{1}$$

is provable in the extended theory.

There is one potential worry about the presence of nonlogical schemata – such as the induction axiom schema of **PA** – that we want to address:

First off, while logical axioms are "good" over any first order language – they are "universal" – nonlogical axioms and schemata on the other hand are specific to a theory and its *basic language*, i.e., the language that existed prior to any extensions by definitions. Thus, the induction schema is an "agent" that yields a specific nonlogical axiom (an *instance* of the schema) for each specific formula, *over the basic language $L_{\mathfrak{N}}$*, that we care to substitute into the metavariable $\mathscr{A}$.

There is no a priori promise that the schema "works" whenever we replace the syntactic variable $\mathscr{A}$ by a formula, say "$\mathscr{B}$", over a language extension *obtained by definitions*. By "works", of course, we mean that the produced schema instance is a theorem in the extended theory.

Well, does it "work"? Indeed it does; for let us look at the formula

$$\mathscr{B}[x \leftarrow 0] \wedge (\forall x)(\mathscr{B} \rightarrow \mathscr{B}[x \leftarrow Sx]) \rightarrow \mathscr{B} \tag{2}$$

where the particular formula $\mathscr{B}$ *may* contain defined symbols. Following the technique of symbol elimination (cf. Remark I.7.4(a), p. 120) – eliminating at the atomic formula level – we obtain the following version of (2), *in the basic language* of **PA**. This translated version has exactly the same form as (2) (i.e., **Ind**), namely

$$\mathscr{B}^*[x \leftarrow 0] \wedge (\forall x)(\mathscr{B}^* \rightarrow \mathscr{B}^*[x \leftarrow Sx]) \rightarrow \mathscr{B}^*$$

Thus – being a schema instance over the basic language – it *is* an axiom of **PA**, and hence also of its extension (by definitions). Now, by (1), the equivalence theorem (Leibniz rule I.4.25) yields the following theorem of the extended theory:

$$(\mathscr{B}[x \leftarrow 0] \wedge (\forall x)(\mathscr{B} \rightarrow \mathscr{B}[x \leftarrow Sx]) \rightarrow \mathscr{B})$$

$$\leftrightarrow$$

$$(\mathscr{B}^*[x \leftarrow 0] \wedge (\forall x)(\mathscr{B}^* \rightarrow \mathscr{B}^*[x \leftarrow Sx]) \rightarrow \mathscr{B}^*)$$

Hence (2) is a theorem of the extended theory.   □

II.1.3 Remark (The Induction "Rule"). In practice, instead of **Ind**, we usually employ the (derived) *rule of inference* **Ind'** that we obtain from **Ind** by invoking modus ponens and the duo I.4.7–I.4.8:[†]

$$\mathscr{A}[x \leftarrow 0], \mathscr{A} \rightarrow \mathscr{A}[x \leftarrow Sx] \vdash \mathscr{A} \qquad (\mathbf{Ind'})$$

The rule is normally applied as follows: We ascertain that the premises apply by

(1) proving $\mathscr{A}[0]$ (this part is called the *basis* of the induction, just as in the informal case over $\mathbb{N}$), and

(2) adding the induction hypothesis (I.H.) $\mathscr{A}$ to the axioms, treating the free variables of $\mathscr{A}$ as new constants[‡] until we can prove $\mathscr{A}[Sx]$.

We then have a proof of $\mathscr{A} \rightarrow \mathscr{A}[Sx]$ by the deduction theorem.

Ind' now allows us to conclude that $\mathscr{A}$ has been proved by induction on x.

What is interesting is that **Ind'** implies **Ind**; thus the two are equivalent. What *makes* this interesting is that while the deduction theorem readily yields **Ind** from **Ind'**, it does so under the restriction that the free variables in $\mathscr{A}[0]$ and $(\forall x)(\mathscr{A} \rightarrow \mathscr{A}[Sx])$ must be treated as new constants (be "frozen"). We can do without the deduction theorem and without the restriction.

Let us then fix a formula $\mathscr{A}$ and prove **Ind** assuming **Ind'** (see, e.g., Shoenfield (1967)).

We let[§]

$$\mathscr{B} \equiv \mathscr{A}[0] \wedge (\forall x)(\mathscr{A} \rightarrow \mathscr{A}[Sx]) \rightarrow \mathscr{A}$$

To prove $\mathscr{B}$, using **Ind'**, we need to prove

$$\mathscr{B}[0] \qquad (1)$$

[†] Throughout this chapter, the symbol $\vdash$ implies a subscript, **PA**, or some extension thereof, unless something else is clear from the context.

[‡] That is, disallowing universal quantification over, or substitution in the variables. Of course, existential quantification is always possible by **Ax2**.

[§] Here "$\equiv$" means equality of strings; cf. I.1.4.

and

$$\mathscr{B} \to \mathscr{B}[Sx] \tag{2}$$

Now, (1) is a tautology, while (2) is tautologically equivalent to

$$(\mathscr{A}[0] \land (\forall x)(\mathscr{A} \to \mathscr{A}[Sx]) \land \neg \mathscr{A}) \lor$$
$$\neg \mathscr{A}[0] \lor \neg(\forall x)(\mathscr{A} \to \mathscr{A}[Sx]) \lor \mathscr{A}[Sx] \tag{3}$$

In turn, (3) – after distributing $\lor$ over $\land$ – is seen to be tautologically equivalent to

$$\mathscr{A}[0] \to (\forall x)(\mathscr{A} \to \mathscr{A}[Sx]) \to \mathscr{A} \to \mathscr{A}[Sx]$$

which is provable by tautological implication and specialization (I.4.7). □

For the next little while we will be rediscovering arithmetic in the formal setting of **PA**. In the process, new predicate and function symbols will be introduced to the language (with their attendant axioms – as in Section I.7).

II.1.4 Definition. We introduce the predicate $\leq$ by $x \leq y \leftrightarrow x < y \lor x = y$.
□

 For the balance of the section $\vdash$ means $\vdash_{\mathbf{PA}}$ unless noted otherwise.

Of course, in the those instances where we add axioms (in order to argue by the deduction theorem, or by auxiliary constant, or by cases, etc.) by a sentence such as "Let $\mathscr{A}$ ..." or "Add $\mathscr{A}$ ...", "$\vdash$" will mean provability in the augmented theory **PA** $+ \mathscr{A}$.

II.1.5 Lemma. $\vdash 0 \leq x$.

Proof. We use induction on x.[†] For convenience, we let $\mathscr{A} \equiv 0 \leq x$.

Basis. $\vdash \mathscr{A}[0]$, since $0 = 0 \models_{\text{Taut}} \mathscr{A}[0]$ by II.1.4.

I.H. Add $\mathscr{A}$.[‡]

Now, $\mathscr{A}[Sx] \equiv 0 < Sx \lor 0 = Sx$; thus, by $<$ 2 (p. 175) and the Leibniz rule (I.4.25, p. 51), $\vdash \mathscr{A}[Sx] \leftrightarrow \mathscr{A} \lor 0 = Sx$. We are done by $\models_{\text{Taut}}$ and I.H. □

II.1.6 Lemma (Transitivity of $<$). $\vdash x < y \land y < z \to x < z$.

[†] This, usually, means that we are using the induction rule, **Ind**$'$, rather than the induction schema itself.

[‡] Instead of "add", we often say "let". Either means that we are about to invoke the deduction theorem, and we are adding a new nonlogical axiom, with all its free variables "frozen".

Proof. Induction on z. Let[†] $\mathscr{A}[z] \equiv x < y \wedge y < z \to x < z$.

  Basis. $\vdash \mathscr{A}[0]$ from < 1 and $\models_{\text{Taut}}$.

  I.H. Add $\mathscr{A}$.

  Add $x < y \wedge y < Sz$ (to prove $x < Sz$). This yields (by < 2, the Leibniz rule, and $\models_{\text{Taut}}$)

$$x < y \wedge y < z \vee x < y \wedge y = z \tag{1}$$

We also have (the I.H.) $\mathscr{A}$:

$$x < y \wedge y < z \to x < z$$

and (Leibniz equality axiom)

$$x < y \wedge y = z \to x < z$$

Thus, using $\models_{\text{Taut}}$, $x < z$ follows from (1). This last conclusion, by $\models_{\text{Taut}}$ and < 2, yields $x < Sz$. $\square$

II.1.7 Corollary (Irreflexivity of <). $\vdash \neg x < x$.

Proof. Induction on x.

  Basis. By < 1.

  I.H. Add $\neg x < x$.

  We want to deduce $\neg Sx < Sx$. Arguing by contradiction, we add $Sx < Sx$, that is, via < 2,

$$\vdash Sx < x \vee Sx = x \tag{1}$$

Now, $\vdash x < Sx$ by < 2, the axiom $\vdash x = x$, and $\models_{\text{Taut}}$. Thus, using (1),

$$\vdash x < Sx \wedge (Sx < x \vee Sx = x)$$

which yields $x < x$ by II.1.6 and the Leibniz equality axiom (**Ax4** of $\wedge$), contradicting the I.H. $\square$

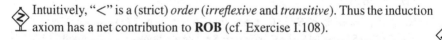 Intuitively, "$<$" is a (strict) *order* (*irreflexive* and *transitive*). Thus the induction axiom has a net contribution to **ROB** (cf. Exercise I.108).

II.1.8 Lemma (Antisymmetry of ≤). $\vdash x \leq y \wedge y \leq x \to x = y$.

Proof. Assume the hypothesis

$$x \leq y \wedge y \leq x$$

[†] That is, "z" is our favourite free variable in the formula – see I.1.11, p. 18.

By II.1.4 this is tautologically equivalent to

$$\neg(x < y \wedge y < x) \to \neg(x < y \wedge x = y)$$
$$\to \neg(y < x \wedge x = y) \to x = y$$

Since $\vdash \neg(x < y \wedge y < x)$ by transitivity and irreflexivity (e.g., use proof by contradiction for this sub-claim) while each of

$$\neg(x < y \wedge x = y)$$

and

$$\neg(y < x \wedge x = y)$$

are theorems by the Leibniz axiom and irreflexivity, $x = y$ follows by modus ponens. $\square$

II.1.9 Lemma. $\vdash x < y \to Sx < Sy$.

Proof. Induction on y.

   Basis. $\vdash x < 0 \to Sx < S0$, by < 1 and $\models_{\text{Taut}}$.

   I.H. Add $x < y \to Sx < Sy$.

   Add $x < Sy$ towards proving $Sx < SSy$.

   Hence $\vdash x < y \vee x = y$, by < 2. By I.H. and the Leibniz equality axiom,

$$\vdash Sx < Sy \vee Sx = Sy$$

Therefore $\vdash Sx < SSy$ by < 2. $\square$

II.1.10 Corollary. $\vdash x < y \leftrightarrow Sx \leq y$.

Proof. $\to$: $\vdash Sx \leq y \leftrightarrow Sx < Sy$, by < 2.

   $\leftarrow$: By $\vdash x < Sx$ and transitivity. $\square$

II.1.11 Proposition. *Axiom < 3 is redundant in the presence of the induction axiom.*

Proof. Exercise II.1. $\square$

 We give notice that in what follows we will often use $x > y$ to mean $y < x$.

We now state without proof some "standard" properties of $+$ and $\times$. The usual proof tool here will be induction. The lemma below will need a double induction, that is, on both x and y.[†]

[†] Actually, if you prove associativity first and also the theorem $S0 + x = Sx$, then you can prove commutativity with a single induction, after you have proved $0 + x = x$.

II.1.12 Lemma. $\vdash x + y = y + x$.

Proof. Exercise II.3.   □

II.1.13 Lemma. $\vdash x + (y + z) = (x + y) + z$.

Proof. Exercise II.4.   □

II.1.14 Lemma. $\vdash x \times y = y \times x$.

Proof. Exercise II.5.   □

II.1.15 Lemma. $\vdash x \times (y \times z) = (x \times y) \times z$.

Proof. Exercise II.6.   □

II.1.16 Lemma. $\vdash x \times (y + z) = (x \times y) + (x \times z)$.

Proof. Exercise II.7.   □

We adopt the usual priorities of arithmetic operations; thus, instead of

$$x \times (y + z) = (x \times y) + (x \times z)$$

we will normally use the *argot*

$$x \times (y + z) = x \times y + x \times z$$

We will adopt one more useful abbreviation (*argot*): From now on we will write xy for $x \times y$ (implied multiplication notation). Moreover we will often take advantage of properties such as commutativity without notice, for example, writing $x + y$ for $y + x$.

II.1.17 Lemma. $\vdash x < y \rightarrow x + z < y + z$.

Proof. Induction on z.

   Basis. From $+1$.

   I.H. Add $x < y \rightarrow x + z < y + z$.

   Let $x < y$. By I.H., $\vdash x + z < y + z$. By II.1.9, $\vdash S(x + z) < S(y + z)$. By $+2$, $\vdash x + Sz < y + Sz$.   □

II.1.18 Corollary. $\vdash x + z < y + z \rightarrow x < y$.

Proof. Let $x + z < y + z$. Add $\neg x < y$, which, by < 3, tautologically implies

$$y < x \lor y = x \tag{1}$$

Since $\vdash y < x \rightarrow y + z < x + z$ (II.1.17) and $\vdash y = x \rightarrow y + z = x + z$ (Leibniz axiom), $\models_{\text{Taut}}$ and (1) yield

$$\vdash y + z \leq x + z$$

Along with the original assumption and transitivity, we have just contradicted irreflexivity. $\qquad\square$

II.1.19 Corollary. $\vdash z > 0 \rightarrow Sx \leq x + z$.

Proof. We have $\vdash 0 < z \rightarrow 0 + x < z + x$. Commutativity of $+$, and $+1$, lead to the claim. $\qquad\square$

II.1.20 Lemma. $\vdash z > 0 \land x < y \rightarrow xz < yz$.

Proof. Induction on z.

Basis. By < 1.

I.H. Add $z > 0 \land x < y \rightarrow xz < yz$.

Let now

$$x < y \tag{1}$$

As $\vdash 0 < Sz$ anyhow (< 2 and II.1.5), we embark on proving

$$xSz < ySz$$

By $\times 2$ (and the Leibniz axiom, twice) the above is provably equivalent to

$$x + xz < y + yz \tag{2}$$

so we will just prove (2).

By II.1.5 there are two cases for z.[†]

Case $z = 0$. Thus, by the Leibniz axiom and $\times 1$,

$$\vdash xz = 0$$

and

$$\vdash yz = 0$$

[†] Cf. p. 51.

By (1), II.1.17, and the Leibniz axiom,[†]

$$\vdash x + xz < y + yz$$

Case $z > 0$. First off, by I.H. and II.1.17,

$$\vdash y + xz < y + yz$$

Also, by II.1.17,

$$\vdash x + xz < y + xz$$

These last two and transitivity (II.1.6) yield

$$\vdash x + xz < y + yz \qquad\qquad \square$$

II.1.21 Corollary (Cancellation Laws).

$$\vdash x + z = y + z \rightarrow x = y$$

$$\vdash z > 0 \wedge xz = yz \rightarrow x = y$$

Proof. By < 3, II.1.7, II.1.17, and II.1.20. $\qquad\qquad \square$

II.1.22 Corollary.

$$\vdash x < y \wedge z < w \rightarrow x + z < y + w$$

$$\vdash x < y \wedge z < w \rightarrow xz < yw$$

Proof. By II.1.6, II.1.17, and II.1.20. $\qquad\qquad \square$

II.1.23 Theorem (Course-of-Values Induction). *For any formula $\mathscr{A}$, the following is provable in* **PA**:

$$(\forall x)((\forall z < x)\mathscr{A}[z] \rightarrow \mathscr{A}) \rightarrow (\forall x)\mathscr{A} \qquad (1)$$

It goes without saying that z is a *new* variable. Such annoying (for being obvious) qualifications we omit as a matter of policy.

In practice, the schema (1) is employed in conjunction with the deduction theorem, as follows: One proves $\mathscr{A}$ with frozen free variables, on the I.H. that

---

[†] This "$\vdash$" is effected in the extension of **PA** that contains the I.H. and also $x < y$ and $z = 0$.

"$(\forall z < x)\mathscr{A}[z]$ is true" (i.e., with the help of the *new axiom* $(\forall z < x)\mathscr{A}[z]$).
This is all one has to do, since then $(\forall z < x)\mathscr{A}[z] \to \mathscr{A}$ (deduction theorem)
and hence (generalization) $(\forall x)((\forall z < x)\mathscr{A}[z] \to \mathscr{A})$.
    By (1), $(\forall x)\mathscr{A}$ now follows.

Proof. To prove (1), we let the name $\mathscr{B}$ (or $\mathscr{B}[x]$, since we are interested in
x) stand for $(\forall z < x)\mathscr{A}[z]$, that is,

$$\mathscr{B} \equiv (\forall z)(z < x \to \mathscr{A}[z])$$

Proving (1) via the deduction theorem dictates that we next assume (1)'s
hypothesis, that is, we add the axiom

$$(\forall x)(\mathscr{B} \to \mathscr{A})[\vec{c}\,]$$

where $\vec{c}$ are *new* distinct constants, substituted into *all* the free variables of
$(\forall x)(\mathscr{B} \to \mathscr{A})$. The above yields

$$\vdash \mathscr{B}[x,\vec{c}\,] \to \mathscr{A}[x,\vec{c}\,] \tag{2}$$

Our next (subsidiary) task is to establish

$$\vdash (\forall x)\mathscr{B}[x,\vec{c}\,] \tag{3}$$

by induction on x.
    Basis. $\vdash \mathscr{B}[0,\vec{c}\,]$ by $\models_{\text{Taut}}$ and < 1.
    I.H. Add $\mathscr{B}[x,\vec{c}\,]$, with frozen x, of course.[†] By (2),[‡]

$$\vdash \mathscr{A}[x,\vec{c}\,] \tag{4}$$

Now, $\mathscr{B}[Sx,\vec{c}\,] \equiv (\forall z < Sx)\mathscr{A}[z,\vec{c}\,]$; thus

$$\mathscr{B}[Sx,\vec{c}\,]$$

$\leftrightarrow \Big($by the Leibniz rule, $\models_{\text{Taut}}$ and $< 2 - \forall$ distributes over $\wedge$ (Exercise I.23)$\Big)$

$$(\forall z)(z < x \to \mathscr{A}[z,\vec{c}\,]) \wedge (\forall z)(z = x \to \mathscr{A}[z,\vec{c}\,])$$

$\leftrightarrow \Big(\text{I.7.2,}\Big)$

$$\mathscr{B}[x,\vec{c}\,] \wedge \mathscr{A}[x,\vec{c}\,]$$

[†] A constant. But we will not bother to name it anything like "c'''".
[‡] x is still frozen.

Pause. We applied the style of "equational proofs" or "calculational proofs" [†] immediately above (chain of equivalences). The "$\leftrightarrow$" is used *conjunctionally*, that is, "$\vdash \mathscr{D}_1 \leftrightarrow D_2 \leftrightarrow \ldots \leftrightarrow D_n$" means "$\vdash (D_1 \leftrightarrow D_2) \wedge \cdots \wedge (D_{n-1} \leftrightarrow D_n)$"; hence, by tautological implication, $\vdash D_1 \leftrightarrow D_n$.

By (4) and the I.H. we have proved $\mathscr{B}[Sx, \vec{c}]$. Hence our induction has concluded: We now have (3).

By (2), (3), $\forall$-monotonicity (I.4.24), and modus ponens we infer $(\forall x)\mathscr{A}[x, \vec{c}]$; hence, by the deduction theorem,

$$(\forall x)(\mathscr{B} \rightarrow \mathscr{A})[\vec{c}] \rightarrow (\forall x)\mathscr{A}[x, \vec{c}]$$

Applying the theorem on constants, we get (1). □

We have applied quite a bit of pedantry above during the application of the deduction theorem, invoking the theorem on constants explicitly. This made it easier to keep track of which variables were frozen, and when.

II.1.24 Corollary (The Least Principle). *The following is provable in* **PA** *for any* $\mathscr{A}$:

$$(\exists x)\mathscr{A} \rightarrow (\exists x)(\mathscr{A} \wedge (\forall z < x) \neg \mathscr{A}[z]) \qquad (LP)$$

Proof. The course-of-values induction schema applied to $\neg\mathscr{A}$ is provably equivalent to (LP) above. □

We now have enough tools to formalize *unbounded search*, (μy), in **PA** (cf. I.8.8, p. 127).

Suppose that we have[‡]

$$\vdash (\exists x)\mathscr{A} \qquad (E)$$

Informally this says that for all values of the free variables there is a (corresponding value) x that makes $\mathscr{A}$ true. In view of the least principle, we must then be able to define a "total function on the natural numbers" which, for each input, returns the *smallest* x that "works". *Formally* this "function" will be a *function letter* – introduced into the theory (**PA**) by an appropriate definition

[†] Cf. Dijkstra and Scholten (1990), Gries and Scheider (1994), Tourlakis (2000a, 2000b. 2001b).
[‡] Reminder: We continue using "$\vdash$" for "$\vdash_{\mathscr{T}}$", where $\mathscr{T}$ is **PA**, possibly extended by definitions.

(Section I.7) – whose natural interpretation in $\mathfrak{N}$ will be the total function we have just described. The formal details are as follows:

Let $\mathscr{B}$ stand for

$$\mathscr{A} \wedge (\forall z < x) \neg \mathscr{A}[z] \qquad (B)$$

By the least principle and (E) (existence condition),

$$\vdash (\exists x) \mathscr{B} \qquad (1)$$

Pause. By $\exists$-*monotonicity* (I.4.23), (1) implies (E), since $\vdash \mathscr{B} \to \mathscr{A}$. Thus (1) and (E) are provably equivalent in **PA**, by II.1.24.

We next show that the "$\exists$" in (1) is really "$\exists!$". To this end, we prove

$$\mathscr{B}[x] \wedge \mathscr{B}[y] \to x = y \qquad (2)$$

Add $\mathscr{B}[x]$ and $\mathscr{B}[y]$ (with frozen free variables), that is, add

$$\mathscr{A}[x] \wedge (\forall z < x) \neg \mathscr{A}[z]$$

and

$$\mathscr{A}[y] \wedge (\forall z < y) \neg \mathscr{A}[z]$$

These entail

$$\mathscr{A}[x] \qquad (3)$$
$$z < x \to \neg \mathscr{A}[z] \qquad (4)$$
$$\mathscr{A}[y] \qquad (5)$$
$$z < y \to \neg \mathscr{A}[z] \qquad (6)$$

We will now show that adding

$$x < y \vee y < x \qquad (7)$$

will lead to a contradiction, therefore establishing (by < 3)

$$x = y$$

Having added (7) as an axiom, we now have two cases to consider:

Case $x < y$. Then (6) yields $\neg \mathscr{A}[x]$ contradicting (3).
Case $y < x$. Then (4) yields $\neg \mathscr{A}[y]$, contradicting (5).
We have established (2).
We can now introduce a *new* function symbol, f, in $L_{\mathfrak{A}}$ by the axiom

$$f\vec{y} = x \leftrightarrow \mathscr{B} \qquad (f)$$

where the list $\vec{y}, x$ contains all the free variables of $\mathscr{B}$ – and perhaps others.

Pause. Is "and perhaps others" right? Should it not be "exactly the free variables of $\mathscr{B}$"?

The "enabling necessary condition" for (f) is, of course, (1) – or, equivalently, (E) – above. We will always speak of (E) as the *enabling existence condition*.

The following alternative notation to (f) above is due to Hilbert and Bernays (1968) and better captures the intuitive meaning of the whole process we have just described:

$$f\vec{y} = (\mu x)\mathscr{A} \qquad\qquad (f')$$

or, using the Whitehead-Russell "ι",

$$(\mu x)\mathscr{A} \overset{\text{def}}{=} (\iota x)\mathscr{B} \qquad\qquad (f'')$$

Thus, we can always introduce a new function symbol f in **PA** by the explicit definition (f') as long as we can prove (E) above.

Note that, by I.7.1 and I.7.3, such extensions of **PA** are conservative.

We can also say that we have introduced a μ-*term*, $(\mu x)\mathscr{A}$, if we want to suppress the details of introducing a new function letter, etc.

Axiom (f) yields at once[†]

$$\vdash \mathscr{B}[f\vec{y}] \qquad\qquad (f^{(3)})$$

therefore, by (B), < 3, and $\models_{\text{Taut}}$,

$$\vdash \mathscr{A}[f\vec{y}] \qquad\qquad (f^{(4)})$$

and

$$\vdash \mathscr{A}[z] \rightarrow f\vec{y} \leq z \qquad\qquad (f^{(5)})$$

Here "μ" is the formal counterpart of the "μ" (unbounded search) of nonformalized recursion theory (I.8.8). We have restricted its application in the formal theory so that functions defined as in (f') are total (due to (E)).

We will also need to formalize primitive recursion, that is, to show that given any function symbols g and h of $L_{\mathfrak{A}}$, of arities $n + 2$ and n respectively, we can introduce a new function symbol f of arity $n + 1$ satisfying the (defining) axioms

$$f0\vec{y}_n = h\vec{y}_n$$
$$fSx\vec{y}_n = gx\vec{y}_n fx\vec{y}_n$$

[†] Where the machinery for "$\vdash$" includes the defining axiom (f).

Needless to stress that $x, \vec{y}_n$ are free variables. Note that the pair of equations above generalizes the manner in which $+$ and $\times$ were introduced as primeval symbols in **ROB** and **PA**.

More "user-friendly" (*argot*) notation for the above recurrence equations is

$$f(0, \vec{y}_n) = h(\vec{y}_n)$$
$$f(Sx, \vec{y}_n) = g(x, \vec{y}_n, f(x, \vec{y}_n))$$

To be able to handle primitive recursion we need to strengthen our grasp of "arithmetic" in **PA**, by developing a few more tools.

First off, for any term t we may introduce a new function symbol f by a "definition" (axiom)

$$f\vec{y} = t \tag{8}$$

where $\vec{y}$ contains all the free variables of t (but may contain additional free variables).

(8) is our *preferred* short notation for *introducing* f. The long notation is by quoting

$$\vdash (\exists x) x = t \tag{8'}$$

and[†]

$$f\vec{y} = (\mu x) x = t \tag{8''}$$

Of course, the enabling condition (8') is satisfied, by logical axiom **Ax2**.

We next introduce the (formal) *characteristic function* of a formula, a formal counterpart of the characteristic function of a relation (I.8.16).

Let $\mathscr{A}$ be any formula, and $\vec{y}_n$ the list of its free variables. We introduce a new n-ary function symbol $\chi_{\mathscr{A}}$ by the explicit definition

$$\chi_{\mathscr{A}} \vec{y}_n = (\mu x)(\mathscr{A} \wedge x = 0 \vee \neg \mathscr{A} \wedge x = \tilde{1}) \tag{C}$$

As always, we must be satisfied that the enabling condition

$$\vdash (\exists x)(\mathscr{A} \wedge x = 0 \vee \neg \mathscr{A} \wedge x = \tilde{1}) \tag{C'}$$

holds. Well, since $\vdash \mathscr{A} \vee \neg \mathscr{A}$, we may use proof by cases.

Case $\mathscr{A}$.[‡] Now, $\vdash \mathscr{A} \wedge 0 = 0 \vee \neg \mathscr{A} \wedge 0 = \tilde{1}$, thus (C') follows.

Case $\neg \mathscr{A}$. This time $\vdash \mathscr{A} \wedge \tilde{1} = 0 \vee \neg \mathscr{A} \wedge \tilde{1} = \tilde{1}$, thus (C') follows once more.

[†] I shall eventually stop issuing annoyingly obvious reminders such as: "x is *not*, of course, chosen among the variables of t".

[‡] Cf. p. 51.

II.1.25 Remark (Disclaimer). *"[T]he* (formal) characteristic function of a formula" (emphasis added) above are strong words indeed. Once a $\chi_{\mathscr{A}}$ has been introduced as above, one may subsequently introduce a *new* function symbol f by the explicit definition $f\vec{y}_n = \chi_{\mathscr{A}}\,\vec{y}_n$, and this too satisfies (C) and its corollaries (9) and (10) below. For example, $\vdash \mathscr{A} \leftrightarrow f\vec{y}_n = 0$.

Thus, "the" characteristic function symbol is in fact not unique.[†] Nevertheless, we may occasionally allow the tongue to slip. The reader is hereby forewarned.   □

II.1.26 Remark (About "$\vdash$", Again: Recursive Extensions of PA). From now on, proofs take place in an (unspecified) extension of **PA** effected by a *finite sequence* of μ-definitions or[‡] definitions of new predicate symbols. To be exact, we work in a theory **PA′** defined as follows: There is a sequence of theories $\mathfrak{T}_i$, for $i = 0, \ldots, n$ each over a language $L_{\mathfrak{A}_i}$, such that

(i) $L_{\mathfrak{A}_0} = L_{\mathfrak{N}}$, $\mathfrak{T}_0 = $ **PA**, $\mathfrak{T}_n = $ **PA′**, and
(ii) for each $i = 0, \ldots, n - 1$, $\mathfrak{T}_{i+1}$ is obtained from $\mathfrak{T}_i$ by
    (a) adding a single new function symbol f to $L_{\mathfrak{A}_i}$, to obtain $L_{\mathfrak{A}_{i+1}}$, and adding the axiom

$$f(\vec{y}) = (\mu x)\mathscr{A} \qquad (F)$$

    to $\mathfrak{T}_i$, having shown first that $\vdash_{\mathfrak{T}_i} (\exists x)\mathscr{A}$, or
    (b) adding a single new n-ary predicate symbol P to $L_{\mathfrak{A}_i}$, to obtain $L_{\mathfrak{A}_{i+1}}$, and adding the axiom[§]

$$P\vec{x}_n \leftrightarrow \mathscr{A}(\vec{x}_n) \qquad (P)$$

    to $\mathfrak{T}_i$.

We will restrict from now on the form of function and predicate definitions (F) and (P) above, so that in each case the formula $\mathscr{A}$ is in $\boldsymbol{\Delta_0}(L_{\mathfrak{A}})$ (see I.9.27), where $L_{\mathfrak{A}}$ is the language to which the new symbol is being added.

Under the above restriction on $\mathscr{A}$, we call any extension **PA′** of **PA**, as described in (i)–(ii) above, a *recursive extension* (Shoenfield (1967), Schwichtenberg (1978)).[¶]   □

[†] Its interpretation, or "extension", in the standard structure *is* unique, of course.
[‡] Inclusively speaking.
[§] Such a definition effected the introduction of "$\leq$" in II.1.4.
[¶] Actually, those authors require $\mathscr{A}$ to be an *open formula*. For the purposes of Theorem II.4.12 later on, the two formulations make no difference.

One can now show that

$$\vdash \mathscr{A} \leftrightarrow \chi_{\mathscr{A}} \vec{y}_n = 0 \tag{9}$$

and

$$\vdash \neg \mathscr{A} \leftrightarrow \chi_{\mathscr{A}} \vec{y}_n = \widetilde{1} \tag{10}$$

For (9), add $\mathscr{A}$,[†] and prove

$$\chi_{\mathscr{A}} \vec{y}_n = 0 \tag{11}$$

We have (cf. $(f^{(4)})$, p. 218)

$$\vdash \mathscr{A} \wedge \chi_{\mathscr{A}} \vec{y}_n = 0 \vee \neg \mathscr{A} \wedge \chi_{\mathscr{A}} \vec{y}_n = \widetilde{1}$$

Since

$$\mathscr{A}, \ \mathscr{A} \wedge \chi_{\mathscr{A}} \vec{y}_n = 0 \vee \neg \mathscr{A} \wedge \chi_{\mathscr{A}} \vec{y}_n = \widetilde{1} \models_{\text{Taut}} \chi_{\mathscr{A}} \vec{y}_n = 0$$

(11) follows. In short, we have the $\rightarrow$-direction of (9). Similarly, one obtains the $\rightarrow$-direction of (10).

A by-product of all this, in view of tautological implication and $\vdash \mathscr{A} \vee \neg \mathscr{A}$, is

$$\vdash \chi_{\mathscr{A}} \vec{y}_n = 0 \vee \chi_{\mathscr{A}} \vec{y}_n = \widetilde{1}$$

The latter yields the $\leftarrow$-directions of (9) and (10) via proof by cases: Say, we work under the case $\chi_{\mathscr{A}} \vec{y}_n = 0$, with frozen $\vec{y}_n$. Add $\neg \mathscr{A}$. By the $\rightarrow$-half of (10),

$$\vdash \chi_{\mathscr{A}} \vec{y}_n = \widetilde{1}$$

hence, using the assumption, we obtain $\vdash 0 = \widetilde{1}$ which contradicts axiom **S1**. Thus $\vdash \mathscr{A}$, and hence the $\leftarrow$-half of (9) is proved. One handles (10) similarly.

Definition by cases. We want to legitimize definitions of new function symbols f such as

$$f \vec{x}_n = \begin{cases} t_1 & \text{if } \mathscr{A}_1 \\ \vdots & \vdots \\ t_k & \text{if } \mathscr{A}_k \end{cases} \tag{12}$$

where

$$\vdash \mathscr{A}_1 \vee \cdots \vee \mathscr{A}_k \tag{13}$$

$$\vdash \neg(\mathscr{A}_i \wedge \mathscr{A}_j) \qquad \text{for all } i \neq j \tag{14}$$

[†] With frozen $\vec{y}_n$, of course.

and $\vec{x}_n$ is the list of all free variables in the right hand side of (12). Our understanding of the informal notation (12) is that

$$\vdash \mathscr{A}_i \rightarrow f\vec{x}_n = t_i \tag{15}$$

holds for $i = 1, \ldots, k$. We (formally) achieve this as follows:

Let $\chi_i \vec{x}_n$ be the[†] *characteristic term* of $\mathscr{A}_1 \vee \cdots \vee \widehat{\mathscr{A}_i} \vee \cdots \vee \mathscr{A}_k$, where $\widehat{\mathscr{A}_i}$ means that $\mathscr{A}_i$ is missing from the disjunction. Then the definition below, in the style of (8) (p. 219), is what we want (cf. I.8.27):

$$f\vec{x}_n = t_1\chi_1\vec{x}_n + \cdots + t_k\chi_k\vec{x}_n \tag{16}$$

Indeed, $\vdash \mathscr{A}_i \rightarrow \neg(\mathscr{A}_1 \vee \cdots \vee \widehat{\mathscr{A}_i} \vee \cdots \vee \mathscr{A}_k)$, by (14) and $\models_{\textbf{Taut}}$; hence

$$\vdash \mathscr{A}_i \rightarrow \chi_i\vec{x}_n = \tilde{1} \tag{17}$$

On the other hand,[‡]

$$\vdash \mathscr{A}_i \rightarrow \chi_j\vec{x}_n = 0 \qquad \text{for } j \neq i \tag{18}$$

Elementary properties[§] of $+$ and $\times$ now yield (15).

We next define the formal *proper subtraction* function(-symbol), δ. Informally,[¶] $\delta(x, y)$ stands for $x \dot{-} y$. In fact, its natural interpretation – as it is suggested by the right hand side of (19) below – in $\mathfrak{N}' = (\mathbb{N}; 0; S, +, \times; <; \ldots)$ *is* $x \dot{-} y$.

We set, exactly as in informal recursion theory

$$\delta(x, y) = (\mu z)(x = y + z \vee x < y) \tag{19}$$

To allow (19) stand, one must show that

$$\vdash (\exists z)(x = y + z \vee x < y) \tag{20}$$

By < 3, one applies proof by cases. If $x < y$ is the hypothesis, then, by **Ax2** and modus ponens, (20) holds.

---

† For the use of the definite article here – and in similar contexts in the future – see the disclaimer II.1.25, p. 220

‡ Since $\vdash \mathscr{A}_i \rightarrow \mathscr{A}_1 \vee \cdots \vee \widehat{\mathscr{A}_j} \vee \cdots \vee \mathscr{A}_k$ for $j \neq i$.

§ $+1$, $\times 1$ and $\times 2$, associativity, commutativity.

¶ When it comes to formal counterparts of known number-theoretic functions we will abuse the formal notation a bit, opting for *argot* in the interest of readability. Here we wrote "$\delta(x, y)$" rather than the nondescript "$f_i^2 xy$", where f_i^2 is the first unused (so far) function symbol of arity 2. See also p. 166.

For the remaining case we do induction on x to prove

$$\vdash x \geq y \rightarrow (\exists z)x = y + z \tag{21}$$

Pause. What happened to "$\vee\, x < y$"?

Basis. For $x = 0$, if $y = 0$ (cases!), then $z = 0$ works (via **Ax2**) by $+\mathbf{1}$. Otherwise ($y \neq 0$), and we are done by $< \mathbf{1}$.

For the induction step we want

$$\vdash Sx \geq y \rightarrow (\exists z)Sx = y + z \tag{21$'$}$$

Let $Sx \geq y$. This entails two cases:

If $Sx = y$, then $z = 0$ works.
Finally, say $Sx > y$. By $< \mathbf{2}$, we have $x \geq y$. By I.H.,

$$\vdash (\exists z)x = y + z \tag{21$''$}$$

Add $x = y + a$ to the hypotheses, where a is a new constant.[†] By $+\mathbf{2}$, we have $\vdash Sx = y + Sa$; hence $\vdash (\exists z)Sx = y + z$, and we are done proving (20).

We have at once

II.1.27 Lemma. $\vdash x < y \rightarrow \delta(x, y) = 0$, *and* $\vdash x \geq y \rightarrow x = y + \delta(x, y)$.

Also,

II.1.28 Lemma. *For any z,* $\vdash z\delta(x, y) = \delta(zx, zy)$.

Proof. If $x < y$, then $zx < zy$, and hence $\vdash \delta(x, y) = 0$ and $\vdash \delta(zx, zy) = 0$.
If $x \geq y$, then $zx \geq zy$ using II.1.20; hence (II.1.27)

$$\vdash x = y + \delta(x, y) \tag{i}$$

and

$$\vdash zx = zy + \delta(zx, zy) \tag{ii}$$

(*i*) yields

$$\vdash zx = zy + z\delta(x, y)$$

using distribution of $\times$ over $+$ (II.1.16), and therefore we are done by (*ii*) and II.1.21. $\qquad\square$

[†] Or, more colloquially, "let $z = a$ work in (21$''$)". We are using *proof by auxiliary constant*.

Now that we have mastered addition, subtraction, and multiplication, we turn our attention to division. First we formalize remainders and quotients.

We start from the trivial observation

$$\vdash (\exists z)(\exists w)x = yw + z \tag{1}$$

(1) follows from $\vdash x = y0 + x$ ($+\mathbf{1}$ and $\times\mathbf{1}$).

Thus we may introduce a new function (of arity 2), the *remainder function* "r", by the following μ-definition:[†]

$$r(x, y) = (\mu z)(\exists w)x = yw + z \tag{rem}$$

We have at once $((f^{(4)})$, p. 218)

$$\vdash (\exists w)x = yw + r(x, y) \tag{EQ}$$

Therefore we can introduce quotients, $q(x, y)$, via the definition of the 2-ary function symbol "q" below:

$$q(x, y) = (\mu w)x = yw + r(x, y) \tag{Q}$$

To improve readability we will usually denote the term $q(x, y)$ by $\lfloor x/y \rfloor$ or

$$\left\lfloor \frac{x}{y} \right\rfloor$$

(note the boldface variables x and y, which distinguish the formal case from the informal one), and we will hardly need to refer to the symbol "q" again.

One more application of $(f^{(4)})$ – to (Q) – yields

$$\vdash x = y \left\lfloor \frac{x}{y} \right\rfloor + r(x, y) \tag{Euc}$$

 The enabling condition (EQ) for (Q) yields

$$\vdash (\exists w)x = 0 \cdot w + r(x, 0)$$

Hence the interesting

$$\vdash x = r(x, 0)$$

Since $\vdash x = 0 \cdot 0 + r(x, 0)$, (Q) and $(f^{(5)})$ (p. 218) yield $\lfloor x/0 \rfloor \leq 0$. Hence, by II.1.5,

$$\vdash \left\lfloor \frac{x}{0} \right\rfloor = 0$$

[†] We write "$r(x, y)$" rather than "rxy" in the interest of user-friendliness of notation.

Adding the usual assumption $y > 0$, we can guarantee the "standard" properties of quotient and remainder, namely the inequality[†] $r(x, y) < y$ and the uniqueness of quotient and remainder.

To see this, add $y \leq r(x, y)$. Thus (II.1.27)

$$\vdash r(x, y) = y + t \tag{2}$$

where we have set $t = \delta(r(x, y), y)$ for convenience. By (*Euc*) and $\times 2$ (invoking associativity *tacitly*, as is usual)

$$\vdash x = yS\left\lfloor \frac{x}{y} \right\rfloor + t$$

Hence,

$$\vdash (\exists w)x = yw + t$$

from which, along with (*rem*) and $(f^{(5)})$ (p. 218),

$$\vdash r(x, y) \leq t$$

Since also $\vdash r(x, y) \geq t$ by (2) and II.1.17, we get $\vdash r(x, y) = t$ (II.1.8); hence $\vdash y = 0$ from (2) and cancellation (II.1.21). By the deduction theorem we have derived

$$\vdash y \leq r(x, y) \rightarrow y = 0$$

or, better still (via < 3, irreflexivity, and II.1.5), the contrapositive

$$\vdash y > 0 \rightarrow r(x, y) < y$$

We next turn to uniqueness. Let[‡] then

$$z < y \wedge (\exists w)x = yw + z \tag{3}$$

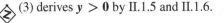 (3) derives $y > 0$ by II.1.5 and II.1.6.

By $(f^{(5)})$ and (*rem*), $\vdash r(x, y) \leq z$. Since we want "=" here, we add

$$r(x, y) < z \tag{4}$$

to obtain a contradiction. Reasoning by *auxiliary constant*, we also add

$$x = yq + z \tag{5}$$

[†] The "$0 \leq r(x, y)$" part is trivial from $\vdash 0 \leq x$ (II.1.5) and substitution.
[‡] Cf. p. 209 for our frequent use of the *argot* "let".

for some new constant q. Let now $s = \delta(z, r(x, y))$. By (4) and II.1.27,

$$\vdash z = r(x, y) + s \wedge s > 0 \tag{6}$$

Thus (5) and *(Euc)* yield (via II.1.21) $\vdash y \lfloor x/y \rfloor = yq + s$; hence (definition of δ and II.1.28)

$$\vdash y\delta\left(\left\lfloor \frac{x}{y} \right\rfloor, q\right) = s$$

We conclude that $\vdash y \le s$ (why?), which, along with $\vdash z < y$, from (3), and $\vdash s \le z$, from (6) and II.1.17, yields (transitivity of $<$) the contradiction $\vdash y < y$. Thus,

$$\vdash z < y \wedge (\exists w)x = yw + z \rightarrow z = r(x, y)$$

It is a trivial matter (not pursued here) now to obtain

$$\vdash z < y \wedge x = yw + z \rightarrow w = \left\lfloor \frac{x}{y} \right\rfloor \wedge z = r(x, y) \tag{UN}$$

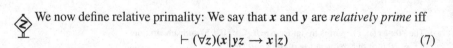

In *(rem)* (p. 224) the formula to the right of (μz) is provably equivalent to the Δ_0-formula $(\exists w)_{<x}x = yw + z$ (why?). Thus the latter could have been used in lieu of the original.

Therefore the addition of r and its defining axiom to the theory results to a *recursive extension* (such as we promised all our extensions by definitions to be [p. 220]).

We next introduce the divisibility predicate "$|$" by the axiom *(D)* below:

$$x|y \leftrightarrow r(y, x) = 0 \tag{D}$$

Once again, the use of boldface variables will signify that we are in the formal domain.

II.1.29 Lemma. $\vdash y|x \leftrightarrow (\exists z)yz = x$.

Proof. By tautological implication, we have two directions to deal with:

$\rightarrow$:   It follows by *(Euc)* above, $+\mathbf{1}$, and **Ax2**.

$\leftarrow$:   By the deduction theorem; assume $(\exists z)yz = x$ towards proving $y|x$. Thus $\vdash (\exists z)yz + 0 = x$ by $+\mathbf{1}$. By *(rem)* and $(f^{(5)})$, we have $\vdash r(x, y) \le 0$. Done, by antisymmetry and II.1.5. $\square$

We now define relative primality: We say that x and y are *relatively prime* iff

$$\vdash (\forall z)(x|yz \rightarrow x|z) \tag{7}$$

We introduce the *metanotation* $RP(x, y)$ to stand for $(\forall z)(x|yz \to x|z)$. Thus, we may write (7) as $\vdash RP(x, y)$.

We emphasize that "RP" is *not* introduced as a predicate into the language,[†] rather it is introduced as a metamathematical abbreviation – that is, we write "$RP(x, y)$" simply to avoid writing the formula in (7).

The technical reason is that we only introduce predicates if we are prepared to give a $\mathbf{\Delta_0}$-formula to the right of "$\leftrightarrow$" in their definition (since we want to only consider recursive extensions[‡] of **PA**).

II.1.30 Remark. Strictly speaking, RP as (informally) defined gives relative primality correctly, as we understand it intuitively, *only when both numbers are nonzero*. It has some pathologies: e.g, the counterintuitive

$$\vdash RP(0, \widetilde{2})$$

Indeed, to prove $(\forall z)(0 \,|\, \widetilde{2}z \to 0\,|\,z)$, strip it of the quantifier and assume

$$r(\widetilde{2}z, 0) = 0$$

Hence (p. 224)

$$\vdash \widetilde{2}z = 0$$

Thus ($\vdash \widetilde{2} > 0$, $\vdash 0z = 0$, and II.1.21) $\vdash z = 0$; hence

$$\vdash 0\,|\,z$$

Similarly, one can prove $\vdash x > 0 \to RP(0, x)$.

Of course, in "reality" (i.e., informally), 0 and 2 are *not* relatively prime, since their greatest common divisor is 2. □

II.1.31 Exercise. Show that $\vdash RP(x, y) \to x > 0 \lor y > 0$. □

II.1.32 Lemma. $\vdash (\exists z)(\exists w)\delta(xz, yw) = \widetilde{1} \to RP(x, y)$.

Proof. Assume the hypothesis

$$(\exists z)(\exists w)\delta(xz, yw) = \widetilde{1}$$

[†] That would have been through a definition like $RP(x, y) \leftrightarrow (\forall z)(x|yz \to x|z)$.
[‡] Cf. p. 220.

Let $\delta(xa, yb) = \tilde{1}$, where a and b are new constants. Add also

$$x \mid yz \tag{8}$$

towards proving $x \mid z$.

Now $\vdash z\delta(xa, yb) = z$;[†] hence

$$\vdash \delta(zxa, zyb) = z \tag{9}$$

by II.1.28. Setting $c = \lfloor zy/x \rfloor$ for convenience,[‡] we get $\vdash xc = zy$ from (8). Thus, by (9), $\vdash \delta(zxa, xcb) = z$; hence (again via II.1.28)

$$\vdash x\delta(za, cb) = z$$

That is, $\vdash x \mid z$, and an application of the deduction theorem followed by generalization yields $(\forall z)(x \mid yz \rightarrow x \mid z)$, i.e., $RP(x, y)$.   □

We have already remarked that we will be using associativity and commutativity of "$+$" and "$\times$" without notice.

II.1.33 Lemma. $\vdash x > 0 \rightarrow RP(x, y) \rightarrow RP(y, x)$.

Proof. The case $y = 0$ is trivial by Remark II.1.30. Thus, take now the case $y > 0$.

We add the assumptions

$$x > 0$$

$$RP(x, y)$$

and

$$y \mid xz \tag{10}$$

towards proving $y \mid z$. By (10),

$$\vdash xz = ya \tag{11}$$

where $a = \lfloor xz/y \rfloor$. Thus $x \mid ya$; hence $x \mid a$ by $RP(x, y)$.

Write then $a = xq$ ($q = \lfloor a/x \rfloor$), from which and (11) $\vdash xz = yxq$. Thus, $\vdash z = yq$, by II.1.21 and our first hypothesis, which proves $y \mid z$.   □

[†] Assuming you believe that $\vdash x = x\tilde{1}$.

[‡] Another aspect of convenience is to invoke commutativity or associativity of either $+$ or $\times$ tacitly.

Thus, what we have explicitly derived under the second case above (via the deduction theorem) was

$$y > 0 \vdash x > 0 \rightarrow RP(x, y) \rightarrow y|xz \rightarrow y|z$$

Hence we also derived (by $\forall$-introduction) what we really wanted,

$$y > 0 \vdash x > 0 \rightarrow RP(x, y) \rightarrow (\forall z)(y|xz \rightarrow y|z)$$

The moral is that even "formal", but "practical", proofs often omit the obvious. While we could just as easily have incorporated these couple of lines in the proof above, we are going to practise this shortening of proofs again and again. Hence the need for this comment.

II.1.34 Lemma. $\vdash k \mid p \rightarrow RP(S(ip), S((i + k)p))$.

Proof. The case $p = 0$ being trivial – $\vdash RP(S0, S0)$ – we argue the case $p > 0$. Add $k \mid p$ towards proving

$$RP(S(ip), S((i + k)p)) \qquad (i)$$

Thus,

$$p = ka \qquad (ii)$$

where $a = \lfloor p/k \rfloor$. By II.1.32

$$\vdash RP(S(ip), p) \qquad (iii)$$

and

$$\vdash RP(S(ip), k) \qquad (iv)$$

each because $\vdash \delta(S(ip), ip) = \widetilde{1}$. Add $S(ip) \mid zS((i + k)p)$ towards proving $\vdash S(ip) \mid z$.

By hypothesis, $\vdash S(ip) \mid zkp$ (fill in the missing steps); hence $\vdash S(ip) \mid zp$ by (iv). Then $\vdash S(ip) \mid z$ by (iii). $\qquad\qquad\square$

We now embark on introducing coding of sequences formally.[†]

The formal counterpart of a sequence $a_0, a_1, \ldots, a_n$ of (variable) length $= n+1$ is a term $t(n, \vec{x})$,[‡] where the parenthesis notation lists all free variables of t. We may also simply write $t[n]$ (see p. 18).

[†] This will be a "careful" repetition of the definition of Gödel's β-function ((2) of p. 159).

[‡] "n" is here a variable, *not* a numeral. That is why we wrote "n" rather than "$\widetilde{n}$".

We introduce the *maximum* of the first $n + 1$ members of a sequence

$$\max_{i \leq n}(t(i, \vec{x})) = (\mu z)((\forall i)_{\leq n} z \geq t(i, \vec{x})) \qquad (M)$$

To legitimize (M) we need to establish

$$\vdash (\exists z)(\forall i)(i \leq n \rightarrow z \geq t(i, \vec{x})) \qquad (M')$$

We prove (M') by induction on n.

For $n = 0$ (M') follows immediately (do $[z \leftarrow t(0, \vec{x})]$ and apply **Ax2**).

Add now (M') for frozen n and $\vec{x}$, and show that

$$\vdash (\exists z)(\forall i)(i \leq Sn \rightarrow z \geq t(i, \vec{x})) \qquad (M'')$$

Let a satisfy (M') (where in the latter n and $\vec{x}$ are still frozen).

That is, formally, add a new constant symbol a, and the new axiom

$$(\forall i)(i \leq n \rightarrow a \geq t(i, \vec{x})) \qquad (M^{(3)})$$

It follows (specialization) that

$$\vdash i \leq n \rightarrow a \geq t(i, \vec{x}) \qquad (1)$$

Since $\vdash a + t(Sn, \vec{x}) \geq a$ and $\vdash a + t(Sn, \vec{x}) \geq t(Sn, \vec{x})$, proof by cases, (1), and $\vdash i \leq Sn \leftrightarrow i \leq n \vee i = Sn$ yield[†]

$$\vdash i \leq Sn \rightarrow a + t(Sn, \vec{x}) \geq t(i, \vec{x}) \qquad (2)$$

Hence (generalization)

$$\vdash (\forall i)(i \leq Sn \rightarrow a + t(Sn, \vec{x}) \geq t(i, \vec{x}))$$

and via **Ax2**

$$\vdash (\exists z)(\forall i)(i \leq Sn \rightarrow z \geq t(i, \vec{x}))$$

This is (M''). Now the deduction theorem confirms the induction step. $\qquad \square$

We next introduce the *least common multiple* of the first $n + 1$ members of a "*positive* sequence":

$$lcm_{i \leq n}(St(i, \vec{x})) = (\mu z)(z > 0 \wedge (\forall i)_{\leq n} St(i, \vec{x}) \,|\, z) \qquad (LCM)$$

To legitimize (LCM) we need to establish

$$\vdash (\exists z)(z > 0 \wedge (\forall i)(i \leq n \rightarrow St(i, \vec{x}) \,|\, z)) \qquad (LCM')$$

[†] With help from the logical $i = Sn \rightarrow t(i, \vec{x}) = t(Sn, \vec{x})$.

One can easily prove (LCM') by induction on n. In outline, a z that works for $n = 0$ is $St(0, \vec{x})$. If now a (auxiliary constant!) works for n (the latter frozen along with $\vec{x}$), then $a\,St(Sn, \vec{x})$ works for Sn. The details are left to the reader.

 The positive sequence above has members $St(i, \vec{x})$, "indexed" by i, where t is some term.

This representation stems from the fact that $\vdash t > 0 \rightarrow t = S\delta(t, \widetilde{1})$.

Alternatively, we could have opted to μ-define *lcm with a condition*: that $(\forall i \leq n)t(i, \vec{x}) > 0$. This approach complicates the defining formula as we try to make the μ-defined object total (when interpreted in the standard structure).

Axiom *(LCM)* implies at once ($f^{(4)}$ and $f^{(5)}$ of p. 218)

$$\vdash (z > 0 \wedge (\forall i)_{\leq n} St(i, \vec{x}) \,|\, z) \rightarrow \underset{i \leq n}{lcm}(St(i, \vec{x})) \leq z \tag{1}$$

$$\vdash \underset{i \leq n}{lcm}(St(i, \vec{x})) > 0 \tag{2}$$

and[†]

$$\vdash (\forall i)_{\leq n}\Big(St(i, \vec{x}) \,|\, \underset{i \leq n}{lcm}(St(i, \vec{x})) \Big) \tag{3}$$

(1) can be sharpened to

$$\vdash (z > 0 \wedge (\forall i)_{\leq n} St(i, \vec{x}) \,|\, z) \rightarrow \underset{i \leq n}{lcm}(St(i, \vec{x})) \,|\, z \tag{1'}$$

Indeed, assume the left hand side of "$\rightarrow$" in $(1')$ and also let

$$r > 0 \wedge z = \underset{i \leq n}{lcm}(St(i, \vec{x}))q + r \tag{4}$$

where the terms r and q are the unique (by (2)) remainder and quotient of the division $z / lcm_{i \leq n}(St(i, \vec{x}))$ respectively. That is, we adopt the negation of the right hand side of "$\rightarrow$" in $(1')$ and hope for a contradiction.

Well, by (3), (4) and our additional assumptions immediately above,

$$\vdash r > 0 \wedge (\forall i)_{\leq n} St(i, \vec{x}) \,|\, r$$

hence

$$\vdash \underset{i \leq n}{lcm}(St(i, \vec{x})) \leq r$$

by (1), contradicting the remainder inequality (by (2) the divisor in (4) is positive). This establishes that $r > 0$ is untenable and proves $(1')$.

We now revisit Lemma I.9.6 (actually proving a bit more).

[†] The big brackets are superfluous but they improve readability.

II.1.35 Lemma. *Let t and s be terms. Then*

$$\vdash s > S0 \wedge (\forall i)_{\leq n} RP(s, St[i]) \rightarrow RP(s, \underset{i \leq n}{lcm}(St[i]))$$

Proof. Let

$$s > S0 \wedge (\forall i)_{\leq n} RP(s, St[i]) \tag{5}$$

Let also (we are implicitly using II.1.33) $lcm_{i \leq n}(St[i]) \mid sz$. By (3) and (5)

$$\vdash (\forall i)_{\leq n} St[i] \mid z$$

Taking cases, if $z = 0$ then

$$\vdash \underset{i \leq n}{lcm}(St[i]) \mid z \tag{6}$$

anyway. If $z > 0$ then (6) is again obtained, this time by $(1')$.   □

The following is the formalized Lemma I.9.6.

II.1.36 Corollary. *Let t and s be terms. Then*

$$\vdash s > S0 \wedge (\forall i)_{\leq n} RP(s, St[i]) \rightarrow \neg s \mid \underset{i \leq n}{lcm}(St[i])$$

Proof. Exercise II.11.   □

II.2. A Formal β-Function

The following steps formalize those taken starting with p. 156. Note that c, p, and q below are just convenient (metamathematical) abbreviations of the respective right hand sides.

Let $t(n, \vec{x})$ be a term, and set

$$c = \underset{i \leq n}{\max}(St[i]) \tag{C}$$

We next let p be the lcm of the sequence $S0, \ldots, Sc$ (informally, $1, \ldots, c+1$). Thus we set[†]

$$p = \underset{i \leq c}{lcm}(Si) \tag{P}$$

Finally, define the term q by the explicit definition[‡]

$$q = \underset{i \leq n}{lcm}(S(p\,St[i])) \tag{Q}$$

[†] That is, p stands for $s[c]$, where $s[n]$ abbreviates $lcm_{i \leq n}(Si)$, n being a variable.
[‡] Of course, a more user-friendly way to write "$S(p\,St[i])$" is "$1 + p(1 + t[i])$".

By (P) and (2) and (3) (p. 231) above,

$$\vdash p > 0 \wedge (\forall i)_{\leq_c} Si \mid p \qquad (P')$$

We can now derive

$$\vdash y \leq c \rightarrow (\forall i)(i \leq n \rightarrow \neg y = t[i]) \\ \rightarrow (\forall i)(i \leq n \rightarrow RP(S(pSy), S(pSt[i]))) \qquad (6)$$

To see why (6) holds, add the assumptions $y \leq c$ and $i \leq n \rightarrow \neg y = t[i]$. We now try to prove

$$i \leq n \rightarrow RP(S(pSy), S(pSt[i]))$$

So add $i \leq n$. This yields $\neg y = t[i]$, which splits into two cases by < 3, namely, $y > t[i]$ and $y < t[i]$.

We will consider first the case $y > t[i]$. Set $k = \delta(y, t[i])$ for convenience.

Now $\vdash 0 < k \wedge k \leq y$ hence also $\vdash k \leq c$ by assumption and transitivity. Thus $\vdash k \mid p$ by (P') and Exercise II.10.

Moreover (by II.1.27) $\vdash y = t[i] + k$, hence (via $+2$) $\vdash Sy = St[i] + k$. Thus II.1.34 yields

$$\vdash RP(S(pSy), S(pSt[i]))$$

The other case, $y < t[i]$, is handled entirely similarly with slightly different start-up details: This time we set $k = \delta(t[i], y)$.

Now $\vdash 0 < k \wedge k \leq t[i]$; hence also $\vdash k < c$.

Why? Well, $\vdash i \leq n \rightarrow St[i] \leq c$ by (C) and (M) (p. 230) via $(f^{(4)})$ (p. 218). Now the assumption $i \leq n$ yields $\vdash St[i] \leq c$, and transitivity does the rest.

Thus $\vdash k \mid p$ by (P'), and one continues the proof as in the previous case: By II.1.27 $\vdash t[i] = y + k$; hence (via $+2$) $\vdash St[i] = Sy + k$. Thus II.1.34 yields

$$\vdash RP(S(pSy), S(pSt[i]))$$

once more.

At this point we have derived (by the deduction theorem)

$$y \leq c \vdash \\ (i \leq n \rightarrow \neg y = t[i]) \rightarrow (i \leq n \rightarrow RP(S(pSy), S(pSt[i]))) \qquad (7)$$

Hence, by ∀-monotonicity (I.4.24),

$$y \leq c \vdash \\ (\forall i \leq n)(\neg y = t[i]) \rightarrow (\forall i \leq n)RP(S(pSy), S(pSt[i])) \qquad (7')$$

(6) now follows by the deduction theorem.

We immediately derive from (6), (Q), and II.1.36 that

$$\vdash y \leq c \rightarrow (\forall i \leq n)(\neg y = t[i]) \rightarrow \neg S(pSy)\,|\,q \qquad (8)$$

Hence, by tautological implication,

$$\vdash y \leq c \rightarrow S(pSy)\,|\,q \rightarrow (\exists i \leq n)y = t[i] \qquad (8')$$

Thus, *informally speaking*, q "codes" the *unordered* set of all "objects"

$$T = \{S(pSt[i]) : i \leq n\}$$

in the sense that if x is in T, then $x\,|\,q$, and, conversely, if $S(pSy)\,|\,q$ – where $y \leq c$ – then $S(pSy)$ is in T. By coding "position information", i, along with the term $t[i]$, we can retrieve from q the ith *sequence member* $t[i]$.

To this end, we define three new function symbols, J, K, L, of arities 2, 1, and 1 respectively:

$$J(x,y) = (x + y)^2 + x \qquad (J)$$

where "$(x + y)^2$" is an abbreviation for "$(x + y) \times (x + y)$",

$$Kz = (\mu x)(x = Sz \vee (\exists y)_{\leq z}J(x,y) = z) \qquad (K)$$

$$Lz = (\mu y)(y = Sz \vee (\exists x)_{\leq z}J(x,y) = z) \qquad (L)$$

J, K, L are the formal counterparts of J, K, L of p. 156.

To legitimize (K) and (L) one needs to show

$$\vdash (\exists x)(x = Sz \vee (\exists y)_{\leq z}J(x,y) = z) \qquad (K')$$

and

$$\vdash (\exists y)(y = Sz \vee (\exists x)_{\leq z}J(x,y) = z) \qquad (L')$$

They are both trivial, since $\vdash Sz = Sz$.

II.2.1 Lemma. $\vdash J(x,y) = J(a,b) \rightarrow x = a \wedge y = b$.

Proof. A straightforward adaptation of the argument following $(*)$ on p. 156. $\square$

II.2.2 Lemma. $\vdash KJ(a,b) = a$ and $\vdash LJ(a,b) = b$.

Proof. We just prove the first contention, the proof of the second being entirely analogous.

First, it is a trivial matter to prove $\vdash x \leq J(x,y)$ and $\vdash y \leq J(x,y)$ (Exercise II.12). Now $\vdash b \leq J(a,b) \wedge J(a,b) = J(a,b)$; hence

$$\vdash a = SJ(a,b) \vee (\exists y)(y \leq J(a,b) \wedge J(a,y) = J(a,b)) \tag{1}$$

By (K), (1) above, and $(f^{(5)})$ (p. 218) we have

$$\vdash KJ(a,b) \leq a \tag{2}$$

while (K) and $(f^{(4)})$ (p. 218) yield

$$\vdash KJ(a,b) = SJ(a,b)$$
$$\vee (\exists y)(y \leq J(a,b) \wedge J(KJ(a,b),y) = J(a,b)) \tag{3}$$

Since $KJ(a,b) = SJ(a,b)$ is untenable by (2), we get

$$\vdash (\exists y)(y \leq J(a,b) \wedge J(KJ(a,b),y) = J(a,b)) \tag{4}$$

Let $c \leq J(a,b) \wedge J(KJ(a,b),c) = J(a,b)$, where c is a new constant. By II.2.1, $\vdash KJ(a,b) = a$. □

To conclude our coding, whose description we launched with the ⬦-sign on p. 232, let finally $a[n]$ be a term.

We code the *sequence* $a(n,\vec{x})$, for $i \leq n$, by following the above steps, letting first t of the previous discussion be an *abbreviation* of the specific term below:

$$t(i,\vec{x}) \stackrel{\text{def}}{=} J(i,a(i,\vec{x})), \qquad \text{where } i \text{ and } \vec{x} \text{ are distinct variables} \tag{T}$$

Thus, by (8′) (p. 234) and substitution, we have

$$\vdash J(i,m) \leq c \wedge S(pSJ(i,m)) \mid q \rightarrow$$
$$(\exists j \leq n)J(i,m) = J(j,a[j])$$

which (by II.2.1†) yields

$$\vdash J(i,m) \leq c \wedge S(pSJ(i,m)) \mid q \rightarrow m = a[i] \wedge i \leq n \tag{5}$$

This motivates the definition (where d is intended to receive the "value"‡ $J(c,q)$)

II.2.3 Definition (The Formal β).

$$\beta(d,i) = (\mu m)(m = d \vee S(pSJ(i,m)) \mid Ld) \tag{B}$$

---

† $\vdash (\exists j)(j \leq n \wedge J(i,m) = J(j,a[j])) \rightarrow m = a[i] \wedge i \leq n$. To see this, assume hypothesis and use a new constant b to eliminate $(\exists j)$.

‡ Of course, regardless of intentions, the letter d in the definition (B) is just a variable, like i,m.

The letter p in (B) is an abbreviation for the term $lcm_{j \leq Kd}(Sj)$ (see (P), p. 232).

That (B) is a legitimate definition, that is,

$$\vdash (\exists m)(m = d \vee S(pSJ(i,m)) \,|\, Ld) \qquad (B')$$

follows from $\vdash x = x$.

II.2.4 Proposition.

(i) $\vdash \beta(x,i) \leq x$. *Moreover,*
(ii) $\vdash \beta(x,i) < x \leftrightarrow (\exists m)(S(pSJ(i,m)) \,|\, Lx)$, *where* $p = lcm_{i \leq Kx}(Si)$.

Proof. (i) is immediate from (B), $\vdash x = x$ and $(f^{(5)})$ (p. 218).

(ii): The $\rightarrow$-part is immediate from (B) and $(f^{(4)})$ (p. 218).

As for $\leftarrow$, assume $(\exists m)(S(pSJ(i,m)) \,|\, Lx)$.

Let $S(pSJ(i,r)) \,|\, Lx$, where r is a new constant. Hence $\vdash S(pSJ(i,r)) \leq Lx$. We also have $\vdash Lx \leq Sx$ by (L) (p. 234) and $(f^{(5)})$ (p. 218); thus $\vdash pSJ(i,r) \leq x$ by transitivity and II.1.9 (contrapositive). But $\vdash r < pSJ(i,r)$ by $\vdash y \leq J(x,y)$ (Exercise II.12); hence $\vdash r < x$. Since $\vdash \beta(x,i) \leq r$ by (B) and $(f^{(5)})$ (p. 218), we are done. $\qquad \square$

All this work yields the "obvious":

II.2.5 Theorem. *For any term* $a(i,\vec{x})$,

$$\vdash (\forall x_1) \ldots (\forall x_m)(\forall n)(\exists z)(\forall i)(i \leq n \rightarrow \beta(z,i) = a(i,\vec{x})) \qquad (6)$$

where m *is the length of* $\vec{x}$.

Proof. We prove instead

$$\vdash (\exists z)(\forall i)(i \leq n \rightarrow \beta(z,i) = a(i,\vec{x}))$$

The proof constructs a "z that works" (and then invokes **Ax2**). To this end, we let t be that in (T), and in turn, let c, p, q stand for the terms in the right hand sides of (C), (P), and (Q) respectively (p. 232). Setting for convenience $d = J(c,q)$, we are reduced to proving

$$i \leq n \rightarrow \beta(d,i) = a(i,\vec{x}) \qquad (7)$$

Thus we add the assumption $i \leq n$. We know

$$\vdash J(i, a(i, \vec{x})) \leq c, \qquad \text{by } (C), (M) \text{ on p. 230, and } (f^{(4)}) \text{ (p. 218)}$$

$$\vdash S(pSJ(i, a(i, \vec{x}))) \mid q, \qquad \text{by } (Q) \text{ and } (3) \text{ on p. 231}$$

Or, using the abbreviation "d" and II.2.2

$$\vdash J(i, a(i, \vec{x})) \leq Kd \qquad (8)$$

$$\vdash S(pSJ(i, a(i, \vec{x}))) \mid Ld \qquad (9)$$

Thus,

$$\vdash (\exists m)(S(pSJ(i, m)) \mid Ld)$$

The above existential theorem and II.2.4(ii) imply

$$\vdash \beta(d, i) < d \qquad (10)$$

so that (B) (p. 235) and $(f^{(4)})$ – through $\vdash \neg \beta(d, i) = d$, by (10) – yield

$$\vdash S(pSJ(i, \beta(d, i))) \mid Ld \qquad (11)$$

By (9), (B), and $(f^{(5)})$ (p. 218), $\vdash \beta(d, i) \leq a(i, \vec{x})$; hence, since J is increasing in each argument (do you believe this?), (8) implies

$$\vdash J(i, \beta(d, i)) \leq Kd$$

Combining the immediately above with (11), we obtain

$$\vdash J(i, \beta(d, i)) \leq Kd \wedge S(pSJ(i, \beta(d, i))) \mid Ld$$

Now (5) on p. 235 yields

$$\vdash \beta(d, i) = a(i, \vec{x})$$

By the deduction theorem, we now have (7). □

II.2.6 Corollary. *For any term $a(i, \vec{x})$,*

$$\vdash (\forall x_1) \dots (\forall x_m)(\forall n)(\exists z)(\forall i)_{i \leq n}(\beta(z, i) < z \wedge \beta(z, i) = a(i, \vec{x}))$$

where m is the length of $\vec{x}$.

Proof. By (10) of the previous proof. □

II.2.7 Example (Some Pathologies). (1) By II.2.4 (*i*) we get $\vdash \beta(0, i) = 0$ (using II.1.5 and II.1.8). Thus, if we introduce an 1-ary function letter f by the explicit definition $fn = 0$, then $\vdash i \leq n \rightarrow \beta(0, i) = fi$. It follows, according to (6) of II.2.5, that 0 "codes" the sequence of the first n members of the term fi – for any n "value".

(2) Next, "compute" $\beta(\widetilde{3}, i)$. Now, $\vdash K\widetilde{3} = \widetilde{4}$ and $\vdash L\widetilde{3} = \widetilde{4}$ (why?). Since $\vdash p = \widetilde{60}$ (why?), we get

$$\vdash \neg S(pSJ(i, m)) \mid L\widetilde{3}$$

since $\vdash S(pSJ(i, m)) \geq \widetilde{61}$ (why?). By II.2.4(*ii*), $\vdash \beta(\widetilde{3}, i) = \widetilde{3}$.[†]

Thus, if we have a function symbol g with a definition $gn = \widetilde{3}$, then a possible proof of

$$\vdash (\exists z)(\forall i)(i \leq w \rightarrow \beta(z, i) = gi)$$

starts with "take z to be $\widetilde{3}$". An alternative "value" for z is the "d" constructed in the proof of II.2.5, adapted to the term gn. We may call the latter z-"value" the "intended one" or the "natural one".

Clearly, "intended" or not, any z that works in (6) of II.2.5 is an in principle acceptable coding of the first "n members" of a term a.

(3) Finally, let us compute $\beta(\widetilde{2}, i)$. Now, $\vdash K\widetilde{2} = \widetilde{1}$ and $\vdash L\widetilde{2} = 0$. Also $\vdash p = \widetilde{2}$. It follows that

$$\vdash \beta(\widetilde{2}, i) = 0$$

since

$$\vdash S(pSJ(i, 0)) \mid 0$$

Thus, if f is introduced as in part (1) of this example by $fn = 0$, then

$$\vdash (\exists z)(\forall i)(i \leq w \rightarrow \beta(z, i) = fi)$$

can be proved by letting z be $\widetilde{2}$, or 0, or by invoking the construction carried out in the proof of II.2.5.[‡] In particular, β *is not 1-1 in its first argument.*

Part (3) of this example shows that an x that passes the test "$\beta(x, i) < x$" is *not* necessarily the d computed in the standard manner as in the proof of II.2.5 – i.e., we cannot expect $\vdash x = d$. After all, $\vdash \beta(\widetilde{2}, i) < \widetilde{2}$.   □

[†] This is related to the example $\beta(\widetilde{4}, i) = \widetilde{4}$ of II.2.14.
[‡] If the latter construction is followed, then $\vdash Lz > 0$, of course.

We are all set to introduce the formal counterparts of $\langle \dots \rangle$, Seq, lh, $*$, $(z)_i$ of p. 165.

II.2.8 Definition (Bold $\langle \dots \rangle$). For any term t and any variable w not free in t, we denote by $\langle t[i] : i < w \rangle$ (or, sloppily, $\langle t[0], \dots, t[w-1] \rangle$) the μ-term

$$(\mu z)(\beta(z, 0) = w \wedge (\forall i)_{<w}(\beta(z, Si) = St[i])) \qquad (FC)$$

$\square$

II.2.9 Proposition. *The μ-term in* (FC) *can be formally introduced.*

Proof. We want

$$\vdash (\exists z)(\beta(z, 0) = w \wedge (\forall i)_{<w}(\beta(z, Si) = St[i])) \qquad (FC')$$

Let $a[w, n]$ abbreviate the term defined by cases below (see p. 221):

$$a[w, n] = \begin{cases} w & \text{if } n = 0 \\ St[\delta(n, S0)] & \text{if } n > 0 \end{cases} \qquad (*)$$

$(*)$ yields ((15) on p. 221)

$$\vdash n = 0 \rightarrow a[w, n] = w$$

and hence (**Ax4**)

$$\vdash a[w, 0] = w \qquad (1)$$

but also

$$\vdash n > 0 \rightarrow a[w, n] = St[\delta(n, S0)]$$

Hence (by $\vdash Sn > 0$ and modus ponens)

$$\vdash a[w, Sn] = St[\delta(Sn, S0)]$$

or, using $\vdash \delta(Sn, S0) = n$ (do you believe this?)

$$\vdash a[w, Sn] = St[n] \qquad (2)$$

Now, by Theorem II.2.5,

$$\vdash (\exists z)(\forall i)(i \leq w \rightarrow \beta(z, i) = a[w, i])$$

In view of the above existential statement, we introduce a new constant c and the assumption

$$(\forall i)(i \leq w \rightarrow \beta(c,i) = a[w,i])$$

By specialization we obtain from the above

$$\vdash 0 \leq w \rightarrow \beta(c,0) = a[w,0]$$

that is,

$$\vdash \beta(c,0) = a[w,0] \tag{3}$$

by II.1.5, and

$$\vdash Si \leq w \rightarrow \beta(c,Si) = a[w,Si]$$

that is,

$$\vdash i < w \rightarrow \beta(c,Si) = a[w,Si] \tag{4}$$

in view of II.1.10. Putting the generalization of (4) together (conjunction) with (3), via (1) and (2), yields

$$\vdash \beta(c,0) = w \wedge (\forall i)_{<w}(\beta(c,Si) = St[i])$$

from which **Ax2** yields (FC'). $\qquad\square$

II.2.10 Definition. We introduce the functions "lh" and "$(\dots)_{\dots}$" by

$$
\begin{aligned}
lh(z) &= \beta(z,0) \\
(z)_i &= \delta(\beta(z,Si),S0)
\end{aligned}
\tag{$**$}
$$

$\qquad\square$

In the second definition in the group $(**)$ above we have introduced a new 2-ary function symbol called, let us say, f, by $fzi = \delta(\beta(z,Si),S0)$ – in informal lightface notation, $\beta(z,i+1) \dot{-} 1$ – and then agreed to denote the term "fzi" by $(z)_i$.[†]

II.2.11 Proposition. *If we let* $b = \langle t[i] : i < x \rangle$, *then we can obtain*

(1) $\vdash lh(b) = x$,

(2) $\vdash (\forall i)_{<x}(b)_i = t[i]$, *or, equivalently,* $\vdash (\forall i)_{<lh(b)}(b)_i = t[i]$, *and*

(3) $\vdash z < b \rightarrow (\neg lh(z) = x \vee (\exists i)_{<x} \neg(z)_i = t[i])$.

[†] Note how this closely parallels the "$(z)_i$" of the prime-power coding. We had set there (p. 137) $(z)_i = \exp(i,z) \dot{-} 1$.

Proof. (1): By $(f^{(4)})$ and (FC) (p. 239),

$$\vdash \beta(b, 0) = x$$

We conclude using II.2.10.

(2): By $(f^{(4)})$ and (FC),

$$\vdash i < x \rightarrow \beta(b, Si) = St[i]$$

We conclude using $\vdash \delta(St[i], S0) = t[i]$ and II.2.10.

(3): To prove the last contention we invoke $(f^{(3)})$, p. 218, in connection with (FC). It yields

$$\vdash (\forall z)_{<b} \neg(\beta(z, 0) = x \wedge (\forall i)(i < x \rightarrow \beta(z, Si) = St[i]))$$

Hence, using II.2.10, the Leibniz rule, and specialization,

$$\vdash z < b \rightarrow (\neg lh(z) = x \vee (\exists i)_{<x} \neg(z)_i = t[i]) \qquad \qquad \square$$

Item (3) above suggests how to test a number for being a sequence code or not. We define (exactly as on p. 165, but in boldface)

II.2.12 Definition. We introduce the unary predicate **Seq** by

$$\textbf{Seq}(z) \leftrightarrow (\forall i)_{<lh(z)} \beta(z, Si) > 0 \wedge$$
$$(\forall x)_{<z} (lh(x) \neq lh(z) \vee (\exists i)_{<lh(z)} (z)_i \neq (x)_i) \qquad \square$$

The first conjunct above tests that we did not forget to add 1 to the sequence members before coding. The second conjunct says that our code z is *minimum*, because, for any smaller "number" x, whatever sequence the latter may code ("minimally" or not) cannot be the same sequence as the one that z (minimally) codes.

II.2.13 Proposition. $\vdash \textbf{Seq}(z) \rightarrow (\forall i)_{<lh(z)} z > (z)_i$.

Proof. Assume the hypothesis. Specialization and II.2.12 yield

$$\vdash i < lh(z) \rightarrow \beta(z, Si) > 0 \tag{1}$$

By II.1.10, $\vdash \beta(z, Si) > 0 \rightarrow \beta(z, Si) \geq S0$; thus, by II.1.27 and II.2.10, (1) yields

$$\vdash i < lh(z) \rightarrow \beta(z, Si) = (z)_i + S0$$

By $+2$ and $+1$ we now get

$$\vdash i < lh(z) \rightarrow \beta(z, Si) = S((z)_i)$$

which, by II.2.4(i), rests our case.   □

II.2.14 Remark. The inequality

$$(z)_i < z \qquad\qquad (i)$$

is very important when it comes to doing induction on sequence codes and (soon) on Gödel numbers.

By Example II.2.7 we see that unless we do something about it, we are *not* justified in expecting $\vdash \beta(z, i) < z$, in general. What we did to ensure that induction on codes *is* possible was to add 1 to each sequence member $t[i]$, a simple trick that was already employed in the prime-power coding (p. 136), although for different reasons there.[†] This device ensures (i) by II.2.13.

Another way to ensure (i) is to invoke Corollary II.2.6 and modify (*FC*) to read instead

$$(\mu z)(\beta(z, 0) = w \wedge w < z \wedge (\forall i)_{<w}(\beta(z, Si) < z \wedge \beta(z, Si) = t[i]))$$

Note that we did not need to "add 1" to t above.

We prefer our first solution (Definition II.2.8), if nothing else, because it allows **0** to code the "empty sequence" (see II.2.15 below), a fact that is intuitively pleasing.

This is a good place to mention that while our "β" (in either the bold or the lightface version) is, essentially, that in Shoenfield (1967), we had to tweak the latter somewhat (especially the derived "$\langle \ldots \rangle$") to get it to be induction-friendly (in particular, to have (i) above).

The version in Shoenfield (1967, (5), p. 116) is

$$\begin{aligned}
\widetilde{\beta}(a, i) = (\mu x)(x = a\ \vee \\
(\exists y)_{<a}(\exists z)_{<a}(a = J'(y, z) \wedge S(zSJ'(x, i)) \,|\, y))
\end{aligned} \qquad (Sh)$$

where[‡]

$$J'(y, z) = (y + z)^2 + y + \widetilde{1}$$

Since, intuitively speaking, 4 is not in the range of J', i.e., formally,

$$\vdash \neg(\exists y)(\exists z)J'(y, z) = \widetilde{4}$$

[†] Namely, to enable us to *implicitly* store in the code the length of the coded sequence.
[‡] The two "J"'s, the one employed here and Shoenfield's J', trivially satisfy $\vdash J'(y, z) = SJ(y, z)$.

we have

$$\vdash \tilde{\beta}(\tilde{4}, i) = \tilde{4}$$

because the search $(\exists y)_{<a}(\exists z)_{<a}(\dots)$ in the defining axiom (Sh) fails. This invalidates (7), p. 116 of Shoenfield (1967),[†] on which (i) hinges in that book.[‡] □

II.2.15 Example. Since $\vdash 0 = \beta(0, 0)$ by II.2.7, we have $\vdash lh(0) = 0$, but also, by < 1, that 0 is a *minimum code* of a sequence. Indeed, $\vdash Seq(0)$ by < 1.

Since the sequence in question has 0 length, it is called the *empty sequence*. We often write the *minimum code*, (FC), of the empty sequence as "$\langle\,\rangle$", that is,

$$\vdash \langle\,\rangle = 0 \qquad\qquad \square$$

To conclude with the introduction of our formal coding tools we will also need a formal concatenation function.

II.2.16 Definition (Concatenation). "$*$" – a 2-ary function symbol – is introduced via the μ-term below (denoted by "$x * y$"):

$$x * y = (\mu z)(\beta(z, 0) = lh(x) + lh(y) \wedge$$
$$(\forall i)_{<lh(x)}\beta(z, Si) = S((x)_i) \wedge \qquad (***)$$
$$(\forall i)_{<lh(y)}\beta(z, Si + lh(x)) = S((y)_i))$$

$$\square$$

The legitimacy of $(***)$ relies on

II.2.17 Proposition.

$$\vdash (\exists w)(\beta(w, 0) = lh(x) + lh(y) \wedge$$
$$(\forall i)_{<lh(x)}\beta(w, Si) = S((x)_i) \wedge \qquad (1)$$
$$(\forall i)_{<lh(y)}\beta(w, Si + lh(x)) = S((y)_i))$$

Since the x and y are free variables, (1) says, informally, that for all "values" of x and y a w that "works" exists. That is, the natural interpretation of $*$ over $\mathfrak{N}$ is *total*. In other words, just like the "$*$" that we saw earlier on, which was based on prime power coding (see I.8.30, p. 136), this new "$*$" makes sense regardless of whether or not its arguments are minimum codes according to (FC).

---

† That $\vdash \neg a = 0 \rightarrow \tilde{\beta}(a, i) < a$.

‡ (8) in *loc. cit.*, p. 117 [deduced from (7) in *loc. cit.*]: $\vdash a \neq 0 \rightarrow lh(a) < a \wedge (a)_i < a$.

Proof. To see why (1) holds, we introduce the 3-ary function symbol a by

$$a(n,x,y) = \begin{cases} lh(x) + lh(y) & \text{if } n = 0 \\ S((x)_{\delta(n,S0)}) & \text{if } 0 < n \wedge n \leq lh(x) \\ S((y)_{\delta(n,S(lh(x)))}) & \text{if } lh(x) < n \end{cases} \qquad (A)$$

By II.2.5,

$$\vdash (\exists w)(\forall i)(i \leq lh(x) + lh(y) \rightarrow \beta(w,i) = a(i,x,y))$$

Let us add a new constant, z, and the assumption

$$(\forall i)(i \leq lh(x) + lh(y) \rightarrow \beta(z,i) = a(i,x,y)) \qquad (2)$$

We now show that we can prove (1) from (2).

> It may be that it is *not* the case that $\vdash Seq(z)$. This is just fine, as the proof needs no such assumption.

We verify each conjunct of (1) separately:

- Since (A) yields (p. 221) $\vdash n = 0 \rightarrow a(n,x,y) = lh(x) + lh(y)$ and hence

$$\vdash a(0,x,y) = lh(x) + lh(y)$$

  (2), specialization, and II.1.5 yield

$$\vdash \beta(z,0) = lh(x) + lh(y) \qquad (3)$$

- Next we prove

$$\vdash i < lh(x) \rightarrow \beta(z,Si) = S((x)_i) \qquad (4)$$

  By II.1.10 this amounts to proving

$$\vdash Si \leq lh(x) \rightarrow \beta(z,Si) = S((x)_i) \qquad (4')$$

  Now, since $\vdash lh(x) \leq lh(x) + lh(y)$, (2) yields

$$\vdash Si \leq lh(x) \rightarrow \beta(z,Si) = a(Si,x,y)$$

  Moreover, (A) and $\vdash 0 < Si$ yield (p. 221)

$$\vdash Si \leq lh(x) \rightarrow a(Si,x,y) = S((x)_{\delta(Si,S0)})$$

  and since $\vdash \delta(Si,S0) = i$, we get $(4')$ above.

- Finally, we want to verify

$$\vdash i < lh(y) \rightarrow \beta(z, Si + lh(x)) = S((y)_i) \qquad (5)$$

which amounts to

$$\vdash Si \le lh(y) \rightarrow \beta(z, Si + lh(x)) = S((y)_i) \tag{5'}$$

using II.1.10. Add $Si \le lh(y)$. Hence $\vdash Si + lh(x) \le lh(x) + lh(y)$ by II.1.17. Thus, by (2),

$$\vdash \beta(z, Si + lh(x)) = a(Si + lh(x), x, y) \tag{6}$$

By (A),

$$\vdash lh(x) < Si + lh(x)$$
$$\rightarrow a(Si + lh(x), x, y) = S((y)_{\delta(Si+lh(x),S(lh(x)))}) \tag{7}$$

Hence, noting that

$$\vdash lh(x) < Si + lh(x) \qquad \text{(by II.1.17)}$$

$$\vdash Si + lh(x) = i + S(lh(x))$$

and

$$\vdash \delta(i + S(lh(x)), S(lh(x))) = i \qquad \text{(by II.1.27)}$$

(6) and (7) yield

$$\vdash \beta(z, Si + lh(x)) = S((y)_i)$$

which by the deduction theorem yields (5′) and hence (5).

Putting now the conjuncts (3)–(5) together and applying **Ax2** yields (1). □

 II.2.18 Example. We verify that for any terms t and s

$$\vdash \langle t[i] : i < n \rangle * \langle s[i] : i < m \rangle = \langle a[i] : i < n + m \rangle$$

where

$$a[i] = \begin{cases} t[i] & \text{if } i < n \\ s[\delta(i,n)] & \text{if } n \le i \wedge i < n + m \\ 0 & \text{otherwise} \end{cases}$$

Setting $x = \langle t[i] : i < n \rangle$ and $y = \langle s[i] : i < m \rangle$ for convenience, we get from II.2.11

$$\vdash lh(x) = n$$

$$\vdash lh(y) = m$$

$$\vdash i < n \rightarrow (x)_i = t[i]$$

and

$$\vdash i < m \rightarrow (y)_i = s[i]$$

Pause. (Important.) None of the above four facts need the assertion that x and y are minimum "$\langle \ldots \rangle$-codes". We have invoked above only the part of II.2.11 that does *not* rely on minimality. Only the last claim in II.2.11 does.

Thus, by II.2.16,

$$\vdash x * y = (\mu z)(lh(z) = n + m \wedge \\ (\forall i)_{<n} \beta(z, Si) = St[i] \wedge \\ (\forall i)_{<m} \beta(z, Si + n) = Ss[i]) \qquad (8)$$

On the other hand,

$$\vdash \langle a[i] : i < n + m \rangle = (\mu w)(lh(w) = n + m \wedge \\ (\forall i)_{<n} \beta(w, Si) = St[i] \wedge \\ (\forall i)_{<m} \beta(w, Si + n) = Ss[i]) \qquad (9)$$

by the way a was defined (noting that $\vdash n \leq i + n$ and $\vdash \delta(i + n, n) = i$). By (8) and (9), we are done.

Pause. Parting comment: From our earlier Pause, we see that even in the case when x and y are *not* the minimum $\langle \ldots \rangle$-codes for the sequences $t[i] : i < n$ and $s[i] : i < m$ respectively, but nevertheless happen to satisfy the following,

$$\vdash lh(x) = n$$

$$\vdash lh(y) = m$$

$$\vdash i < n \rightarrow (x)_i = t[i]$$

and

$$\vdash i < m \rightarrow (y)_i = s[i]$$

Then the work immediately above still establishes

$$\vdash x * y = \langle a[i] : i < n + m \rangle \qquad \square$$

II.2.19 Proposition. *If we let b abbreviate $\langle t[i] : i < x \rangle$, then we have*

$$\vdash Seq(b)$$

Proof. Immediate from II.2.8, II.2.11, and II.2.12. $\qquad \square$

II.2.20 Exercise. $\vdash Seq(x * y)$. $\qquad \square$

II.2.21 Example. We introduce the abbreviation t below:

$$t = (\mu z)(Seq(z) \wedge$$
$$lh(z) = lh(x) + lh(y) \wedge$$
$$(\forall i)_{<lh(x)}(z)_i = (x)_i \wedge \tag{1}$$
$$(\forall i)_{<lh(y)}(z)_{i+lh(x)} = (y)_i)$$

and prove

$$\vdash t = x * y$$

By the definition of "$*$" (II.2.16) and $(f^{(4)})$ (p. 218) – using

$$\vdash x = Sy \rightarrow \delta(x, S0) = y$$

(by II.1.27) and II.2.10 to remove instances of β – we obtain

$$\vdash lh(x * y) = lh(x) + lh(y) \wedge$$
$$(\forall i)_{<lh(x)}(x * y)_i = (x)_i \wedge \tag{2}$$
$$(\forall i)_{<lh(y)}(x * y)_{i+lh(x)} = (y)_i$$

Since also $\vdash Seq(x * y)$, by II.2.20 we get

$$\vdash t \leq x * y \tag{3}$$

by (2), (1), and $(f^{(5)})$ (p. 218).

To get the converse inequality, we use (1) and $(f^{(4)})$ to get

$$\vdash Seq(t) \wedge$$
$$lh(t) = lh(x) + lh(y) \wedge$$
$$(\forall i)_{<lh(x)}(t)_i = (x)_i \wedge \tag{4}$$
$$(\forall i)_{<lh(y)}(t)_{i+lh(x)} = (y)_i$$

The first conjunct of (4) and II.2.12 (via II.1.27) yield,[†] from the remaining conjuncts of (4),

$$\vdash \beta(t, 0) = lh(x) + lh(y) \wedge$$
$$(\forall i)_{<lh(x)}\beta(t, Si) = S((x)_i) \wedge$$
$$(\forall i)_{<lh(y)}\beta(t, Si + lh(x)) = S((y)_i)$$

By the definition of "$*$" (II.2.16) and $(f^{(5)})$,

$$\vdash x * y \leq t$$

which along with (3) and II.1.8 rests the case. $\square$

[†] More expansively, $\vdash (\forall i)_{<lh(t)}\beta(t, Si) \geq S(0)$ by II.2.12, which implies (by II.1.27 and II.2.10) $\vdash (\forall i)_{<lh(t)}\beta(t, Si) = (t)_i + S0$, but $\vdash z + S0 = Sz$.

II.2.22 Example. By II.2.18,

$$\vdash 0 * \langle t[i] : i < x \rangle = \langle t[i] : i < x \rangle$$

and

$$\vdash \langle t[i] : i < x \rangle * 0 = \langle t[i] : i < x \rangle$$

In particular, $\vdash 0 * \langle\,\rangle = \langle\,\rangle$ (cf. II.2.15), i.e., $\vdash 0 * 0 = 0$. Note however that if a is the term that denotes the *natural* coding of the empty sequence, that is, $\vdash \beta(a, 0) = 0$ and $\vdash 0 < a$ (note that $\vdash \neg Seq(a)$), then

$$\vdash 0 * a = 0$$

since $\vdash Seq(0 * a)$, and therefore

$$\vdash \neg 0 * a = a \qquad \qquad \square$$

Thus we have *two* ways to "β-code" sequences $t[i]$, for $i < x$. One is to pick *any* c that satisfies $\vdash \beta(c, 0) = x$ and $\vdash (\forall i)_{<x}\beta(c, i) = t[i]$. The other is to employ *the* minimum "$\langle \dots \rangle$-code", $b = \langle t[i] : i < x \rangle$. By II.2.19, b (but not necessarily c) "satisfies" *Seq*.

II.3. Formal Primitive Recursion

II.3.1 Theorem (Primitive Recursive Definitions). *Let h and g be n-ary and $(n + 2)$-ary function symbols. Then we may introduce a new $(n + 1)$-ary function symbol, f, such that*

$$\vdash f(0, \vec{y}) = h(\vec{y})$$

and

$$\vdash f(Sx, \vec{y}) = g(x, \vec{y}, f(x, \vec{y}))$$

Proof. We prove

$$\vdash (\exists z)\Big(Seq(z) \wedge lh(z) = Sx \wedge (z)_0 = h(\vec{y}) \tag{1}$$
$$\wedge (\forall i)_{<x}((z)_{Si} = g(i, \vec{y}, (z)_i))\Big)$$

This is done by formal induction on x.

For $x = 0$, taking $z = \langle h(\vec{y}) \rangle$ works (this invokes < 1 and **Ax2**).

Assume now (1) for frozen variables (the I.H.). For the induction step we want to show

$$\vdash (\exists z)\Big(Seq(z) \wedge lh(z) = SSx \wedge (z)_0 = h(\vec{y}) \tag{2}$$
$$\wedge (\forall i)_{<Sx}((z)_{Si} = g(i, \vec{y}, (z)_i)) \Big)$$

To this end, let a be a new constant, and add the assumption (invoking I.4.27 and (1))

$$\vdash Seq(a) \wedge lh(a) = Sx \wedge (a)_0 = h(\vec{y}) \tag{3}$$
$$\wedge (\forall i)_{<x}((a)_{Si} = g(i, \vec{y}, (a)_i))$$

Set now

$$b = a * \langle g(x, \vec{y}, (a)_x) \rangle$$

By II.2.21 and $(f^{(4)})$ (p. 218)

$$\vdash Seq(b) \wedge$$
$$lh(b) = SSx \wedge$$
$$(\forall i)_{<Sx}(b)_i = (a)_i \wedge \tag{4}$$
$$(b)_{Sx} = g(x, \vec{y}, (a)_x)$$

the last conjunct being distilled from

$$\vdash i < S0 \rightarrow (b)_{i+Sx} = g(x, \vec{y}, (a)_x)$$

and $\vdash i < S0 \leftrightarrow i = 0$.

Thus, using (3) and the Leibniz axiom (**Ax4**),

$$\vdash Seq(b) \wedge$$
$$lh(b) = SSx \wedge$$
$$(b)_0 = h(\vec{y}) \wedge$$
$$(\forall i)_{<x}(b)_{Si} = g(i, \vec{y}, (b)_i) \wedge$$
$$(b)_{Sx} = g(x, \vec{y}, (b)_x) \qquad (\text{since } \vdash (b)_x = (a)_x \text{ by } (4))$$

In short, using $(\text{I}.7.2) \vdash (b)_{Sx} = g(x, \vec{y}, (b)_x) \leftrightarrow (\forall i)_{i=x}(b)_{Si} = g(i, \vec{y}, (b)_i)$, $\forall$-distribution over $\wedge$, and < 2, we have

$$\vdash Seq(b) \wedge lh(b) = SSx \wedge (b)_0 = h(\vec{y}) \wedge (\forall i)_{<Sx}(b)_{Si} = g(i, \vec{y}, (b)_i)$$

This proves (2) by **Ax2** and concludes the inductive proof of (1). We can now let

$$F(x, \vec{y}) = (\mu z)\Big(Seq(z) \wedge lh(z) = Sx \wedge$$
$$(z)_0 = h(\vec{y}) \wedge (\forall i)_{<x}((z)_{Si} = g(i, \vec{y}, (z)_i)) \Big)$$

where F is a new function symbol.

By $(f^{(4)})$ (p. 218),

$$\vdash (F(x, \vec{y}))_0 = h(\vec{y})$$
$$\vdash i < x \rightarrow (F(x, \vec{y}))_{Si} = g(i, \vec{y}, (F(x, \vec{y}))_i) \tag{5}$$

The first part of (5) yields (by substitution)

$$\vdash (F(0, \vec{y}))_0 = h(\vec{y}) \tag{6}$$

The second yields (by substitution $[x \leftarrow Si]$)

$$\vdash (F(Si, \vec{y}))_{Si} = g(i, \vec{y}, (F(Si, \vec{y}))_i) \tag{7}$$

We now claim that

$$\vdash (F(Si, \vec{y}))_i = (F(i, \vec{y}))_i \tag{8}$$

It is convenient to prove a bit more, by induction on i, namely, that

$$\vdash (\forall x)(i < x \rightarrow (F(i, \vec{y}))_i = (F(x, \vec{y}))_i) \tag{9}$$

The case for $i = 0$ follows from (6) and the first of (5).

We now assume (9) as our I.H., *freezing the free variables* (i and $\vec{y}$).

For our induction step we need to prove

$$(\forall x)(Si < x \rightarrow (F(Si, \vec{y}))_{Si} = (F(x, \vec{y}))_{Si})$$

but we prove instead

$$Si < x \rightarrow (F(Si, \vec{y}))_{Si} = (F(x, \vec{y}))_{Si}$$

To this end, we assume $Si < x$ – freezing x – and proceed to verify

$$\vdash (F(Si, \vec{y}))_{Si} = (F(x, \vec{y}))_{Si} \tag{10}$$

Well, since also $\vdash i < x$ and $\vdash i < Si$, the I.H. (9) yields by specialization

$$\vdash (F(i, \vec{y}))_i = (F(x, \vec{y}))_i$$

and

$$\vdash (F(i, \vec{y}))_i = (F(Si, \vec{y}))_i$$

Therefore,

$$\vdash g(i, \vec{y}, (F(Si, \vec{y}))_i) = g(i, \vec{y}, (F(x, \vec{y}))_i)$$

which, by the second part of (5) and (7), verifies (10).

Having thus established (9) and therefore (8), (7) becomes

$$\vdash (F(Si, \vec{y}))_{Si} = g(i, \vec{y}, (F(i, \vec{y}))_i)$$

All this proves that if f is a new function symbol introduced by

$$f(x, \vec{y}) = (F(x, \vec{y}))_x$$

then the two formulas stated in the theorem are proved. $\square$

II.3.2 Exercise. Verify that if h and g are μ-defined from $\Delta_0(L_{\mathfrak{A}})$ formulas, then so is f. $\square$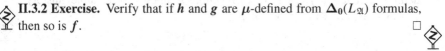

II.3.3 Theorem (Course-of-Values Recursion). *Let h and g be function symbols of arities n and $n + 2$ respectively. Then we may introduce a new function symbol H of arity $n + 1$ such that*

$$\vdash (H(0, \vec{y}))_0 = h(\vec{y})$$
$$\vdash (H(Sx, \vec{y}))_{Sx} = g(x, \vec{y}, H(x, \vec{y})) \tag{1}$$
$$\vdash Seq(H(x, \vec{y})) \wedge lh(H(x, \vec{y})) = Sx$$

Proof. Invoking II.3.1, we introduce a new function symbol H by the primitive recursion below, and show that it works:

$$H(0, \vec{y}) = \langle h(\vec{y}) \rangle$$
$$H(Sx, \vec{y}) = H(x, \vec{y}) * \langle g(x, \vec{y}, H(x, \vec{y})) \rangle \tag{2}$$

The third formula in the group (1) is proved by induction on x. The basis is immediate from the basis of the recursion (2) (note that a "$\langle \ldots \rangle$" term satisfies *Seq* by II.2.19).

Assume now the contention for frozen x. By II.2.21 and the second formula in group (2),

$$\vdash lh(H(Sx, \vec{y})) = SSx$$

and

$$\vdash Seq(H(Sx, \vec{y}))$$

This concludes the induction. The other two contentions in the theorem are direct consequences of the two recurrence equations (2). $\square$

II.3.4 Remark. Introducing yet another function symbol, f, by

$$f(x, \vec{y}) = (H(x, \vec{y}))_x$$

yields at once

$$\vdash f(0, \vec{y}) = h(\vec{y})$$
$$\vdash f(Sx, \vec{y}) = g(x, \vec{y}, H(x, \vec{y}))$$

This is the standard way that course-of-values recursion is presented, i.e., defining a function f from known functions h and g and from the "history", $H(x, \vec{y}) = \langle f(i, \vec{y}) : i \leq x \rangle$, of f. □

We recall the concept of a *term-definable* (total) function (p. 181). We can now prove

II.3.5 Theorem. *Every primitive recursive function is term-definable in (some recursive extension of)* **PA**.

Proof. The proof is by (informal) induction on $\mathfrak{PR}$. For the basis, we already know that the initial functions are term-definable in **ROB**, a subtheory of **PA**.

Let now $f, g_1, \ldots, g_n$ be defined by the terms $\boldsymbol{f}, \boldsymbol{g_1}, \ldots, \boldsymbol{g_n}$. We will argue that the term $\boldsymbol{f}(\boldsymbol{g_1}(\vec{\boldsymbol{y}}_m), \ldots, \boldsymbol{g_n}(\vec{\boldsymbol{y}}_m))$ defines $\lambda \vec{y}_m . f(g_1(\vec{y}_m), \ldots, g_n(\vec{y}_m))$.

So let $f(g_1(\vec{a}_m), \ldots, g_n(\vec{a}_m)) = b$ be true. Then, for appropriate $\vec{c}_n$,

$$f(c_1, \ldots, c_n) = b$$
$$g_1(\vec{a}_m) = c_1$$
$$\vdots$$
$$g_n(\vec{a}_m) = c_n$$

By the induction hypothesis,

$$\vdash \boldsymbol{f}(\widetilde{c_1}, \ldots, \widetilde{c_n}) = \widetilde{b}$$
$$\vdash \boldsymbol{g_1}(\widetilde{a_1}, \ldots, \widetilde{a_m}) = \widetilde{c_1}$$
$$\vdots$$
$$\vdash \boldsymbol{g_n}(\widetilde{a_1}, \ldots, \widetilde{a_m}) = \widetilde{c_n}$$

Hence (via **Ax4**)

$$\vdash \boldsymbol{f}(\boldsymbol{g_1}(\widetilde{a_1}, \ldots, \widetilde{a_m}), \ldots, \boldsymbol{g_n}(\widetilde{a_1}, \ldots, \widetilde{a_m})) = \widetilde{b}$$

Let finally h and g be term-defined by $\boldsymbol{h}$ and $\boldsymbol{g}$ respectively, and let f be given by the schema below (for all $x, \vec{y}$):

$$f(0, \vec{y}) = h(\vec{y})$$
$$f(x + 1, \vec{y}) = g(x, \vec{y}, f(x, \vec{y}))$$

We verify that the term $f(x, \vec{y})$, where f is introduced by formal primitive recursion below, defines f:

$$f(0, \vec{y}) = h(\vec{y})$$
$$f(x + 1, \vec{y}) = g(x, \vec{y}, f(x, \vec{y})) \tag{1}$$

To this end, we prove by (informal) induction on a that[†]

$$f(a, \vec{b}) = c \quad \text{implies} \quad \vdash f(\tilde{a}, \tilde{\vec{b}}) = \tilde{c} \tag{2}$$

Let $a = 0$. Then $f(0, \vec{b}) = c$ entails $h(\vec{b}) = c$; hence $\vdash h(\tilde{\vec{b}}) = \tilde{c}$ by the I.H. (of the $\mathfrak{PR}$-induction).

By the first equation of group (1), $\vdash f(0, \tilde{\vec{b}}) = \tilde{c}$, which settles the basis of the a-induction. Now fix a, and take (2) as the I.H. of the a-induction. We will argue the case of (2) when a is replaced by $a + 1$.

Let $f(a + 1, \vec{b}) = c$. Then $g(a, \vec{b}, d) = c$, where $f(a, \vec{b}) = d$. By the I.H. of the $\mathfrak{PR}$-induction and a-induction, i.e., (2),

$$\vdash f(\tilde{a}, \tilde{\vec{b}}) = \tilde{d}$$

and

$$\vdash g(\tilde{a}, \tilde{\vec{b}}, \tilde{d}) = \tilde{c}$$

Thus, by **Ax4** and the second equation of group (1),

$$\vdash f(S\tilde{a}, \tilde{\vec{b}}) = \tilde{c} \qquad \square$$

The above suggests the following definition.

II.3.6 Definition (Formal Primitive Recursive Functions). A defined function symbol f is *primitive recursive* iff there is a sequence of function symbols $g_1, \ldots, g_p$ such that $g_p \equiv f$ and, for all $i = 1, \ldots, p$, g_i has been introduced by a *primitive recursive definition*

$$g_i(0, v_1, \ldots, v_{n-1}) = g_j(v_1, \ldots, v_{n-1})$$
$$g_i(Sv_0, v_1, \ldots, v_{n-1}) = g_k(v_0, v_1, \ldots, v_{n-1}, g_i(v_0, \ldots, v_{n-1}))$$

where $j < i$ and $k < i$ and the g_j and g_k have arities $n - 1$ and $n + 1$ respectively, or by *composition*, that is, an explicit definition such as

$$g_i(v_0, \ldots, v_{n-1}) = g_{j_0}(g_{j_1}(v_0, \ldots, v_{n-1}), \ldots, g_{j_r}(v_0, \ldots, v_{n-1}))$$

[†] $\tilde{\vec{b}}$ means $\tilde{b}_1, \tilde{b}_2, \ldots$.

254   *II. The Second Incompleteness Theorem*

using, again, *previous* function symbols (of the correct arities) in the sequence (i.e., $j_m < i$, $m = 0, \ldots, r$). Otherwise, g_i is one of S, Z (*zero function symbol*), or U_i^n ($n > 0$, $1 \leq i \leq n$) (*projection function symbols*), where the latter two symbols are introduced by the defining axioms

$$Z(v_0) = 0$$

and

$$U_i^n(v_0, \ldots, v_{n-1}) = v_{i-1} \qquad \text{for each positive } n \text{ and } 1 \leq i \leq n \text{ in } \mathbb{N}.$$

A sequence such as $g_1, \ldots, g_p$ is a *formal (primitive recursive) derivation* of g_p.

A *term* is *primitive recursive* iff it is defined from $\mathbf{0}$, variables, and primitive recursive function symbols according to the standard definition of "term".

A *predicate* P is primitive recursive iff there is a primitive recursive function symbol f of the same arity as P such that $P(\vec{x}) \leftrightarrow f(\vec{x}) = \mathbf{0}$ is provable.

By a slip of the tongue, we may say that P is primitive recursive iff its formal characteristic function, χ_P, is. (See however II.1.25, p. 220.) □

II.3.7 Remark. In order that the primitive recursive derivations do not monopolize our supply of function symbols, we can easily arrange that primitive recursive function symbols are chosen from an appropriate *subsequence* of the "standard function symbol sequence f_i^n" of p. 166. We fix the following very convenient scheme that is informed by the table on p. 138.

Let a be the *largest f-index* used to allocate all the finitely many function symbols required by $L_{\mathfrak{N}}$ (three) and the finitely many required for the introduction of β and the $\langle \ldots \rangle$ coding.

We let, for convenience, $b = a + 1$, and we allocate the formal primitive recursive function symbols from among the members of the subsequence $(f_{bk}^n)_{k \geq 0}$, being very particular about the k-value (k-*code*) chosen (p. 138):

(1) $k = \langle 0, 1, 0 \rangle$ is used for Z ($n = 1$, of course).
(2) $k = \langle 0, 1, 1 \rangle$ is used for S.[†]
(3) $k = \langle 0, n, i, 2 \rangle$ is used for U_i^n. That is, U_i^n is allocated as

$$f_{b\langle 0,n,i,2 \rangle}^n$$

[†] It does no harm that S is already allocated as f_0^1. After all, every "real" primitive recursive function – i.e., function viewed *extensionally* as a set of input-output pairs – has infinitely many different derivations, and hence infinitely many function symbols allocated to it.

(4) $k = \langle 1, m, f, g_1, \ldots, g_n \rangle$ is used if f_{bk}^m is allocated to denote the result of composition from function symbols (already allocated) with k-codes equal to $f, g_1, \ldots, g_n$. Of these, the symbol with code f must have arity n; all the others, arity m.

(5) $k = \langle 2, n + 1, h, g \rangle$ is used if f_{bk}^{n+1} is allocated to denote the result of primitive recursion from function symbols (already allocated) with k-codes h, g. Of these the first must have arity n, the second $n + 2$.

This allocation scheme still leaves an infinite supply of unused (so far) symbols to be used for future extensions of the language.

As a parting comment we note that the seemingly contrived allocation scheme above forces the set of k-codes to be primitive recursive. (See Exercise II.13). □

By the proof of II.3.5, having the formal and informal derivations of the boldface f and the lightface f "track each other" in the obvious way – i.e., assembling the boldface version exactly as the lightface version was assembled – we obtain $f^{\mathfrak{N}'} = f$, and

II.3.8 Theorem. *If f is a formal primitive recursive function of arity n (in $L_{\mathfrak{N}'}$), then, for all $\vec{a}_n$ in $\mathbb{N}^n$,*

$$\models_{\mathfrak{N}'} f(\widetilde{\vec{a}}_n) = \widetilde{b} \quad implies \quad \vdash f(\widetilde{\vec{a}}_n) = \widetilde{b}$$

II.3.9 Corollary. *The "implies" in II.3.8 can be strengthened to "iff". That is, every informal primitive recursive function is strongly term-definable (p. 187), indeed by a primitive recursive term.*

Proof. Assume $\vdash f(\widetilde{\vec{a}}_n) = \widetilde{b}$. Writing f for $f^{\mathfrak{N}'}$, assume $f(\vec{a}_n) = c \neq b$. Then also $\vdash f(\widetilde{\vec{a}}_n) = \widetilde{c}$; hence $\vdash \widetilde{b} = \widetilde{c}$. Now, $b \neq c$ yields (already in **ROB**) $\vdash \neg \widetilde{b} = \widetilde{c}$, contradicting consistency of **PA**'. □

 We are reminded that **ROB** and **PA** are consistent (since they each have a model), a fact that we often use implicitly.[†]

It is trivial then that each recursive extension **PA**' is also consistent, for such an extension is conservative.

[†] We base this assertion, of course, on the existence of a standard model $\mathfrak{N}$, a fact that provides a non-constructive proof of consistency. Constructive proofs of the consistency of **ROB** and **PA** are also known. See for example Shoenfield (1967), Schütte (1977).

II.3.10 Corollary. *For every closed primitive recursive term t there is a unique $n \in \mathbb{N}$ such that $\vdash t = \tilde{n}$.*

Proof. Existence is by II.3.8 (use $n = t^{\mathfrak{N}'}$). Uniqueness follows from definability of $x = y$ in **ROB** (I.9.39, p. 181) and hence in any extension **PA'**. □

Pause. Is it true that *every* closed term in a recursive extension of **PA** is provably equal to a numeral?

II.3.11 Corollary. *Every primitive recursive relation is strongly definable by some formal primitive recursive predicate.*

II.4. The Boldface Δ and Σ

II.4.1 Definition. Let $L_{\mathfrak{A}}$ be a language of arithmetic. The symbol $\Sigma_1(L_{\mathfrak{A}})$ denotes the set of formulas $\{(\exists x)\mathscr{A} : \mathscr{A} \in \Delta_0(L_{\mathfrak{A}})\}$.

If the language is $L_{\mathfrak{N}}$, then we simply write Σ_1. □

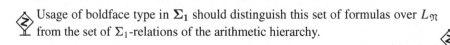 Usage of boldface type in Σ_1 should distinguish this set of formulas over $L_{\mathfrak{N}}$ from the set of Σ_1-relations of the arithmetic hierarchy.

We also define variants of Δ_0 and Σ_1 above.

II.4.2 Definition. Let $L_{\mathfrak{A}}$ be a language of arithmetic. The symbol $\Delta_0^+(L_{\mathfrak{A}})$ denotes the smallest set of formulas over $L_{\mathfrak{A}}$ that includes the atomic formulas, but also the negations of atomic formulas, and moreover satisfies the closure conditions: If $\mathscr{A}$ and $\mathscr{B}$ are in $\Delta_0^+(L_{\mathfrak{A}})$, then so are $\mathscr{A} \vee \mathscr{B}$, $\mathscr{A} \wedge \mathscr{B}$, $(\exists x)_{<t}\mathscr{A}$, and $(\forall x)_{<t}\mathscr{A}$ (where we require that the variable x not occur in the term t).

If the language is $L_{\mathfrak{N}}$, then we simply write Δ_0^+. □

II.4.3 Definition. Let $L_{\mathfrak{A}}$ be a language of arithmetic. The symbol $\Delta_0'(L_{\mathfrak{A}})$ denotes the smallest set of formulas over $L_{\mathfrak{A}}$ that includes the *restricted atomic formulas* (defined below) – and their negations – and moreover satisfies the closure conditions: If $\mathscr{A}$ and $\mathscr{B}$ are in $\Delta_0'(L_{\mathfrak{A}})$, then so are $\mathscr{A} \vee \mathscr{B}$, $\mathscr{A} \wedge \mathscr{B}$, $(\exists x)_{<y}\mathscr{A}$ and $(\forall x)_{<y}\mathscr{A}$ (where $x \not\equiv y$).

Correspondingly, the symbol $\Sigma_1'(L_{\mathfrak{A}})$ denotes the set of formulas $\{(\exists x)\mathscr{A} : \mathscr{A} \in \Delta_0'(L_{\mathfrak{A}})\}$.

If the language is $L_{\mathfrak{N}}$, then we simply write Δ_0' and Σ_1'.

Now, the *restricted atomic formulas* over $L_{\mathfrak{A}}$ are $x = y$, $0 = y$, $f\vec{x}_n = y$, and $P\vec{x}_n$ for all n-ary function (f) and predicate (P) symbols. □

The restriction on the atomic formulas above was to have function and predicate letters act on variables (rather than arbitrary terms). Similarly, we have used $(\exists x)_{<y}$ and $(\forall x)_{<y}$ rather than $(\exists x)_{<t}$ and $(\forall x)_{<t}$ in II.4.3.

The superscript "+" in II.4.2 is indicative of the (explicit) presence of only *positive* closure operations ($\neg$ does not participate in the definition). The same is true of the $\Delta_0'(L_{\mathfrak{A}})$ formulas.

It turns out that both $\Delta_0^+(L_{\mathfrak{A}})$ and $\Delta_0'(L_{\mathfrak{A}})$ formulas are closed under negation (in the definition of these sets of formulas the application of "$\neg$" has been pushed as far to the right as possible). We prove this contention below. The introduction of $\Delta_0^+(L_{\mathfrak{A}})$ and $\Delta_0'(L_{\mathfrak{A}})$ is only offered for convenience (proof of II.4.12 below). Neither symbol is standard in the literature.

II.4.4 Lemma. *For any $\mathscr{A} \in \Delta_0^+(L_{\mathfrak{A}})$ (respectively, $\mathscr{A} \in \Delta_0'(L_{\mathfrak{A}})$) there is a $\mathscr{B} \in \Delta_0^+(L_{\mathfrak{A}})$ (respectively, $\mathscr{B} \in \Delta_0'(L_{\mathfrak{A}})$) such that $\vdash \neg\mathscr{A} \leftrightarrow \mathscr{B}$, where "$\vdash$" denotes logical (pure) provability.*

Proof. We do induction on $\Delta_0^+(L_{\mathfrak{A}})$ (respectively, $\Delta_0'(L_{\mathfrak{A}})$; the induction variable is $\mathscr{A}$).

Basis. If $\mathscr{A}$ is atomic (or restricted atomic), then we are done at once. If it is a negated atomic (or negated restricted atomic) formula, then we are done by $\models_{\text{Taut}} \mathscr{B} \leftrightarrow \neg\neg\mathscr{B}$.

Now $\mathscr{A}$ can have the following forms (if not atomic or negated atomic) by II.4.2–II.4.3:

(i) $\mathscr{A} \equiv \mathscr{B} \vee \mathscr{C}$. Then $\vdash \neg\mathscr{A} \leftrightarrow \neg\mathscr{B} \wedge \neg\mathscr{C}$, and we are done by the I.H. via the Leibniz rule.

(ii) $\mathscr{A} \equiv \mathscr{B} \wedge \mathscr{C}$. Then $\vdash \neg\mathscr{A} \leftrightarrow \neg\mathscr{B} \vee \neg\mathscr{C}$, and we are done by the I.H. via the Leibniz rule.

(iii) $\mathscr{A} \equiv (\exists x)_{<t}\mathscr{B}$. Then $\vdash \neg\mathscr{A} \leftrightarrow (\forall x)_{<t}\neg\mathscr{B}$, and we are done by the I.H. via the Leibniz rule.

(iv) $\mathscr{A} \equiv (\forall x)_{<t}\mathscr{B}$. Then $\vdash \neg\mathscr{A} \leftrightarrow (\exists x)_{<t}\neg\mathscr{B}$, and we are done by the I.H. via the Leibniz rule. $\square$

II.4.5 Corollary. *For any $\mathscr{A} \in \Delta_0(L_{\mathfrak{A}})$ there is a $\mathscr{B} \in \Delta_0^+(L_{\mathfrak{A}})$ such that we can prove $\mathscr{A} \leftrightarrow \mathscr{B}$ without nonlogical axioms.*

Conversely, every $\mathscr{B} \in \Delta_0^+(L_{\mathfrak{A}})$ is a formula of $\Delta_0(L_{\mathfrak{A}})$.

II.4.6 Lemma. *Let $\mathscr{A}$ and $\mathscr{B}$ be in $\Sigma_1(L_{\mathfrak{A}})$ (respectively, $\Sigma_1'(L_{\mathfrak{A}})$). Then each of the following is provably equivalent in $\mathbf{PA}'$ to a formula in $\Sigma_1(L_{\mathfrak{A}})$*

(*respectively,* $\Sigma_1'(L_{\mathfrak{A}})$):

 (*i*) $\mathscr{A} \vee \mathscr{B}$
 (*ii*) $\mathscr{A} \wedge \mathscr{B}$
 (*iii*) $(\exists x).\mathscr{A}$
 (*iv*) $(\exists x)_{<z}\mathscr{A}$
 (*v*) $(\forall x)_{<z}\mathscr{A}.$

By **PA**' *over* $L_{\mathfrak{A}}$, *above, we understand an extension by definitions of* **PA** *over* $L_{\mathfrak{N}}$. *(In this connection cf. II.1.2, p. 207).*

Proof. The proof is a straightforward formalization of the techniques employed in the proof of I.8.48. The only case that presents some interest is (*v*), and we give a proof here. The proof hinges on the fact[†]

$$\vdash (\forall x)_{<z}(\exists y).\mathscr{B} \leftrightarrow (\exists w)(\forall x)_{<z}(\exists y)_{<w}\mathscr{B}$$

where w is a *new* variable.

The $\leftarrow$-direction of the above being trivial, we just prove

$$\vdash (\forall x)_{<z}(\exists y).\mathscr{B} \rightarrow (\exists w)(\forall x)_{<z}(\exists y)_{<w}\mathscr{B} \tag{1}$$

by induction on z.[‡]

The basis, $z = \mathbf{0}$ is settled by $< \mathbf{1}$. Take now (1), with frozen variables, as the I.H. Add the assumption

$$(\forall x)_{<Sz}(\exists y).\mathscr{B} \tag{2}$$

We want

$$\vdash (\exists w)(\forall x)_{<Sz}(\exists y)_{<w}\mathscr{B} \tag{3}$$

By (2) and specialization,

$$\vdash x < z \vee x = z \rightarrow (\exists y).\mathscr{B} \tag{4}$$

Since $\vdash x < z \rightarrow x < z \vee x = z$ and $\vdash x = z \rightarrow x < z \vee x = z$, (4) yields

$$\vdash x < z \rightarrow (\exists y).\mathscr{B} \tag{5}$$

[†] If we set $\mathscr{A} \equiv (\exists y).\mathscr{B}$, where $\mathscr{B} \in \Delta_0(L_{\mathfrak{A}})$, then $(\exists w)(\forall x)_{<z}(\exists y)_{<w}\mathscr{B}$ is the $\Sigma_1(L_{\mathfrak{A}})$ formula we want in order to establish (*v*).

[‡] The reader who has had some axiomatic set theory will notice the remarkable similarity of (1) with the axiom of *collection* (cf. volume 2, Chapter III). Indeed, (1) interprets collection, if we interpret the basic predicate "$\in$" of set theory as the predicate "$<$" of arithmetic.

and

$$\vdash x = z \rightarrow (\exists y).\mathscr{B} \tag{6}$$

By (5) and the I.H. we obtain

$$\vdash (\exists w)(\forall x)_{<z}(\exists y)_{<w}\mathscr{B} \tag{7}$$

By (6) we obtain $\vdash (\exists y).\mathscr{B}[x \leftarrow z]$; hence

$$\vdash (\exists w)(\exists y)_{<w}\mathscr{B}[x \leftarrow z] \tag{8}$$

Pause. Why is (8) true? Well, it follows immediately provided we believe

$$\vdash (\exists y).\mathscr{B} \rightarrow (\exists w)(\exists y)_{<w}\mathscr{B}$$

To establish the above let $(\exists y).\mathscr{B}[y]$. Add now $\mathscr{B}[c]$, where c is a new constant. Then $\vdash c < Sc \wedge \mathscr{B}[c]$; hence $\vdash (\exists y)(y < Sc \wedge \mathscr{B})$ (by **Ax2**) and thus $\vdash (\exists w)(\exists y)(y < w \wedge \mathscr{B})$ (by **Ax2** again).

Now, arguing by auxiliary constant once more, relying on (7) and (8), we add new constants a and b and the assumptions

$$(\forall x)_{<z}(\exists y)_{<a}\mathscr{B} \tag{7$'$}$$

and

$$(\exists y)_{<b}\mathscr{B}[x \leftarrow z] \tag{8$'$}$$

By the Leibniz axiom (**Ax4**) and tautological implication, (8$'$) yields

$$\vdash x = z \rightarrow (\exists y)_{<b}\mathscr{B} \tag{9}$$

On the other hand, we obtain from (7$'$)

$$\vdash x < z \rightarrow (\exists y)_{<a}\mathscr{B} \tag{10}$$

Now set $c = S(a + b)$. Proof by cases from (9) and (10) yields[†]

$$\vdash x < z \vee x = z \rightarrow (\exists y)_{<c}\mathscr{B}$$

Hence

$$\vdash (\forall x)_{x<Sz}(\exists y)_{<c}\mathscr{B}$$

By **Ax2**, (3) follows. $\square$

[†] Via the obvious $\vdash (\exists y)_{<a}\mathscr{B} \rightarrow (\exists y)_{<c}\mathscr{B}$ and $\vdash (\exists y)_{<b}\mathscr{B} \rightarrow (\exists y)_{<c}\mathscr{B}$, obtained from II.1.17, tautological implication, and $\exists$-monotonicity (I.4.23).

II.4.7 Lemma. *For any* $\mathscr{A} \in \Delta_0(L_\mathfrak{A})$ *(respectively,* $\mathscr{A} \in \Delta_0'(L_\mathfrak{A})$*) there is a provably equivalent – in pure logic –* $\mathscr{B} \in \Sigma_1(L_\mathfrak{A})$ *(respectively,* $\mathscr{B} \in \Sigma_1'(L_\mathfrak{A})$*).*

Proof. Let x be a variable that is not free in $\mathscr{A}$. Then

(a) $(\exists x)\mathscr{A} \in \Sigma_1(L_\mathfrak{A})$ (respectively, $(\exists x)\mathscr{A} \in \Sigma_1'(L_\mathfrak{A})$), and
(b) $\vdash (\exists x)\mathscr{A} \leftrightarrow \mathscr{A}$. □

II.4.8 Lemma. *For any* $\mathscr{A} \in \Delta_0(L_\mathfrak{A})$ *there is a* $\mathscr{B} \in \Sigma_1'(L_\mathfrak{A})$ *such that*

$$\vdash_{\mathbf{PA'}} \mathscr{A} \leftrightarrow \mathscr{B}$$

 The absence of a prime from $\Delta_0(L_\mathfrak{A})$ is intentional. **PA′** is as in II.4.6

Proof. In view of II.4.5, we do induction on the definition of $\Delta_0^+(L_\mathfrak{A})$ (induction variable is $\mathscr{A}$).

Basis. *Atomic formulas of type* $t = s$: We first look at the subcase $t = y$. We embark on an informal induction on the formation of t.

For the basis we have two cases: $t \equiv x$ and $t \equiv 0$. Both lead to restricted atomic formulas (II.4.3), and we are done by II.4.7. If now $t \equiv ft_1 \ldots t_n$, then (one point rule, I.7.2)

$$\vdash ft_1 \ldots t_n = y \leftrightarrow$$
$$(\exists x_1) \ldots (\exists x_n)(\underbrace{t_1 = x_1}_{\text{I.H.}} \wedge \ldots \wedge \underbrace{t_n = x_n}_{\text{I.H.}} \wedge \underbrace{fx_1 \ldots x_n = y}_{\Delta_0'(L_\mathfrak{A})})$$

where the x_i are new variables. By the I.H. on terms, the Leibniz rule, and II.4.6–II.4.7 we are done.

We can now conclude the $t = s$ case:

$$\vdash t = s \leftrightarrow (\exists y)(t = y \wedge s = y), \quad \text{where } y \text{ is a new variable.}$$

Basis. *Atomic formulas of type* $Pt_1 \ldots t_n$: Done by

$$\vdash Pt_1 \ldots t_n \leftrightarrow$$
$$(\exists x_1) \ldots (\exists x_n)(t_1 = x_1 \wedge \ldots \wedge t_n = x_n \wedge \underbrace{Px_1 \ldots x_n}_{\Delta_0'(L_\mathfrak{A})})$$

Basis. *Negated atomic formulas of type* $\neg t = s$:

$$\vdash \neg t = s \leftrightarrow (\exists x)(\exists y)(t = x \wedge s = y \wedge \underbrace{\neg x = y}_{\Delta_0'(L_\mathfrak{A})})$$

Basis. *Negated atomic formulas of type* $\neg Pt_1 \dots t_n$:

$$\vdash \neg Pt_1 \dots t_n \leftrightarrow$$

$$(\exists x_1) \dots (\exists x_n)(t_1 = x_1 \wedge \dots \wedge t_n = x_n \wedge \underbrace{\neg P x_1 \dots x_n}_{\Delta_0'(L_{\mathfrak{A}})})$$

The induction steps are for $\vee, \wedge, (\exists x)_{<t}, (\forall x)_{<t}$ and follow at once from II.4.6 (i), (ii), (iii), (iv), and (v), and $(\exists x)_{<t}\mathcal{A} \leftrightarrow (\exists z)(z = t \wedge (\exists x)_{<z}\mathcal{A})$. $\square$

II.4.9 Corollary. *For any* $\mathcal{A} \in \Sigma_1(L_{\mathfrak{A}})$ *there is a provably (in* **PA**$'$*) equivalent* $\mathcal{B} \in \Sigma_1'(L_{\mathfrak{A}})$.

Proof. II.4.8 and II.4.6(iii). $\square$

II.4.10 Lemma. *Let* **PA**$'$ *over* $L_{\mathfrak{A}'}$ *be a* recursive extension *of* **PA**. *Then for each* $\mathcal{A} \in \Sigma_1(L_{\mathfrak{A}'})$ *there is a formula* $\mathcal{B} \in \Sigma_1(L_{\mathfrak{N}})$ *such that* $\vdash_{\mathbf{PA}'} \mathcal{A} \leftrightarrow \mathcal{B}$.

Proof. It suffices to prove that if $L_{\mathfrak{A}'}$ is obtained from $L_{\mathfrak{A}}$ by the addition of either a single function or a single predicate symbol, and the defining axiom was added to a recursive extension $\mathfrak{T}$ (over $L_{\mathfrak{A}}$) of **PA** yielding **PA**$'$, then for each $\mathcal{A} \in \Sigma_1(L_{\mathfrak{A}'})$ there is a formula $\mathcal{B} \in \Sigma_1(L_{\mathfrak{A}})$ such that $\vdash_{\mathbf{PA}'} \mathcal{A} \leftrightarrow \mathcal{B}$.

In view of II.4.9 and II.4.6(iii) it suffices to prove this latter claim just for all $\mathcal{A} \in \Delta_0'(L_{\mathfrak{A}'})$. We do induction on the definition of $\Delta_0'(L_{\mathfrak{A}'})$ (induction variable: $\mathcal{A}$).

Basis. *Restricted atomic cases and their negations.* All cases are trivial (II.4.7), except $f\vec{x}_n = y$, $\neg f\vec{x}_n = y$, $P\vec{x}_n$, and $\neg P\vec{x}_n$, when f, or P, is the new symbol.

Say we have the case $\neg f\vec{x}_n = y$, where

$$f\vec{x}_n = (\mu z)\mathcal{B}$$

and $\mathcal{B} \in \Delta_0(L_{\mathfrak{A}})$.

 Note the absence of a prime from $\mathfrak{A}$ in $\Delta_0(L_{\mathfrak{A}})$. "$\vdash$" below is "$\vdash_{\mathbf{PA}'}$".

Thus,

$$\vdash f\vec{x}_n = y \leftrightarrow \mathcal{B}[y] \wedge (\forall z)_{<y}\neg\mathcal{B}[z]$$

and therefore

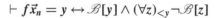
$$\vdash \neg f\vec{x}_n = y \leftrightarrow (\exists w)(\neg y = w \wedge \mathcal{B}[w] \wedge (\forall z)_{<w}\neg\mathcal{B}[z])$$

The right hand side of "$\leftrightarrow$" above is in $\Sigma_1(L_{\mathfrak{A}})$.

If on the other hand P is the new symbol, then we have

$$P\vec{x}_n \leftrightarrow \mathscr{B}$$

and $\mathscr{B} \in \Delta_0(L_{\mathfrak{A}})$, and this rests the case via II.4.7.

The induction steps are (II.4.3) for $\vee$, $\wedge$, $(\exists x)_{<z}$, $(\forall x)_{<z}$ and follow at once from II.4.6 (i), (ii), (iv), and (v) and the standard technique of eliminating defined symbols (at the atomic formula level).   □

II.4.11 Corollary. *Let* **PA**$'$ *over* $L_{\mathfrak{A}'}$ *be a recursive extension of* **PA**. *Then for each* $\mathscr{A} \in \Sigma_1(L_{\mathfrak{A}'})$ *there is a formula* $\mathscr{B} \in \Sigma_1'(L_{\mathfrak{N}})$ *such that* $\vdash_{\mathbf{PA}'} \mathscr{A} \leftrightarrow \mathscr{B}$.

Proof. II.4.10 guarantees a $\mathscr{C} \in \Sigma_1(L_{\mathfrak{N}})$ such that $\vdash_{\mathbf{PA}'} \mathscr{A} \leftrightarrow \mathscr{C}$. Now II.4.9 provides the $\mathscr{B} \in \Sigma_1'(L_{\mathfrak{N}})$ we want (recall that **PA**$'$ extends **PA**).   □

II.4.12 Theorem. *Let* **PA**$'$ *over* $L_{\mathfrak{N}'}$ *be a recursive extension of* **PA**, *and* $\mathscr{A}$ *a sentence in* $\Sigma_1(L_{\mathfrak{N}'})$. *Then* $\models_{\mathfrak{N}'} \mathscr{A}$ *implies that* $\vdash_{\mathbf{PA}'} \mathscr{A}$.

 Or, "any really true sentence of $\Sigma_1(L_{\mathfrak{N}'})$ is provable".

Proof. By II.4.11 there is a sentence $\mathscr{C} \in \Sigma_1'(L_{\mathfrak{N}})$ (note the absence of a prime from $\mathfrak{N}$) such that

$$\vdash_{\mathbf{PA}'} \mathscr{A} \leftrightarrow \mathscr{C} \tag{1}$$

Thus (do you believe this?)

$$\models_{\mathfrak{N}'} \mathscr{A} \leftrightarrow \mathscr{C}$$

Hence

$$\models_{\mathfrak{N}'} \mathscr{C}$$

Now, there is a formula $\mathscr{B}(x) \in \Delta_0'(L_{\mathfrak{N}})$ such that[†]

$$\mathscr{C} \equiv (\exists x).\mathscr{B}(x)$$

Hence

$$\models_{\mathfrak{N}'} (\exists x).\mathscr{B}(x)$$

Therefore, taking reducts,

$$\models_{\mathfrak{N}} (\exists x).\mathscr{B}(x) \tag{2}$$

[†] $\equiv$ means string equality.

By (2), $(\mathscr{B}(\overline{n}))^{\mathfrak{N}} = \mathbf{t}$ for some $n \in \mathbb{N}$; hence (p. 171)

$$(\mathscr{B}(\widetilde{n}))^{\mathfrak{N}} = \mathbf{t} \tag{3}$$

By (3), and an easy induction on the structure of $\mathscr{B}(\widetilde{n})$ (Definition II.4.3), following the steps of the proof of I.9.48 (p. 185),

$$\vdash_{\mathbf{ROB}} \mathscr{B}(\widetilde{n})$$

Hence $\vdash_{\mathbf{PA'}} (\exists x).\mathscr{B}(x)$ by **Ax2**. By (1), $\vdash_{\mathbf{PA'}} \mathscr{A}$.   □

II.4.13 Example. Equipped with II.4.12, one can now readily answer the "why"s that were embedded in Example II.2.7.   □

II.4.14 Remark. What II.4.12 above – and, earlier on, II.3.8 – do for us, practically, is to eliminate the need for *formal* proofs of "true" Σ_1-sentences $\mathscr{A}$, proofs that would be normally carried out in tedious detail within some recursive extension of **PA**. This is achieved by a two-pass process:

(1) We somehow convince ourselves that $\mathscr{A}$ is "true", i.e., true in $\mathfrak{N}$ or in some expansion thereof that accommodates whatever defined symbols are used inside the sentence.

(2) We then invoke (Meta)theorem II.4.12, which guarantees $\vdash \mathscr{A}$, without our having to write down a single line of a formal proof!

The reader may object: Obviously all the work was shifted to item (1). But how does one "prove" informally the "truth" of a sentence? Does it not necessitate as much work – in an informal deduction[†] – as a formal proof does? For example, that $lcm\{2, 3, 4, 5\} = 60$ is definitely obvious, but just proclaiming so does not prove this fact. How do we *know* that it is true?

The explanation is not mathematical, but it rather hinges on the *sociology of informal proofs*. An *informal* proof, viewed as a social activity, ends as soon as a reasonably convincing case has been made. Informal proofs are often sketchy (hence shorter than formal proofs), and the participants (prover and reader) usually agree that a certain level of detail can be left untold,[‡] and are also prepared to accept "the obvious" – the latter being informed by a vast database of "real" mathematical knowledge that goes all the way back to one's primary school years.

[†] Of course, one would never think of establishing truth in mathematical practice purely by semantic means, for this would involve messy infinitary arguments, while deductions, even informal ones, have the advantage of being finitary.

[‡] We are guilty of often leaving details for the reader to work out in our *formal* proofs. This dilutes the formal proof by an informality that we exercise for pedagogical reasons.

Surely sometime in the past we have learnt how to compute the *lcm*, or the gcd, of a set of numbers, or that $13 + 7 = 20$ is "true". We retrieve all this from the "database". Of course, a formal proof in **ROB** of, say, $\widetilde{13} + \widetilde{7} = \widetilde{20}$ is another matter, and is certainly not totally painless, as the reader who has read the lemma on the definability of $x + y = z$ (I.9.41, p. 181) will agree.[†]

Thus, deductions in "real" mathematics need only be as long (i.e., as detailed) as necessary for their *acceptability*. Indeed, one often finds that the level of detail included in various informal proofs of the same result is inversely proportional to the mathematical sophistication of the targeted audiences in each presentation.[‡] By contrast, formal proofs are, by definition (cf. I.3.17, p. 37), audience-independent.

Back to practice: Having performed step (1) above, we can now do one of two things: We can decide not to cheat by using step (2), but write down instead a formal argument that *translates* the informal one of step (1), treating the latter merely as a set of "organizational notes".[§]

Or, we can take the easy way out and invoke step (2). This avenue establishes the formal result entirely metamathematically. We have shown that a formal proof exists, without constructing the proof.

The approach is analogous to that of employing *Church's thesis* to informally "prove" results in recursion theory. This thesis says: "If we have shown by an informal argument that a partial function f is computable *in the intuitive sense* – for example, we have written informal instructions for its computation – then this f is partial recursive, that is, a *formal* algorithm for its computation exists (e.g., an algorithm formalized as a Turing machine, or as a Kleene schemata description)". *We do not need to exhibit this formal algorithm.*

Compare with "If we have shown by an informal argument that a sentence $\mathscr{A}$ among the types allowed in II.4.12 and II.3.8 is true in the standard structure, then there exists a proof for this $\mathscr{A}$ in (some conservative extension of) **PA**".

However, even though Church's thesis and II.4.12 and II.3.8 are *applied* in a similar manner, there is a major difference between them: The former

---

[†] It is normal (sloppy) practice to invoke "general" results where it would have been more appropriate, in order to avoid circularity, not to do so. For example, we may claim that (for any *fixed a* and *b* in $\mathbb{N}$) the "true" sentence $\widetilde{a} + \widetilde{b} = \widetilde{a + b}$ is provable, by invoking II.4.12 or II.3.8. Correct practice would have been to say that the provability of the sentence is due to I.9.41, since the latter was used towards the establishment of II.4.12 and II.3.8. But this puts a heavier burden on memory.

[‡] For example, it is held that part of the reason that Gödel never published a complete proof of his second incompleteness theorem was that the result was readily believed by his targeted audience.

[§] This is entirely analogous to what a computer programmer might do. He would first develop *pseudocode* (an informal program) towards the solution of a problem. He would then translate the pseudocode to a formal *computer program* written in the appropriate *programming language*.

is only a belief based on empirical evidence,[†] while the latter two are meta-theorems.

Unlike "true" existential sentences, "true" *universal*, or Π_1, sentences (of arithmetic) are not necessarily provable, as Gödel's first incompleteness theorem tells us. The Π_1-formulas are those of the form $\neg \mathscr{A}$ where $\mathscr{A}$ is Σ_1.

For example, if T is the *formal* Kleene predicate, then there are infinitely many "true" Π_1-sentences of the form $\neg(\exists y)T(\tilde{a}, \tilde{a}, y)$ that are not provable (cf. (3) in I.9.37). □ ⌽

II.5. Arithmetization

We now resume and conclude the discussion on the arithmetization of formal arithmetic(s) that we have started in Section I.9 (p. 168). The arithmetization is really a package that includes the Gödel numbers of terms and formulas on one hand, and, on the other hand, a *finite* suite of formal *predicates* and *functions* introduced to *test* for properties of Gödel numbers (e.g., testing whether x is a number for a formula, term, variable, etc.).

This package will be totally contained inside an appropriate *recursive extension* of **PA** over the appropriate $L_{\mathfrak{N}'}$ that contains all the tools (such as "$\langle \ldots \rangle$", etc.) needed to form Gödel numbers as on p. 168 and all the additional test predicates and functions (and their defining axioms).

Gödel numbers, "$\ulcorner \ldots \urcorner$", will be certain *closed terms* over $L_{\mathfrak{N}'}$ rather than "real" natural numbers from $\mathbb{N}$. We will rely on the numbering given on p. 168; however, we shall now employ the formal *minimum coding* $\langle \ldots \rangle$ given by *(FC)* on p. 239 and also make the following trivial amendment to notation: Every occurrence of a specific natural number n inside $\langle \ldots \rangle$ brackets is now changed to the term denoted by $\tilde{n}$ (*numeral*).

We next introduce the testing tools, that is, a finite sequence of atomic formulas and terms (introducing appropriate predicate and function symbols in a manner that respects the rules for *recursive extensions*) that enable the theory to "reason about" formulas (using Gödel numbers as aliases of such formulas).

[†] In some contexts – such as when partial function *oracles* are allowed – there is evidence to the contrary: If $L(x, \alpha, y)$ is the relation that says "program x with input the oracle α has a computation of length less than y", then this *is* intuitively computable: Just let program x crank with input α and keep track of the number of steps. If the program halts in fewer than y steps, then stop everything and return "yes"; otherwise, if x has already performed the yth step, stop everything and return "no". Now, if α and β are partial functions, $\alpha \subseteq \beta$ does not guarantee that $L(x, \alpha, y)$ and $L(x, \beta, y)$ yield the same answer, that is, L is *non-monotone*, or *inconsistent*. In most foundations of computability inconsistent relations such as L are not allowed, i.e., L is *not* formally computable in such theories; hence Church's thesis fails *with respect to such theories*. This particular "negative evidence" is eliminated if we use a different foundation of computability that was introduced in Tourlakis (1986, 1996, 2001a). Now L *is* formally computable. See also Kalmár's (1957) objections.

In each case we precede the definition with a comment stating the intended meaning.

$Var(x,i)$ holds if "$x = \ulcorner v_i \urcorner$".

$$Var(x,i) \leftrightarrow Seq(x) \wedge lh(x) = i + \widetilde{4} \wedge (x)_0 = S0$$
$$\wedge (x)_{S0} = \ulcorner (\urcorner \wedge (x)_{i+SSS0} = \ulcorner) \urcorner \qquad\qquad (Var)$$
$$\wedge (\forall z)_{<lh(x)}(S0 < z \wedge z < i + \widetilde{3} \rightarrow (x)_z = \ulcorner \square \urcorner)$$

We prefer to write, say, "$\ldots = \ulcorner (\urcorner$" rather than "$\ldots = \langle 0, \widetilde{9} \rangle$".

$Func(x,i,j)$ holds if "$x = \ulcorner f_i^j \urcorner$".

$$Func(x,i,j) \leftrightarrow Seq(x) \wedge lh(x) = i + j + \widetilde{6} \wedge (x)_0 = \widetilde{2}$$
$$\wedge (x)_{S0} = \ulcorner (\urcorner \wedge (x)_{i+j+\widetilde{5}} = \ulcorner) \urcorner$$
$$\wedge (x)_{i+\widetilde{3}} = \ulcorner \# \urcorner$$
$$\wedge (\forall z)_{<lh(x)}(S0 < z \wedge z < i + j + \widetilde{5} \qquad\qquad (Func)$$
$$\wedge \neg z = i + \widetilde{3}$$
$$\rightarrow (x)_z = \ulcorner \triangle \urcorner)$$

$Pred(x,i,j)$ means "$x = \ulcorner P_i^j \urcorner$".

$$Pred(x,i,j) \leftrightarrow Seq(x) \wedge lh(x) = i + j + \widetilde{6} \wedge (x)_0 = \widetilde{3}$$
$$\wedge (x)_{S0} = \ulcorner (\urcorner \wedge (x)_{i+j+\widetilde{5}} = \ulcorner) \urcorner$$
$$\wedge (x)_{i+\widetilde{3}} = \ulcorner \# \urcorner$$
$$\wedge (\forall z)_{<lh(x)}(S0 < z \wedge z < i + j + \widetilde{5} \qquad\qquad (Pred)$$
$$\wedge \neg z = i + \widetilde{3}$$
$$\rightarrow (x)_z = \ulcorner \bigcirc \urcorner)$$

Before we proceed with terms and other constructs we introduce the lower-case versions of the above predicates:

$$var(x) \quad \leftrightarrow (\exists i)_{\leq x} Var(x,i) \qquad \text{holds if "}x = \ulcorner v_i \urcorner \text{, for some } i\text{"}$$
$$func(x,n) \leftrightarrow (\exists i)_{\leq x} Func(x,i,n) \quad \text{holds if "}x = \ulcorner f_i^n \urcorner \text{, for some } i\text{"}$$
$$pred(x,n) \leftrightarrow (\exists i)_{\leq x} Pred(x,i,n) \quad \text{holds if "}x = \ulcorner P_i^n \urcorner \text{, for some } i\text{"}$$

Term is a new function symbol introduced by course-of-values recursion (Theorem II.3.3, p. 251) below. $Term(\ulcorner t \urcorner) = 0$ means "t is a term".

On the other hand, the statement "$Term(x) = 0$ implies that '$x = \ulcorner t \urcorner$ for some term t'" is *not* a fair description of what "*Term*" does. That is, after establishing that $Term(x) = 0$, we can only infer that x "behaves like" the Gödel number of some term, not that it *is* the Gödel number of some term. See Lemma II.6.24

(p. 297) later on, and the comment following it. If "implies" is not apt, then we cannot say "means" either, for the latter is *argot* for equivalence. We *can* say "holds if", though.

Similar caution must be exercised when interpreting the remaining functions and predicates introduced in this section.

$$
Term(x) = \begin{cases} \mathbf{0} & \text{if } var(x) \vee x = \langle \mathbf{0}, S\mathbf{0}\rangle \\ & \quad \vee\, Seq(x) \wedge (\exists y)_{<x}(lh(x) = Sy \wedge func((x)_0, y) \\ & \quad \wedge (\forall z)_{<y} Term((x)_{Sz}) = \mathbf{0}) \\ S\mathbf{0} & \text{otherwise} \end{cases}
$$

$$(Trm)$$

$AF(\ulcorner \mathscr{A} \urcorner)$ says "$\mathscr{A}$ is atomic".

$$
\begin{aligned}
AF(x) &\leftrightarrow (\exists y)_{<x}(\exists z)_{<x}(Term(y) = \mathbf{0} \wedge Term(z) = \mathbf{0} \\
&\qquad \wedge x = \langle \ulcorner = \urcorner, y, z\rangle) \\
&\qquad \vee Seq(x) \wedge (\exists y)_{<x}(lh(x) = Sy \wedge pred((x)_0, y) \\
&\qquad \wedge (\forall z)_{<y} Term((x)_{Sz}) = \mathbf{0})
\end{aligned}
$$

$$(Af)$$

WFF is a new function symbol introduced by course-of-values recursion below. $WFF(\ulcorner \mathscr{A} \urcorner) = \mathbf{0}$ means "$\mathscr{A} \in$ **Wff**":

$$
WFF(x) = \begin{cases} \mathbf{0} & \text{if } AF(x) \vee Seq(x) \wedge (lh(x) = \tilde{3} \wedge [(x)_0 = \ulcorner \vee \urcorner \\ & \quad \wedge WFF((x)_{S\mathbf{0}}) = \mathbf{0} \wedge WFF((x)_{SS\mathbf{0}}) = \mathbf{0} \\ & \quad \vee (x)_0 = \ulcorner \exists \urcorner \wedge var((x)_{S\mathbf{0}}) \wedge WFF((x)_{SS\mathbf{0}}) = \mathbf{0}] \\ & \quad \vee lh(x) = \tilde{2} \wedge (x)_0 = \ulcorner \neg \urcorner \wedge WFF((x)_{S\mathbf{0}}) = \mathbf{0}) \\ S\mathbf{0} & \text{otherwise} \end{cases}
$$

$$(Wf\!f)$$

We now introduce a new function symbol, $Free$, such that $Free(i, x) = \mathbf{0}$ *intuitively* says "if x is the Gödel number of a term *or* formula E, then v_i occurs free in E" (see I.1.10, p. 18).

$$
Free(i, x) = \begin{cases} \mathbf{0} & \text{if } Var(x, i) \vee \\ & \quad \Big(AF(x) \vee (Term(x) = \mathbf{0} \wedge (\exists y)_{<x} func((x)_0, y)) \Big) \wedge \\ & \quad (\exists y)_{<lh(x)}(0 < y \wedge Free(i, (x)_y) = \mathbf{0}) \vee \\ & \quad WFF(x) = \mathbf{0} \wedge (x)_0 = \ulcorner \neg \urcorner \wedge Free(i, (x)_{S\mathbf{0}}) = \mathbf{0} \vee \\ & \quad WFF(x) = \mathbf{0} \wedge (x)_0 = \ulcorner \vee \urcorner \wedge \\ & \quad (Free(i, (x)_{S\mathbf{0}}) = \mathbf{0} \vee Free(i, (x)_{SS\mathbf{0}}) = \mathbf{0}) \vee \\ & \quad WFF(x) = \mathbf{0} \wedge (x)_0 = \ulcorner \exists \urcorner \wedge \\ & \quad \neg Var((x)_{S\mathbf{0}}, i) \wedge Free(i, (x)_{SS\mathbf{0}}) = \mathbf{0} \\ S\mathbf{0} & \text{otherwise} \end{cases}
$$

$$(Fr)$$

We next introduce Gödel's *substitution function*, a precursor of Kleene's S-m-n functions. We introduce a new function symbol by course-of-values recursion. $Sub(x, y, z)$ will have the following intended effect:

$$Sub(\ulcorner t[v_i]\urcorner, \tilde{i}, \ulcorner s\urcorner) = \ulcorner t[s]\urcorner \quad \text{and} \quad Sub(\ulcorner \mathscr{A}[v_i]\urcorner, \tilde{i}, \ulcorner s\urcorner) = \ulcorner \mathscr{A}[s]\urcorner$$

However, in the latter case, the result will be as described iff s is *substitutable* in v_i. Otherwise the result 0 will be returned (a good choice for "not applicable", since no Gödel number equals 0).

This will make the definition a bit more complicated than usual.[†] Our definition of *Sub* below tracks I.3.11, p. 32.

$$Sub(x, i, z) = \begin{cases} z & \text{if } Var(x, i) \\ x & \text{if } x = \langle 0, S0 \rangle \\ x & \text{if } var(x) \wedge \neg Var(x, i) \\ (\mu w)\Big(Seq(w) \wedge lh(w) = lh(x) \wedge (w)_0 = (x)_0 \wedge \\ \qquad (\forall y)_{<lh(x)}(y > 0 \rightarrow (w)_y = Sub((x)_y, i, z))\Big) \\ \qquad \text{if } (Term(x) = 0 \wedge (\exists y)_{<x} func((x)_0, y)) \vee AF(x) \\ \langle (x)_0, Sub((x)_{S0}, i, z) \rangle \\ \qquad \text{if } WFF(x) = 0 \wedge (x)_0 = \ulcorner \neg \urcorner \wedge \\ \qquad Sub((x)_{S0}, i, z) > 0 \\ \langle (x)_0, Sub((x)_{S0}, i, z), Sub((x)_{SS0}, i, z) \rangle \\ \qquad \text{if } WFF(x) = 0 \wedge (x)_0 = \ulcorner \vee \urcorner \wedge \\ \qquad Sub((x)_{S0}, i, z) > 0 \wedge Sub((x)_{SS0}, i, z) > 0 \\ x \quad \text{if } WFF(x) = 0 \wedge (x)_0 = \ulcorner \exists \urcorner \wedge Var((x)_{S0}, i) \\ \langle (x)_0, (x)_{S0}, Sub((x)_{SS0}, i, z) \rangle \\ \qquad \text{if } WFF(x) = 0 \wedge (x)_0 = \ulcorner \exists \urcorner \wedge \neg Var((x)_{S0}, i) \\ \qquad \wedge Sub((x)_{SS0}, i, z) > 0 \wedge \\ \qquad (\exists j)_{\leq (x)_{S0}}(Var((x)_{S0}, j) \wedge Free(j, z) > 0) \\ 0 \quad \text{otherwise} \end{cases}$$

$$(Sub)$$

A two variable version, the family of the "lowercase sub_i", for $i \in \mathbb{N}$, is also useful:

$$sub_i(x, z) = Sub(x, \tilde{i}, z) \qquad (sub_i)$$

For any terms $t[v_m], s$ and formula $\mathscr{A}[v_m]$, we will (eventually) verify that $\vdash sub_m(\ulcorner t[v_m]\urcorner, \ulcorner s\urcorner) = \ulcorner t[s]\urcorner$ and $\vdash sub_m(\ulcorner \mathscr{A}[v_m]\urcorner, \ulcorner s\urcorner) = \ulcorner \mathscr{A}[s]\urcorner$, assuming that s is substitutable for v_m in $\mathscr{A}$.

[†] In the literature – e.g., Shoenfield (1967), Smoryński (1978, 1985) – one tends to define *Sub* with no regard for substitutability, i.e., pretending that all is O.K. One then tests for substitutability via other predicates or functions that are subsequently introduced.

In the next section we introduce (informally) yet another special case of *sub* that has special significance towards proving the second incompleteness theorem. For now, we continue with the introduction of predicates that "recognize" Gödel numbers of logical axioms.

The introductory remarks to Section II.1 announced that we will be making an amendment to what the set Λ of logical axioms is, in order to make the arithmetization more manageable.

Indeed, while keeping the schemata **Ax2–Ax4** the same, we change group **Ax1** – the "propositional axioms" – to consist of the four schemata below:

(1) $\mathscr{A} \vee \mathscr{A} \to \mathscr{A}$
(2) $\mathscr{A} \to \mathscr{A} \vee \mathscr{B}$
(3) $\mathscr{A} \vee \mathscr{B} \to \mathscr{B} \vee \mathscr{A}$
(4) $(\mathscr{A} \to \mathscr{B}) \to (\mathscr{C} \vee \mathscr{A} \to \mathscr{C} \vee \mathscr{B})$.

Of course, the new axiom group **Ax1** is equivalent with the old one on p. 34, as Exercises I.26–I.41 (pp. 193–195) in Chapter I show.

Thus we introduce predicates $Prop_i$ ($i = 1, 2, 3, 4$) such that $Prop_i(x)$ is intended to hold if x is a Gödel number of a formula belonging to the schema (i), $i = 1, 2, 3, 4$. Here is one case (case (4)); cases (1)–(3) are less work.

$$Prop_4(x) \leftrightarrow (\exists y)_{<x}(\exists z)_{<x}(\exists w)_{<x}\Big(WFF(y) = 0 \wedge WFF(z) = 0 \wedge$$
$$WFF(w) = 0 \wedge$$
$$x = \Big\langle \ulcorner \to \urcorner, \big\langle \ulcorner \to \urcorner, y, z \big\rangle, \big\langle \ulcorner \to \urcorner, \langle \ulcorner \vee \urcorner, w, y \rangle, \langle \ulcorner \vee \urcorner, w, z \rangle \big\rangle \Big\rangle \Big)$$

To simplify notation we have used the abbreviation "$\langle \ulcorner \to \urcorner, a, b \rangle$" for the fully explicit "$\langle \ulcorner \vee \urcorner, \langle \ulcorner \neg \urcorner, a \rangle, b \rangle$". This will recur in our descriptions of the remaining axiom schemata.

We finally introduce

$$Prop(x) \leftrightarrow Prop_1(x) \vee Prop_2(x) \vee Prop_3(x) \vee Prop_4(x) \qquad (Prop)$$

which is intended to hold if x is a Gödel number of an axiom of type **Ax1** (as the **Ax1** group was amended above).

For **Ax2** we introduce *SubAx*.

SubAx(x) holds if $x = \ulcorner \mathscr{A}[z \leftarrow t] \to (\exists z)\mathscr{A} \urcorner$ for some formula $\mathscr{A}$, variable z, and term t that is substitutable for z in $\mathscr{A}$. Thus,

$$SubAx(x) \leftrightarrow (\exists y)_{<x}(\exists z)_{<x}(\exists w)_{<x}(\exists i)_{<x}\Big(WFF(y) = 0 \wedge$$
$$Term(w) = 0 \wedge Var(z, i) \wedge Sub(y, i, w) > 0 \wedge \qquad (SubAx)$$
$$x = \langle \ulcorner \to \urcorner, Sub(y, i, w), \langle \ulcorner \exists \urcorner, z, y \rangle \rangle \Big)$$

Note that $Sub(y,i,w) > 0$ "says" that the term with Gödel number w is substitutable for v_i in the formula with Gödel number y. We will verify this in the next section, however the reader can already intuitively see that this is so from the definition of Sub.

For the axioms **Ax3** we introduce $Id\,Ax$:

$Id\,Ax(x)$ holds if $x = \ulcorner y = y \urcorner$ for some variable y. Thus,

$$Id\,Ax(x) \leftrightarrow Seq(x) \wedge lh(x) = SSS0 \wedge$$
$$(x)_0 = \ulcorner = \urcorner \wedge (x)_{S0} = (x)_{SSS0} \wedge var((x)_{S0}) \tag{IdAx}$$

Finally, for **Ax4** we introduce $Eq\,Ax$. $Eq\,Ax(x)$ holds if $x = \ulcorner t = s \rightarrow (\mathscr{A}[t] \leftrightarrow \mathscr{A}[s]) \urcorner$ for some formula $\mathscr{A}$ and substitutable terms t and s. Thus,

$$Eq\,Ax(x) \leftrightarrow (\exists y)_{<x}(\exists z)_{<x}(\exists w)_{<x}(\exists v)_{<x}(\exists i)_{<x}\Big(WFF(y) = 0 \wedge$$
$$Term(w) = 0 \wedge Term(v) = 0 \wedge Var(z,i) \wedge$$
$$Sub(y,i,w) > 0 \wedge Sub(y,i,v) > 0 \wedge$$
$$x = \langle \ulcorner \rightarrow \urcorner, \langle \ulcorner = \urcorner, w, v \rangle, \langle \ulcorner \leftrightarrow \urcorner, Sub(y,i,w), Sub(y,i,v) \rangle\rangle\Big) \tag{EqAx}$$

Thus, we introduce LA by

$$LA(x) \leftrightarrow Prop(x) \vee Sub\,Ax(x) \vee Id\,Ax(x) \vee Eq\,Ax(x) \tag{LA}$$

$LA(\ulcorner \mathscr{A} \urcorner)$ means "$\mathscr{A} \in \Lambda$".

We have two rules of inference. Thus, we introduce MP by

$$MP(x,y,z) \leftrightarrow WFF(x) = 0 \wedge WFF(z) = 0 \wedge y = \langle \ulcorner \rightarrow \urcorner, x, z \rangle \tag{MP}$$

We also introduce $\exists$-introduction, EI, by

$$EI(x,y) \leftrightarrow (\exists u)_{<x}(\exists v)_{<x}(\exists z)_{<x}(\exists i)_{<x}\big(x = \langle \ulcorner \rightarrow \urcorner, u, v \rangle$$
$$\wedge WFF(u) = 0 \wedge WFF(v) = 0$$
$$\wedge Var(z,i) \wedge Free(i,v) > 0$$
$$\wedge y = \langle \ulcorner \rightarrow \urcorner, \langle \ulcorner \exists \urcorner, z, u \rangle, v \rangle\big) \tag{EI}$$

We will use one more rule of inference in the definition of **Proof** below. This rule, *substitution of terms*, is, of course a *derived* rule (I.4.12, p. 44), but we acknowledge it explicitly for our future convenience (Lemma II.6.27 – via II.6.18 and II.6.20). Thus, we introduce SR. $SR(x,y)$ holds if $x = \ulcorner \mathscr{A} \urcorner$ and $y = \ulcorner \mathscr{A}[v_i \leftarrow t] \urcorner$ for some term t substitutable in v_i:

$$SR(x,y) \leftrightarrow WFF(x) = 0 \wedge WFF(y) = 0 \wedge$$
$$(\exists i)_{\leq x}(\exists z)_{\leq y}(Term(z) = 0 \wedge y = Sub(x,i,z)) \tag{SR}$$

II.5.1 Remark. In introducing *MP*, *EI*, and *SR* above we have been consistent with our definition of inference rules (I.3.15, p. 36) in that the rules *act* on *formulas.*

It is technically feasible to have the rules act on *any* strings over our alphabet instead – e.g., view modus ponens as acting on arbitrary strings $\mathscr{A}$ and $(\mathscr{A} \rightarrow \mathscr{B})$ to produce the string $\mathscr{B}$. Indeed the literature, by and large, takes this point of view for the arithmetized versions above, defining, for example *MP'* just as

$$MP'(x, y, z) \leftrightarrow y = \langle \ulcorner \rightarrow \urcorner, x, z \rangle \qquad (MP')$$

Still, informally speaking, a proof built upon the primed versions of the rules of inference is composed entirely of formulas[†] – as it should be – for these rules produce formulas if their inputs are all formulas, and, of course, a proof starts with formulas (the axioms). □

We are ready to test for proofs. Fix a set of nonlogical axioms Γ.

 Suppose that *we already have* a formula Γ such that $\Gamma(\ulcorner \mathscr{A} \urcorner)$ means "$\mathscr{A} \in \Gamma$".

Then, we may introduce the *informal* abbreviation **Proof** such that **Proof**(x) holds if $x = \langle \ulcorner \mathscr{A}_1 \urcorner, \ulcorner \mathscr{A}_2 \urcorner, \dots \rangle$ where $\mathscr{A}_1, \mathscr{A}_2, \dots$ is some Γ-proof.

$$\begin{aligned}
\textbf{Proof}(x) \text{ stands for } \quad & Seq(x) \wedge lh(x) > 0 \wedge \\
& (\forall i)_{<lh(x)}(LA((x)_i) \vee \Gamma((x)_i) \vee \\
& (\exists j)_{<i}(\exists k)_{<i} MP((x)_j, (x)_k, (x)_i) \vee \qquad (Proof) \\
& (\exists j)_{<i} EI((x)_j, (x)_i) \vee \\
& (\exists j)_{<i} SR((x)_j, (x)_i)))
\end{aligned}$$

 Note that this introduction of **Proof** is *modulo* Γ, but most of the time we will not indicate that dependence explicitly (e.g., as **Proof**$_\Gamma$), since whatever Γ we have in mind will be usually clear from the context.

Gödel's theorem is about theories with a "recognizable" set of axioms, i.e., recursively axiomatized theories. However, it is also applicable if we take the somewhat more general assumption that Γ is a semi-recursive set, which formally means that we want the formula Γ to be Σ_1 over the language where we effect our coding. Informally, this means that even if we cannot recognize the axioms, nevertheless we can form an effective (algorithmic) listing of all of them; Gödel's results still hold for such theories.[‡]

[†] As an easy informal induction on the length of proofs shows.

[‡] It is easy to see that the the set of theorems of a theory with a semi-recursive set of axioms is also semi-recursive by closure of semi-recursive relations under ($\exists y$). This immediately opens any such theory for arithmetic to the incompletableness phenomenon, since the set of all true sentences is not semi-recursive.

The reader will note that **PA** is much "simpler" than semi-recursive. Indeed, it is not hard to show that if $\Gamma = $ **PA**, then the formula Γ is primitive recursive.[†] In this case, $\Gamma(x)$ will be equivalent to a lengthy disjunction of cases. For example, the case corresponding to axiom $+1, z + 0 = z$, is given by the (Δ_0) formula

$$(\exists z)_{<x}\big(var(z) \wedge x = \langle \ulcorner = \urcorner, \langle \ulcorner + \urcorner, z, 0 \rangle, z \rangle\big)$$

We leave the details to the reader (Exercise II.15).

II.6. Derivability Conditions; Fixed Points

In this section we will find a sentence in $L_{\mathfrak{N}}$ that says "I am not a theorem".

We will also show that certain *derivability conditions* (*the Ableitbarkeitsforderungen*,[‡] of Hilbert and Bernays (1968)) hold in some appropriate extension, C, of **PA**. These conditions talk about deducibility (derivability) in appropriate consistent extensions Γ of **PA**. At the end of all this, to prove the second incompleteness theorem becomes a comparatively easy matter.

We begin by assessing the success of our arithmetization. We have made claims throughout Section II.5 about what each introduced predicate or function meant intuitively. We want to make precise and substantiate such claims now, although we will not exhaustively check every predicate and function that we have introduced. Just a few spot checks of the most interesting cases will suffice.

 First off, we fix our attention on an arbitrary *consistent* extension, Γ of **PA**, over some language $L_{\mathfrak{A}}$, where Γ is Σ_1 over the language of C below.

We also fix an *extension by definitions*[§] of **PA**, where all the predicates and functions that we have introduced in our arithmetization along with their defining axioms, reside. We will use the symbol C ("C" for *Coding*) for this specific extension. The language of C will be denoted by $L_{\mathfrak{N}_C}$.

As a matter of fact, in the interest of avoiding circumlocutions – in particular when proving the Hilbert-Bernays **DC 3** later on (II.6.34) – we allow *all the primitive recursive function symbols* (II.3.6, p. 253) to be in $L_{\mathfrak{N}_C}$ and, correspondingly, *all the defining axioms of those symbols* (i.e., through composition, primitive recursion, or, explicitly, as definitions of Z and U_i^n) to be included in C.

[†] Cf. II.3.6, p. 253.

[‡] The *Ableitbarkeitsforderungen* that we employ are close, but not identical, to the originals in Hilbert and Bernays (1968). We have allowed some influence from Löb's modern derivability conditions, which we also state and prove, essentially using the old *Ableitbarkeitsforderungen* as stepping stones in the proof. Once we have Löb's conditions, we employ them to prove the second incompleteness theorem.

[§] Recall that extensions by definitions are conservative.

We note that C is a recursive extension of **PA** in the sense of p. 220, since *Proof* (introduced in the previous section), and *Deriv* and Θ (p. 281 and p. 280 below) are not introduced as *predicates*, but rather are introduced as informal abbreviations.

Note that the languages $L_{\mathfrak{A}}$ and $L_{\mathfrak{M}_C}$ may, *a priori*, extend the language of **PA**, $L_{\mathfrak{M}}$, in unrelated directions, since the former extension accommodates whatever new nonlogical axioms are peculiar to Γ, while the latter simply accommodates symbols (beyond the basic symbols (NL) of p. 166) that are introduced by *definitions* (see, however, II.6.32, p. 303). Thus, Γ may fail to be a conservative extension of **PA**, unlike C.

When we write "$\vdash$" in this section, we mean "$\vdash_C$" (or "$\vdash_C$ *plus some 'temporary' assumptions*", if a proof by deduction theorem, auxiliary constant, etc., has been embarked upon). Exceptions will be noted.

The informal symbols "*Proof*", "*Deriv*", and "Θ" (the latter two to be introduced shortly) will abbreviate "*Proof*$_\Gamma$", "*Deriv*$_\Gamma$", and "Θ_Γ" respectively.

Now (equipped with the tools of Sections I.8–I.9 and II.3.6–II.3.11) it is easy to verify that all the atomic formulas and terms that we have introduced in Section II.5, to "recognize specified sets of Gödel numbers" – *except* for the formula *Proof*, because of our assumptions on Γ – are primitive recursive. For atomic formulas this means that the corresponding characteristic terms[†] are.

Pause. Is that so? How about the definition of *Sub* (p. 268)? It contains unbounded search.

In particular, the lightface versions of each such formula or term (e.g., $Var(x, i), Term(x), WFF(x)$) – i.e., the informal versions obtained by replicating the formal definitions in the metatheory – are in $\mathfrak{PR}_*$ or $\mathfrak{PR}$. Applying Theorem II.3.8 and its Corollary II.3.9, we have that *the boldface version in each case both defines semantically (in $L_{\mathfrak{M}_C}$) and strongly defines (in C) the corresponding lightface version*.

Since every Gödel number t is a closed primitive recursive term, II.3.10 implies that for any such t there is a unique $n \in \mathbb{N}$ such that $\vdash t = \tilde{n}$.

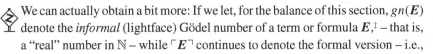 We can actually obtain a bit more: If we let, for the balance of this section, $gn(E)$ denote the *informal* (lightface) Gödel number of a term or formula E,[‡] – that is, a "real" number in $\mathbb{N}$ – while $\ulcorner E \urcorner$ continues to denote the formal version – i.e.,

---

[†] Cf. II.1.25 (p. 220) and p. 222.

[‡] An arbitrary term is generically denoted by t or s in the metalanguage; a formula, by a calligraphic uppercase letter such as $\mathscr{A}$. The "E" here is a compromise notation that captures (in the metalanguage) an arbitrary term *or* formula, that is, an arbitrary well-formed *expression* (string).

a closed term in $L_{\mathfrak{N}_C}$ – then (by II.3.8–II.3.9)

$$\vdash \ulcorner E \urcorner = \widetilde{gn(E)} \tag{1}$$

To see why (1) is true, consider the Gödel numbering that we have developed starting on p. 168. In the interest of clarity of argument, let us introduce the functions f_i ($i = 1, \ldots, 11$) below – as well as their lightface versions, the latter (not shown) by the same (although lightface) defining equations. Note the deliberate choice of *distinct* variables v_i ($i = 1, \ldots, 11$):

$$
\begin{aligned}
f_1(v_1) &= \langle 0, v_1 \rangle \\
f_2(v_2) &= \langle 0, v_2 \rangle \\
f_3(v_3) &= \langle 0, v_3 \rangle \\
f_4(v_4) &= \langle 0, v_4 \rangle \\
f_5(v_5) &= \langle 0, v_5 \rangle \\
f_6(v_6) &= \langle 0, v_6 \rangle \\
f_7(v_7) &= \langle 0, v_7 \rangle \\
f_8(v_8) &= \langle 0, v_8 \rangle \\
f_9(v_9) &= \langle 0, v_9 \rangle \\
f_{10}(v_{10}) &= \langle 0, v_{10} \rangle \\
f_{11}(v_{11}) &= \langle 0, v_{11} \rangle
\end{aligned}
$$

Then any formal (respectively, informal) Gödel number $\ulcorner E \urcorner$ (respectively, $gn(E)$) is a closed instance of $h[v_1, \ldots, v_{11}]$ (respectively, of $h[v_1, \ldots, v_{11}]$) for some appropriate formal (respectively, informal) primitive recursive h (respectively, h). The closed instance is obtained in a specific way: Each v_i (respectively, each v_i) is being replaced by $\widetilde{i}$ (respectively, i).

As we assume that the formal and informal definitions "track each other" – i.e., have identical primitive recursive derivations except for typeface – then

$$h = h^{\mathfrak{N}_C} \tag{1'}$$

Now (1) translates to

$$\vdash h[\widetilde{1}, \ldots, \widetilde{11}] = (h[1, \ldots, 11])^{\sim}$$

where '$(h[1, \ldots, 11])^{\sim}$' denotes that '$\sim$' applies to all of '$h[1, \ldots, 11]$'; which holds by (1') and II.3.8.

II.6.1 A Very Long Example. It is trivial that, for any $i \in \mathbb{N}$, $Var(gn(v_i), i)$ is true.[†] Thus, both

$$\models_{\mathfrak{N}_C} \mathbf{Var}(\widetilde{gn(v_i)}, \widetilde{i})$$

[†] Do you believe this? See II.4.14 for a discussion of such outrageous claims.

and

$$\vdash \textbf{Var}(\widetilde{gn(v_i)}, \tilde{i})$$

hold, for all $i \in \mathbb{N}$.[†] By (1), p. 274, and **Ax4**,

$$\models_{\mathfrak{N}_C} \textbf{Var}(\ulcorner v_i \urcorner, \tilde{i})$$

and

$$\vdash \textbf{Var}(\ulcorner v_i \urcorner, \tilde{i})$$

hold for all $i \in \mathbb{N}$. From the last two and **Ax2** follow, for all $i \in \mathbb{N}$,

$$\models_{\mathfrak{N}_C} \textit{var}(\ulcorner v_i \urcorner)$$

and

$$\vdash \textit{var}(\ulcorner v_i \urcorner)$$

Conversely, if $\vdash \textbf{Var}(\tilde{n}, \tilde{i})$, then $Var(n, i)$ is true by II.3.9, and therefore, trivially, $gn(v_i) = n$ through examination of the lightface definition (Var) (identical to the boldface version given on p. 266). Hence

$$\vdash \widetilde{gn(v_i)} = \tilde{n}$$

Thus (by (1), p. 274)

$$\vdash \ulcorner v_i \urcorner = \tilde{n} \tag{2}$$

In other words, $\vdash \textbf{Var}(\tilde{n}, \tilde{i})$ implies that $\tilde{n}$ is *provably*[‡] *equal* to the (formal) Gödel number of the variable v_i.

Similarly, $\vdash \textit{var}(\tilde{n})$ implies that $\tilde{n}$ is provably equal to the formal Gödel number of some variable.

We next verify that **Term** behaves as intended. Indeed, by *induction on terms*, t, we can show *in the metatheory* that $Term(gn(t)) = 0$ is true. For example, if $t = ft_1 \ldots t_n$, then the I.H. implies that $Term(gn(t_i)) = 0$ is true, for $i = 1, \ldots, n$. Given that $gn(t) = \langle gn(f), gn(t_1), \ldots, gn(t_n) \rangle$, and consulting (Trm),[§] we see that indeed, $Term(gn(t)) = 0$ is true. The basis cases are covered as easily. Thus (II.3.8),

$$\vdash \textbf{Term}(\widetilde{gn(t)}) = 0$$

[†] We recall that the metasymbol $\tilde{i}$ denotes a (formal) term. On the other hand, i in v_i is part of the name.

[‡] In **C**.

[§] This is the formal definition on p. 267. The informal definition is the lightface version of (Trm).

hence, by (1) and **Ax4**

$$\vdash Term(\ulcorner t \urcorner) = 0$$

Conversely, $\vdash Term(\tilde{n}) = 0$ implies that $Term(n) = 0$ is true. Now an informal (i.e., metamathematical) course-of-values induction on n easily shows that $n = gn(t)$ for some term t, and thus we obtain $\vdash \ulcorner t \urcorner = \tilde{n}$, exactly as we have obtained (2). For example, say that $Term(n) = 0$ because the third disjunct[†] in the top case of the definition (Trm) (p. 267) is true, namely,

$$Seq(n) \wedge (\exists y)_{<n}(lh(n) = y + 1 \wedge func((n)_0, y)$$
$$\wedge (\forall z)_{<y} Term((n)_{z+1}) = 0)$$

Let $y = k < n$ be a value that works above, i.e., such that the following is true:

$$Seq(n) \wedge (lh(n) = k + 1 \wedge func((n)_0, k) \wedge (\forall z)_{<k} Term((n)_{z+1}) = 0)$$

By the I.H.[‡] there are terms $t_1, \dots, t_k$ such that $gn(t_z) = (n)_z$ for $z = 1, \dots, k$. Moreover, the truth of $func((n)_0, k)$ implies, analogously to the case of **Var**, that there is a function symbol f, of arity k, such that $gn(f) = (n)_0$. Therefore,

$$n = \langle (n)_0, (n)_1, \dots, (n)_k \rangle$$
$$= \langle gn(f), gn(t_1), \dots, gn(t_k) \rangle$$
$$= gn(ft_1 \dots t_k)$$

We have verified that $\vdash Term(\tilde{n}) = 0$ implies that $\tilde{n}$ is provably equal to the (formal) Gödel number of some term t.

One similarly verifies that **WFF** behaves as intended. We next verify that **Free** and **Sub** behave as intended.

Suppose that v_i appears free in $\mathscr{A}$. Referring to the lightface definition of **Free** (which is structurally identical to (Fr), p. 267) and using induction on $\mathscr{A}$ or t – as the case may be – it is easy to show that $Free(i, gn(\mathscr{A})) = 0$ or (correspondingly) $Free(i, gn(t)) = 0$. Thus, one can prove (by (1) and II.3.8–II.3.10)

$$\vdash Free(\tilde{i}, \ulcorner \mathscr{A} \urcorner) = 0$$

or (correspondingly)

$$\vdash Free(\tilde{i}, \ulcorner t \urcorner) = 0$$

Conversely, assume that $\vdash Free(\tilde{i}, \tilde{n}) = 0$ for some i, n in $\mathbb{N}$. Thus, $Free(i, n) = 0$ is true. A course-of-values induction on n (over $\mathbb{N}$) shows that $n = gn(E)$

[†] A *disjunct* is a member of a disjunction.
[‡] "If $Seq(n)$, then $i < lh(n) \rightarrow (n)_i < n$" is crucial here.

(where E is some term *or* formula) and v_i is free in E. For example, if $Free(i, n) = 0$ is true because the last disjunct in (Fr) (p. 267) obtains, namely,

$$WFF(n) = 0 \wedge (n)_0 = gn(\exists) \wedge \neg Var((n)_1, i) \wedge Free(i, (n)_2) = 0$$

then the behaviour of WFF ensures that, for some formula $\mathscr{A}$,

$$n = gn(\mathscr{A}) \tag{i}$$

Moreover (by (i) and properties of WFF), $n = \langle (n)_0, (n)_1, (n)_2 \rangle$; hence for some $j \in \mathbb{N}$ and formula $\mathscr{B}$,

$$(n)_2 = gn(\mathscr{B}), (n)_1 = gn(v_j) \quad \text{and} \quad \mathscr{A} \equiv ((\exists v j).\mathscr{B})$$

By $\neg Var((n)_1, i)$, we have $i \neq j$. By I.1.10 and the I.H. (by which the conjunct[†] $Free(i, (n)_2) = 0$ ensures that v_i is free in $\mathscr{B}$) it follows that v_i is free in $\mathscr{A}$. Thus, by (1) (p. 274) and (i), $\vdash \ulcorner \mathscr{A} \urcorner = \tilde{n}$ in this case, and hence (by assumption and **Ax4**)

$$\vdash \textbf{\textit{Free}}(\tilde{i}, \ulcorner \mathscr{A} \urcorner) = 0$$

All in all, using (1) for the *if* part,

$$v_i \text{ is free in } t \quad \text{iff} \quad \vdash \textbf{\textit{Free}}(\tilde{i}, \ulcorner t \urcorner) = 0$$
$$v_i \text{ is free in } \mathscr{A} \quad \text{iff} \quad \vdash \textbf{\textit{Free}}(\tilde{i}, \ulcorner \mathscr{A} \urcorner) = 0$$

Turning now to **Sub**, we validate the claims made on p. 268.

Let at first t and s be terms. It is easy (metamathematical induction on t) to see that

$$Sub(gn(t), i, gn(s)) = gn(t[v_i \leftarrow s]) \tag{3}$$

is true. Indeed (sampling the proof), let $t \equiv ft_1 \ldots t_n$. Referring to the definition of Sub (mirrored on p. 268 for the boldface case), we obtain

$$Sub(gn(t), i, gn(s)) = \langle gn(f), Sub(gn(t_1), i, gn(s)), \ldots,$$
$$Sub(gn(t_n), i, gn(s)) \rangle$$

Using the I.H. on t, the above translates to

$$Sub(gn(t), i, gn(s)) = \langle gn(f), gn(t_1[v_i \leftarrow s]), \ldots, gn(t_n[v_i \leftarrow s]) \rangle$$

Since $gn(t[v_i \leftarrow s]) = \langle gn(f), gn(t_1[v_i \leftarrow s]), \ldots, gn(t_n[v_i \leftarrow s]) \rangle$, (3) follows.

[†] A *conjunct* is a member of a conjunction.

Let next $\mathcal{A}$ be a formula, and s be a term *substitutable* for v_i in $\mathcal{A}$. By induction on $\mathcal{A}$ we can show in the metatheory that

$$Sub(gn(\mathcal{A}), i, gn(s)) = gn(\mathcal{A}[v_i \leftarrow s]) \tag{4}$$

is true. Sampling the proof of (4), let us consider an interesting case, namely, $\mathcal{A} \equiv (\exists vj).\mathcal{B}$, where $i \neq j$. By the I.H. on formulas,

$$Sub(gn(\mathcal{B}), i, gn(s)) = gn(\mathcal{B}[v_i \leftarrow s]) \tag{5}$$

By I.3.10–I.3.11, $\mathcal{A}[v_i \leftarrow s] \equiv (\exists v_j)(\mathcal{B}[v_i \leftarrow s])$. Thus,

$$gn(\mathcal{A}[v_i \leftarrow s]) = \langle gn(\exists), gn(v_j), gn(\mathcal{B}[v_i \leftarrow s]) \rangle \tag{6}$$

By the definition of *Sub*,

$$Sub(gn(\mathcal{A}), i, gn(s)) = \langle gn(\exists), gn(v_j), Sub(gn(\mathcal{B}), i, gn(s)) \rangle$$

Thus, (5) and (6), yield (4).

Here is the other interesting case, where $\mathcal{A}$ is a formula, and s is a term that is *not* substitutable for v_i in $\mathcal{A}$. By induction on $\mathcal{A}$ we can show this time that

$$Sub(gn(\mathcal{A}), i, gn(s)) = 0 \tag{7}$$

We just sample the same case as above. First off, for any term *or* formula E, $gn(E) > 0$.

Pause. Do you believe this?

Consider first the subcase where $\mathcal{A} \equiv (\exists v_j).\mathcal{B}$, and s *is* substitutable for v_i in $\mathcal{B}$. By (4), $Sub(gn(\mathcal{B}), i, gn(s)) = gn(\mathcal{B}[v_i \leftarrow s]) > 0$.

Now, by assumption of *non*-substitutability for v_i in $\mathcal{A}$, it must be that v_j is free in s. Thus, the relevant condition for the definition of *Sub* (this is the second from last condition, p. 268) fails, since $Free(j, gn(s)) = 0$. Therefore the definition (of *Sub*) returns 0 (the "otherwise"), and (7) is correct in this subcase.

The remaining subcase is that substitutability of s failed earlier, that is (I.H.), $Sub(gn(\mathcal{B}), i, gn(s)) = 0$. Still, the relevant condition in the definition of *Sub* is the second from last. It fails once more, since the conjunct "$Sub(gn(\mathcal{B}), i, gn(s)) > 0$" is false. Therefore,

$$Sub(gn(\mathcal{A}), i, gn(s)) > 0 \quad \text{iff} \quad \text{the term } s \text{ is substitutable for } v_i \text{ in } \mathcal{A}$$

We translate the above, as well as (3) and (4), to the formal domain using (1) (p. 274) and II.3.8–II.3.9:

$\vdash \boldsymbol{Sub}(\ulcorner \mathscr{A} \urcorner, \tilde{i}, \ulcorner s \urcorner) > 0$ iff   the term s is substitutable for v_i in $\mathscr{A}$

$\vdash \boldsymbol{Sub}(\ulcorner t[v_i] \urcorner, \tilde{i}, \ulcorner s \urcorner) = \ulcorner t[s] \urcorner$

$\vdash \boldsymbol{Sub}(\ulcorner \mathscr{A}[v_i] \urcorner, \tilde{i}, \ulcorner s \urcorner) = \ulcorner \mathscr{A}[s] \urcorner$ if the term s is substitutable for v_i in $\mathscr{A}$

The claims regarding $\boldsymbol{sub_i}$ (p. 268) are trivial consequences of the above.

We finally check $\boldsymbol{Proof}$.

Turning to the lightface version (see p. 271 for the boldface version), let $n = \langle gn(\mathscr{A}_0), \dots, gn(\mathscr{A}_{k-1}) \rangle$, where $k > 0$ and $\mathscr{A}_0, \dots, \mathscr{A}_{k-1}$ is a Γ-proof. Assume that $\Gamma(x)$ (and hence $\Gamma(x)$) is primitive recursive, an assumption that is correct in the case of **PA** (Exercise II.15)

That $Seq(n)$, and $lh(n) = k > 0$, is true outright. To see that $Proof(n)$ is true we now only need to investigate $(n)_i$ for $i < k$ and verify that it satisfies

$$LA((n)_i) \lor \Gamma((n)_i) \lor (\exists j)_{<i} (\exists m)_{<i} MP((n)_j, (n)_m, (n)_i) \lor$$
$$(\exists j)_{<i} EI((n)_j, (n)_i) \lor (\exists j)_{<i} SR((n)_j, (n)_i) \tag{8}$$

Now, $(n)_i = gn(\mathscr{A}_i)$, for $i < k$. The following exhaust all cases:[†]

Case 1. $\mathscr{A}_i \in \Lambda \cup \Gamma$. Then the first or second disjunct of (8) is true.

Case 2. For some $j < i$ and $m < i$, one has $\mathscr{A}_m \equiv \mathscr{A}_j \to \mathscr{A}_i$. Then $MP((n)_j, (n)_m, (n)_i)$ is true.

Case 3. For some $j < i$, one has $\mathscr{A}_j \equiv \mathscr{B} \to \mathscr{C}$, x is not free in $\mathscr{C}$, and $\mathscr{A}_i \equiv (\exists x).\mathscr{B} \to \mathscr{C}$. Then $EI((n)_j, (n)_i)$ is true.

Case 4. For some $j < i$, and some variable x and term t substitutable for x in $\mathscr{A}_j$, one has $\mathscr{A}_i \equiv \mathscr{A}_j[x \leftarrow t]$. Then $SR((n)_j, (n)_i)$ is true.

Thus, (8) is true under all possible cases.

Conversely, one can show by (metamathematical) course-of-values induction on $lh(n)$ that if $Proof(n)$ is true, then there is a Γ-proof $\mathscr{A}_0, \dots, \mathscr{A}_{k-1}$, where $k = lh(n) > 0$, such that $n = \langle gn(\mathscr{A}_0), \dots, gn(\mathscr{A}_{k-1}) \rangle$. Note that truth of $Proof(n)$ guarantees the truth of $Seq(n)$ and $lh(n) > 0$ (definition of $Proof$). Suppose that $k = 1$ (basis). Then the only disjuncts that can be true in (8) are the first two (why?). If the first one holds, then this implies that $(n)_0 = gn(\mathscr{B})$ for some formula $\mathscr{B} \in \Lambda$ (see (LA), p. 270); if the second one holds, then $(n)_0 = gn(\mathscr{B})$ for some formula $\mathscr{B} \in \Gamma$ (hence, in either case, the one-member sequence "$\mathscr{B}$" – which we may want to rename "$\mathscr{A}_0$" – is a Γ-proof). The reader

---

[†] Recall that we have agreed to allow the redundant *substitution* rule (p. 270).

will have no difficulty completing the induction. By the primitive recursiveness of ***Proof***† and II.3.9,

$$Proof(n) \text{ is true } \quad \text{iff} \quad \vdash \textbf{\textit{Proof}}(\tilde{n}) \qquad (9)$$

In particular, if $\mathscr{A}_0, \ldots, \mathscr{A}_{k-1}$ is some Γ-proof and we set

$$n = \langle gn(\mathscr{A}_0), \ldots, gn(\mathscr{A}_{k-1}) \rangle$$

and also set

$$t = \langle \ulcorner \mathscr{A}_0 \urcorner, \ldots, \ulcorner \mathscr{A}_{k-1} \urcorner \rangle$$

– the (formal) Gödel number of the proof – then $\vdash t = \tilde{n}$. Since $Proof(n)$ is true, (9) yields $\vdash \textbf{\textit{Proof}}(\tilde{n})$; hence

$$\vdash \textbf{\textit{Proof}}(t)$$

Now, noting that $\textbf{\textit{Proof}}^{\mathfrak{N}_c} = Proof$, the only-if direction of (9) also holds for our $\Sigma_1 \Gamma$ (by II.4.12). Thus, the claim we made above, "In particular, if $\mathscr{A}_0, \ldots, \mathscr{A}_{k-1}$ is, etc.", still holds *without* the restriction to a primitive recursive Γ. Is the *if* direction still true?   □

II.6.2 Definition (The Provability Predicate). The informal abbreviation Θ introduced below is called the *provability predicate*:

$$\Theta(x) \text{ stands for } (\exists y)(\textbf{\textit{Proof}}(y) \wedge (\exists i)_{<lh(y)} x = (y)_i) \qquad \qquad □$$

II.6.3 Remark. (1) The definition of Θ and the use of the term "provability formula" is modulo a fixed set of nonlogical axioms Γ. We have already fixed attention on such a set, but our attention may shift to other sets, Γ', Γ'', etc., on occasion. All that is required is to observe that:

- Any such primed Γ is a consistent extension of **PA**.
- The corresponding predicate, Γ, is Σ_1 over the language of C, $L_{\mathfrak{N}_c}$, where all the symbols (and more) that we use for Gödel coding belong.

(2) Note that we did not require x to be the last formula in the "proof" y.

---

† We have cheated here and allowed – for the purpose of this informal verification – the "local" assumption that Γ is primitive recursive. Recall that our "global" assumption on Γ – the one operative throughout this section, *outside this example* – is that it is a Σ_1 formula. In this connection the reader should note the concluding sentence of this example.

(3) $\Theta(x)$ is in $\Sigma_1(L_{\mathfrak{N}_c})$.

(4) From the work in the previous example we see that $\Theta(x)$ *intuitively* says that x is the Gödel number of some theorem. More precisely, if $\Gamma \vdash \mathscr{A}$ ($\mathscr{A}$ is over Γ's language) and we set $t = \ulcorner \mathscr{A} \urcorner$ (and also $n = gn(\mathscr{A})$), then $\Theta(\tilde{n})$ is a true $\Sigma_1(L_{\mathfrak{N}_c})$ *sentence*. By II.4.12,

$$\vdash \Theta(\tilde{n})$$

Hence (by $\vdash t = \tilde{n}$)

$$\vdash \Theta(t)$$

The reader is reminded that "$\vdash$" means "$\vdash_C$" unless noted otherwise.

(5) The preceding discussion completes the (sketch of) proof of I.9.33 given on p. 177. That is, that the lightface version of $\Theta(x)$, $\Theta(x)$, is semi-recursive. A number of issues that were then left open (e.g., the recursiveness of Λ and of the rules I_1 and I_2) have now been settled.

(6) It is useful to introduce an informal abbreviation for

$$\textbf{\textit{Proof}}(y) \wedge (\exists i)_{<lh(y)} x = (y)_i$$

We will use the name $\textbf{\textit{Deriv}}$ introduced by

$$\textbf{\textit{Deriv}}(y,x) \text{ stands for } \textbf{\textit{Proof}}(y) \wedge (\exists i)_{<lh(y)} x = (y)_i$$

That is, $\textbf{\textit{Deriv}}(y,x)$ intuitively says that the proof (coded by) y derives the theorem (coded by) x. $\square$

In the following lemmata we "do" some of the metatheory of arithmetic *within* formal arithmetic, having arithmetized the language. For example, the first lemma says that if we substitute a term into some variable of another term, then we obtain a term. The chapter concludes with a proof (second incompleteness theorem) that we *cannot* "do" *all* the metatheory within the theory.

II.6.4 Lemma.

$$\vdash Term(x) = 0 \wedge Term(z) = 0 \to Term(Sub(x,i,z)) = 0$$

Proof. We do this proof in some detail, since in most others in the following sequence of lemmata, we delegate much of the burden of proof to the reader. (Warning: This proof will be rather pedantic.)

We do (formal) course-of-values induction (II.1.23, p. 214) on x, proceeding according to the definition of ***Term*** (p. 267). The latter yields (cf. (9), p. 221)

$$\vdash \mathbf{Term}(x) = 0 \leftrightarrow var(x) \vee x = \langle 0, S0 \rangle$$
$$\vee Seq(x) \wedge (\exists y)_{<x}(lh(x) = Sy \qquad (1)$$
$$\wedge func((x)_0, y) \wedge (\forall z)_{<y} \mathbf{Term}((x)_{Sz}) = 0)$$

Now assume

$$\mathbf{Term}(x) = 0 \quad \text{and} \quad \mathbf{Term}(z) = 0 \qquad (2)$$

and prove

$$\mathbf{Term}(\mathbf{Sub}(x, i, z)) = 0 \qquad (3)$$

We have cases according to (1):

Case of $var(x)$:   Thus (p. 266) $\vdash (\exists j)_{\leq x} Var(x, j)$. We may now add a new constant a and the assumption

$$a \leq x \wedge Var(x, a)$$

The subcase $a = i$ and the definition of ***Sub***, (first case; cf. also the definition by cases in (15), p. 221) yield

$$\vdash \mathbf{Sub}(x, i, z) = z \qquad (4)$$

Similarly, the (only other) subcase, $\neg a = i$, and the definition of ***Sub*** (third case) yield

$$\vdash \mathbf{Sub}(x, i, z) = x \qquad (5)$$

Either of (4) or (5) yields (3) because of (2), and we are done in this case.

Pause. In the last subcase I used (tacitly) the "fact" that

$$\vdash Var(x, a) \rightarrow \neg a = i \rightarrow \neg Var(x, i)$$

Is this indeed a fact? (*Hint. lh.*)

Case of $x = \langle 0, S0 \rangle$:   The definition of ***Sub*** (second case) yields (5), and once again we have derived (3) from (2).

Finally the hard case, that is, that of the third disjunct in (1) above:

$$Seq(x) \wedge (\exists y)_{<x}(lh(x) = Sy$$
$$\wedge func((x)_0, y) \wedge (\forall z)_{<y} \mathbf{Term}((x)_{Sz}) = 0) \qquad (hc)$$

By (*hc*) we have

$$\vdash Seq(x) \tag{6}$$

and we may also introduce a new constant **b** and the assumption

$$b < x \wedge lh(x) = Sb$$
$$\wedge func((x)_0, b) \wedge (\forall z)_{<b} Term((x)_{Sz}) = 0 \tag{7}$$

On the other hand, the definition of **Sub** (fourth case) and (2) (cf. (15), p. 221) yields

$$\vdash Sub(x, i, z) = (\mu w)(Seq(w) \wedge lh(w) = lh(x) \wedge (w)_0 = (x)_0 \wedge$$
$$(\forall y)_{<lh(x)}(y > 0 \rightarrow (w)_y = Sub((x)_y, i, z))) \tag{8}$$

Using the abbreviation $t = Sub(x, i, z)$ for convenience, we get from (8) via $(f^{(4)})$ (p. 218)

$$\vdash Seq(t) \tag{9}$$

$$\vdash lh(t) = lh(x) \tag{10}$$

$$\vdash (t)_0 = (x)_0 \tag{11}$$

and

$$\vdash (\forall y)_{<lh(x)}(y > 0 \rightarrow (t)_y = Sub((x)_y, i, z)) \tag{12}$$

By (6) and II.2.13

$$\vdash y < lh(x) \rightarrow (x)_y < x \tag{13}$$

Now (7) yields

$$\vdash (\forall z)_{<lh(x)}(z > 0 \rightarrow Term((x)_z) = 0)$$

Thus (12) and (13) – via the I.H. – yield

$$\vdash (\forall y)_{<lh(x)}(y > 0 \rightarrow Term((t)_y) = 0) \tag{14}$$

that is, reintroducing **b** (see (7)),

$$\vdash (\forall y)_{<b} Term((t)_{Sy}) = 0 \tag{14'}$$

By (7), (14'), (10), and (11) we now have

$$\vdash lh(t) = Sb$$
$$\wedge func((t)_0, b) \wedge (\forall y)_{<b} Term((t)_{Sy}) = 0$$

284   II. The Second Incompleteness Theorem

Since $\vdash lh(t) \leq t$ by II.2.4(*i*), and therefore $\vdash b < t$ by the first conjunct of the above, we obtain

$$\vdash b < t \wedge lh(t) = Sb$$
$$\wedge func((t)_0, b) \wedge (\forall y)_{<b} Term((t)_{Sy}) = 0$$

Hence, using a new variable z, **Ax2**, and (9),

$$\vdash Seq(t) \wedge (\exists z)_{<t}(lh(t) = Sz$$
$$\wedge func((t)_0, z) \wedge (\forall y)_{<z} Term((t)_{Sy}) = 0) \tag{15}$$

By the definition of **Term**, $\vdash Term(t) = 0$ follows immediately from (15). $\square$

The above result (and the following suite of similar results) is much easier to prove if instead of free variables we have numerals. We just invoke II.3.8–II.3.9. However, we do need the versions with free variables.

The following says that a term is substitutable into any variable of another term.[†]

II.6.5 Corollary.

$$\vdash Term(x) = 0 \wedge Term(z) = 0 \rightarrow Sub(x, i, z) > 0$$

Proof. Assume the hypothesis (to the left of "$\rightarrow$"). Then (lemma above)

$$\vdash Term(Sub(x, i, z)) = 0 \tag{1}$$

Now

$$\vdash Term(x) = 0 \rightarrow Seq(x) \wedge lh(x) > 0$$

by inspection of the cases involved in the definition of **Term** (see (1) in the previous proof). By II.2.13,

$$\vdash Seq(x) \wedge lh(x) > 0 \rightarrow x > (x)_0$$

Thus, by II.1.5 and II.1.6,

$$\vdash Term(x) = 0 \rightarrow x > 0 \tag{2}$$

By substitution from (1), and (2), we have $\vdash Sub(x, i, z) > 0$. $\square$

[†] By a *variable of a term* we understand *any* variable – hence the free i in the corollary statement – not only one that actually occurs in the term. That is, we mean a variable x of t in the sense $t[x]$.

II.6.6 Lemma. $\vdash Term(x) = 0 \wedge Free(i,x) = 0 \rightarrow Sub(x,i,z) \geq z.$

Proof. Assume the hypothesis, i.e.,

$$Term(x) = 0$$

and

$$Free(i,x) = 0$$

We do (formal) course-of-values induction on x. By the definition of *Free* (p. 267), we have just two cases to consider. Here is why, and what:

In principle, we have the *three* cases as in the proof of II.6.4. However, the first case considered in II.6.4 has here only its *first* subcase tenable, namely, where $a = i$. This is because the other subcase – $\neg a = i$ – implies as before $\vdash var(x) \wedge \neg Var(x,i)$. Now this forces $\vdash Free(i,x) = S0$, for if we have $\vdash var(x)$, then $\vdash (x)_0 = \tilde{1}$; hence we cannot have any of $\vdash (\exists y)_{<x} func((x)_0, y),^\dagger \vdash AF(x)$, or $\vdash WFF(x) = 0$.

Thus, (4) of the proof of II.6.4 holds.

The second case in the proof of II.6.4, namely, $x = \langle 0, S0\rangle$ is also untenable as before (see (Fr), p. 267), since here $\vdash (x)_0 = 0$.

This leaves the hard case (hc), which yields (8) of the proof of II.6.4, namely,

$$\vdash Sub(x,i,z) = (\mu w)\Big(Seq(w) \wedge lh(w) = lh(x) \wedge (w)_0 = (x)_0$$
$$\wedge (\forall y)_{<lh(x)}(y > 0 \rightarrow (w)_y = Sub((x)_y,i,z))\Big)$$

Thus, by $(f^{(4)})$ (p. 218), tautological implication, and eliminating $\forall$,

$$\vdash 0 < y \wedge y < lh(x) \rightarrow (Sub(x,i,z))_y = Sub((x)_y,i,z) \tag{1}$$

The definition of *Free* (p. 267) yields

$$\vdash (\exists y)_{<lh(x)}(0 < y \wedge Free(i,(x)_y) = 0)$$

Invoking proof by auxiliary constant, we now add

$$0 < a \wedge a < lh(x) \wedge Free(i,(x)_a) = 0 \tag{2}$$

where a is a new constant. By (2) (first two conjuncts) and (1),

$$\vdash (Sub(x,i,z))_a = Sub((x)_a,i,z) \tag{3}$$

Now, since (cf. (Trm), p. 267)

$$\vdash Term(x) = 0 \rightarrow (\forall z)_{<lh(x)}(0 < z \rightarrow Term((x)_z) = 0)$$

---

† Since $\vdash lh((x)_0) \leq (x)_0$ and (see $(Func)$, p. 266) $\vdash lh((x)_0) \geq \tilde{6}$.

we obtain by hypothesis and specialization

$$\vdash 0 < a \wedge a < lh(x) \to Term((x)_a) = 0$$

Thus

$$\vdash Term((x)_a) = 0 \tag{4}$$

via the first two conjuncts of (2). Using now (2) (last conjunct), (4), and the I.H. – this uses II.2.13 – we get

$$\vdash Sub((x)_a, i, z) \geq z \tag{5}$$

Since $\vdash Seq(Sub(x, i, z))$, II.2.13 yields

$$\vdash Sub(x, i, z) > (Sub(x, i, z))_a$$

By (3) and (5) we now have $\vdash Sub(x, i, z) > z$. □

The next three claims (stated without proof, since their proofs are trivial variations of the preceding three) take care of atomic formulas.

II.6.7 Lemma.

$$\vdash AF(x) \wedge Term(z) = 0 \to AF(Sub(x, i, z)) = 0$$

II.6.8 Corollary.

$$\vdash AF(x) \wedge Term(z) = 0 \to Sub(x, i, z) > 0$$

II.6.9 Lemma. $\vdash AF(x) \wedge Free(i, x) = 0 \to Sub(x, i, z) \geq z.$

For the next result we want to ensure that a substitution actually took place (hence the condition on substitutability, "$Sub(x, i, z) > 0$", which was unnecessary in the cases of term or atomic formula targets).

II.6.10 Proposition.

$$\vdash Term(x) = 0 \vee WFF(x) = 0 \to$$
$$Sub(x, i, z) > 0 \wedge Free(i, x) = 0 \to Sub(x, i, z) \geq z$$

Proof. The case for $Term(x) = 0$ is II.6.6 above. The subcase $AF(x)$ for $WFF(x) = 0$ is II.6.9 (for either of these the assumption $Sub(x, i, z) > 0$ is redundant).

We consider here one among the other subcases of $WFF(x) = 0$. Once again, we employ course-of-values induction on x, legitimized by II.2.13.[†] Add then the assumptions $WFF(x) = 0$, $Sub(x, i, z) > 0$ and $Free(i, x) = 0$.

Subcase. $x = \langle \ulcorner \vee \urcorner, (x)_{S0}, (x)_{SS0} \rangle$. Thus (definition of WFF, p. 267)

$$\vdash WFF((x)_{S0}) = 0$$

and

$$\vdash WFF((x)_{SS0}) = 0$$

By the definition of *Free* (p. 267) we now have

$$\vdash Free(i, (x)_{S0}) = 0 \vee Free(i, (x)_{SS0}) = 0 \tag{1}$$

By definition of *Sub* we obtain

$$\vdash Sub(x, i, z) = \langle \ulcorner \vee \urcorner, Sub((x)_{S0}, i, z), Sub((x)_{SS0}, i, z) \rangle \tag{2}$$

Pause. Wait a minute! The above is so *provided* that

$$\vdash Sub((x)_{S0}, i, z) > 0 \wedge Sub((x)_{SS0}, i, z) > 0$$

Is this satisfied? Why?

By (1), we consider cases. So add $Free(i, (x)_{S0}) = 0$. The I.H. and $\vdash Sub((x)_{S0}, i, z) > 0$ yield

$$\vdash Sub((x)_{S0}, i, z) \geq z$$

and hence

$$\vdash Sub(x, i, z) \geq z$$

exactly as in II.6.6, invoking $\vdash \langle \dots, u, \dots \rangle > u$. The other case works out as well. $\qquad \square$

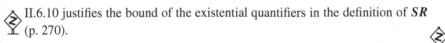 II.6.10 justifies the bound of the existential quantifiers in the definition of SR (p. 270).

II.6.11 Lemma.

$$\vdash WFF(x) = 0 \wedge Term(z) = 0 \wedge Sub(x, i, z) > 0 \rightarrow$$
$$WFF(Sub(x, i, z)) = 0$$

[†] This is the last time we are reminded of the role of II.2.13 in our inductions on Gödel numbers.

Proof. We add the assumptions

$$WFF(x) = 0$$
$$Term(z) = 0$$

and

$$Sub(x, i, z) > 0 \tag{1}$$

and do (formal) course-of-values induction on x, proceeding according to the definition of WFF (p. 267), towards proving

$$\vdash WFF(Sub(x, i, z)) = 0$$

We illustrate what is involved by considering one case, leaving the rest to the reader.

Case "$\neg$". $x = \langle \ulcorner \neg \urcorner, y \rangle$ and $\vdash WFF(y) = 0$, where we have used the abbreviation $y = (x)_{S0}$ for convenience.

We also add the assumption (which we hope to contradict)

$$Sub(y, i, z) = 0$$

Thus,

$$\vdash \neg Sub(y, i, z) > 0$$

and therefore, via tautological implication and **Ax4** (since $\vdash (x)_{S0} = y$),

$$\vdash \neg(WFF(x) = 0 \land (x)_0 = \ulcorner \neg \urcorner \land Sub((x)_{S0}, i, z) > 0)$$

Thus, the "otherwise"[†] in the definition of *Sub* (p. 268) is provable, since $(x)_0 = \ulcorner \neg \urcorner$ is refutable in all the other cases. Therefore, $\vdash Sub(x, i, z) = 0$, contradicting the assumption (1).

We have just established

$$\vdash Sub(y, i, z) > 0$$

Hence also (definition of *Sub*)

$$\vdash Sub(x, i, z) = \langle \ulcorner \neg \urcorner, Sub(y, i, z) \rangle \tag{2}$$

By I.H.,

$$\vdash WFF(Sub(y, i, z)) = 0$$

[†] That is, the conjunction of the negations of all the explicit cases.

Hence, by the definition of $\boldsymbol{WFF}$,

$$\vdash \boldsymbol{WFF}(\langle \ulcorner \neg \urcorner, \boldsymbol{Sub}(y,i,z) \rangle) = 0$$

Thus

$$\vdash \boldsymbol{WFF}(\boldsymbol{Sub}(x,i,z)) = 0$$

by (2) and **Ax4**. □

Pause. Where was the assumption $\boldsymbol{Term}(z) = 0$ used?

II.6.12 Lemma.
$$\vdash (\forall j)\, \boldsymbol{Free}(j,z) > 0 \wedge \boldsymbol{Term}(z) = 0 \wedge$$
$$\boldsymbol{WFF}(x) = 0 \rightarrow \boldsymbol{Sub}(x,i,z) > 0$$

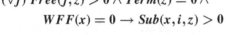 This says that closed terms are always substitutable.

Proof. We assume

$$(\forall j)\boldsymbol{Free}(j,z) > 0$$

$$\boldsymbol{Term}(z) = 0$$

and

$$\boldsymbol{WFF}(x) = 0$$

and do (formal) course-of-values induction on x proceeding according to the definition of $\boldsymbol{WFF}$ to show

$$\vdash \boldsymbol{Sub}(x,i,z) > 0$$

We illustrate what is involved by considering one case, leaving the remaining cases to the reader.

Case "∃". Add $x = \langle \ulcorner \exists \urcorner, y, w \rangle$, $\boldsymbol{WFF}(w) = 0$, and $\boldsymbol{var}(y)$, where we have used the abbreviations $y = (x)_{s0}$ and $w = (x)_{ss0}$ for convenience. The latter expands to (see (var), p. 266)

$$\vdash (\exists j)_{\leq y} \boldsymbol{Var}(y,j) \tag{3}$$

Assume first the interesting subcase.

Subcase. Add $\neg \boldsymbol{Var}(y,i)$. By (3) we may add a new constant $\boldsymbol{a}$ and the assumption

$$a \leq y \wedge \boldsymbol{Var}(y,a) \tag{4}$$

By $\vdash (\forall j) Free(j, z) > 0$

$$\vdash Free(a, z) > 0$$

Hence by (4) and **Ax2**

$$\vdash (\exists j)_{\leq y}(Var(y, j) \wedge Free(j, z) > 0)$$

By the I.H. $\vdash Sub(w, i, z) > 0$; hence (definition of *Sub*)

$$\vdash Sub(x, i, z) = \langle \ulcorner \exists \urcorner, y, Sub(w, i, z) \rangle$$

Thus, $\vdash Sub(x, i, z) > 0$ (e.g., $\vdash Sub(x, i, z) > \ulcorner \exists \urcorner$).

Subcase. Add $Var(y, i)$. Now

$$\vdash Sub(x, i, z) = x$$

Therefore, once more, $\vdash Sub(x, i, z) > 0$. (Why is $\vdash x > 0$?)   □

II.6.13 Lemma. $\vdash Proof(x) \wedge Proof(y) \rightarrow Proof(x * y)$.

Proof. Assume $Proof(x)$ and $Proof(y)$. Thus,

$$
\begin{aligned}
&\vdash Seq(x) \wedge lh(x) > 0 \wedge \\
&(\forall i)_{<lh(x)}(LA((x)_i) \vee \Gamma((x)_i) \vee \\
&(\exists j)_{<i}(\exists k)_{<i} MP((x)_j, (x)_k, (x)_i) \vee \\
&(\exists j)_{<i} EI((x)_j, (x)_i) \vee \\
&(\exists j)_{<i} SR((x)_j, (x)_i))
\end{aligned}
\tag{1}
$$

and

$$
\begin{aligned}
&\vdash Seq(y) \wedge lh(y) > 0 \wedge \\
&(\forall i)_{<lh(y)}(LA((y)_i) \vee \Gamma((y)_i) \vee \\
&(\exists j)_{<i}(\exists k)_{<i} MP((y)_j, (y)_k, (y)_i) \vee \\
&(\exists j)_{<i} EI((y)_j, (y)_i) \vee \\
&(\exists j)_{<i} SR((y)_j, (y)_i))
\end{aligned}
\tag{2}
$$

By II.2.21 (p. 247) we have

$$\vdash Seq(x * y)$$

$$\vdash lh(x * y) = lh(x) + lh(y) \tag{3}$$

Hence

$$\vdash Seq(x * y) \wedge lh(x * y) > 0 \tag{4}$$

(4) settles the first two conjuncts of **Proof**$(x * y)$. We now address the last conjunct, $(\forall i)_{<lh(x*y)}(LA((x * y)_i) \vee \ldots)$. It suffices to drop the quantifier and show

$$
\begin{aligned}
\vdash\ &i < lh(x * y) \rightarrow \\
&LA((x * y)_i) \vee \Gamma((x * y)_i) \vee \\
&(\exists j)_{<i}(\exists k)_{<i} MP((x * y)_j, (x * y)_k, (x * y)_i) \vee \\
&(\exists j)_{<i} EI((x * y)_j, (x * y)_i) \vee \\
&(\exists j)_{<i} SR((x * y)_j, (x * y)_i)
\end{aligned} \tag{5}
$$

We note, again by II.2.21,

$$
\begin{aligned}
\vdash\ &(\forall i)_{<lh(x)}(x * y)_i = (x)_i \wedge \\
&(\forall i)_{<lh(y)}(x * y)_{i+lh(x)} = (y)_i
\end{aligned} \tag{6}
$$

To prove (5), add $i < lh(x * y)$. By (3) and (6) we have two cases.

Case 1. $i < lh(x)$. By (1) and (6) (first conjunct) we obtain (using **Ax4**)

$$
\begin{aligned}
\vdash\ &LA((x * y)_i) \vee \Gamma((x * y)_i) \vee \\
&(\exists j)_{<i}(\exists k)_{<i} MP((x * y)_j, (x * y)_k, (x * y)_i) \vee \\
&(\exists j)_{<i} EI((x * y)_j, (x * y)_i) \vee \\
&(\exists j)_{<i} SR((x * y)_j, (x * y)_i)
\end{aligned} \tag{7}
$$

Case 2. $i \geq lh(x)$. Setting $m = \delta(i, lh(x))$, we get $\vdash m + lh(x) = i$ (cf. II.1.27) and $\vdash m < lh(y)$ (using (3)). By (2),

$$
\begin{aligned}
\vdash\ &LA((y)_m) \vee \Gamma((y)_m) \vee \\
&(\exists j)_{<m}(\exists k)_{<m} MP((y)_j, (y)_k, (y)_m) \vee \\
&(\exists j)_{<m} EI((y)_j, (y)_m) \vee \\
&(\exists j)_{<m} SR((y)_j, (y)_m)
\end{aligned}
$$

Hence, bringing in (6) (second conjunct) via **Ax4** and noting that $\vdash j < m \rightarrow j < lh(y)$, we get

$$
\begin{aligned}
\vdash\ &LA((x * y)_{m+lh(x)}) \vee \Gamma((x * y)_{m+lh(x)}) \vee \\
&(\exists j)_{<m}(\exists k)_{<m} MP((x * y)_{j+lh(x)}, (x * y)_{k+lh(x)}, (x * y)_{m+lh(x)}) \vee \\
&(\exists j)_{<m} EI((x * y)_{j+lh(x)}, (x * y)_{m+lh(x)}) \vee \\
&(\exists j)_{<m} SR((x * y)_{j+lh(x)}, (x * y)_{m+lh(x)})
\end{aligned}
$$

Translating the above (via proof by cases) in terms of i, we obtain (7) once more.

To sample one case, we translate the last disjunct. We claim that

$$\vdash (\exists j)_{<m} SR((x * y)_{j+lh(x)}, (x * y)_{m+lh(x)}) \rightarrow$$
$$(\exists j)_{<i} SR((x * y)_j, (x * y)_i)$$

Assume the hypothesis, and add a new constant a and the assumption

$$a < m \wedge SR((x * y)_{a+lh(x)}, (x * y)_{m+lh(x)})$$

By II.1.17, $\vdash a + lh(x) < m + lh(x)$, that is, $a + lh(x) < i$. The above now yields

$$a + lh(x) < i \wedge SR((x * y)_{a+lh(x)}, (x * y)_i)$$

Hence (**Ax2**)

$$\vdash (\exists j)(j < i \wedge SR((x * y)_j, (x * y)_i))$$

After all this, the deduction theorem establishes (5).   □

II.6.14 Lemma.

$$\vdash Proof(w) \wedge (\exists j)_{<lh(w)}(\exists k)_{<lh(w)} MP((w)_j, (w)_k, x)$$
$$\rightarrow Proof(w * \langle x \rangle)$$

Proof. Assume

$$Proof(w) \tag{8}$$

and

$$(\exists j)_{<lh(w)}(\exists k)_{<lh(w)} MP((w)_j, (w)_k, x) \tag{9}$$

and set $t = w * \langle x \rangle$. Since $\vdash Seq(t) \wedge lh(t) = S(lh(w))$, the first two required conjuncts for $Proof(w * \langle x \rangle)$ are settled. To conclude, we add $i < lh(t)$, that is $(< 2) i < lh(w) \vee i = lh(w)$, and prove

$$\vdash LA((t)_i) \vee \Gamma((t)_i) \vee$$
$$(\exists j)_{<i}(\exists k)_{<i} MP((t)_j, (t)_k, (t)_i) \vee$$
$$(\exists j)_{<i} EI((t)_j, (t)_i) \vee \tag{10}$$
$$(\exists j)_{<i} SR((t)_j, (t)_i)$$

There are two cases:

Case 1. $i < lh(w)$. Thus (cf. (6) (first conjunct) of the previous proof)

$$\vdash (t)_i = (w)_i \tag{11}$$

By (8),

$$\vdash LA((w)_i) \vee \Gamma((w)_i) \vee$$
$$(\exists j)_{<i}(\exists k)_{<i} MP((w)_j, (w)_k, (w)_i) \vee$$
$$(\exists j)_{<i} EI((w)_j, (w)_i) \vee$$
$$(\exists j)_{<i} SR((w)_j, (w)_i)$$

Now (10) follows from the above and (11) via **Ax4**.

Case 2. $i = lh(w)$. Thus, $\vdash (t)_i = x$. By (9),

$$\vdash (\exists j)_{<lh(w)}(\exists k)_{<lh(w)} MP((w)_j, (w)_k, (t)_i)$$

or, using (11),

$$\vdash (\exists j)_{<i}(\exists k)_{<i} MP((t)_j, (t)_k, (t)_i)$$

(10) now follows by tautological implication. □

II.6.15 Lemma.

$$\vdash Proof(w) \wedge (\exists j)_{<lh(w)} SR((w)_j, x) \rightarrow Proof(w * \langle x \rangle)$$

Proof. A trivial rephrasing of the previous proof. □

II.6.16 Exercise. Show by a formal course-of-values induction on i that

$$\vdash i < lh(x) \wedge Proof(x) \rightarrow WFF((x)_i) = 0$$

Moreover, show that this is so regardless of whether one adopts MP, EI, and SR as we did on p. 270 or, instead, one adopts their primed versions (cf. II.5.1, p. 271). □

II.6.17 Corollary.

$$\vdash Deriv(y, u) \wedge Deriv(z, \langle \ulcorner \rightarrow \urcorner, u, x \rangle) \rightarrow Deriv(y * z * \langle x \rangle, x)$$

Note. We have used the abbreviation "$\langle \ulcorner \rightarrow \urcorner, u, x \rangle$" for "$\langle \ulcorner \vee \urcorner, \langle \ulcorner \neg \urcorner, u \rangle, x \rangle$".

Proof. By Lemma II.6.13, $\vdash Proof(y * z)$. The result follows at once from II.6.14 since $\vdash MP(u, \langle \ulcorner \rightarrow \urcorner, u, x \rangle, x)$.

This last claim also uses II.6.16, since the hypothesis (to the left of "$\rightarrow$") yields $\vdash WFF(u) = 0$ and $\vdash WFF(\langle \ulcorner \rightarrow \urcorner, u, x \rangle) = 0$. □

A special case of the above is: *For any formulas $\mathscr{A}$ and $\mathscr{B}$,*

$$\vdash \boldsymbol{Deriv}(y, \ulcorner \mathscr{A} \urcorner) \wedge \boldsymbol{Deriv}(z, \ulcorner \mathscr{A} \rightarrow \mathscr{B} \urcorner)$$
$$\rightarrow \boldsymbol{Deriv}(y * z * \langle \ulcorner \mathscr{B} \urcorner \rangle, \ulcorner \mathscr{B} \urcorner)$$

II.6.18 Corollary.

$$\vdash \boldsymbol{Term}(z) = 0 \wedge \boldsymbol{Sub}(x, i, z) > 0 \rightarrow$$
$$\boldsymbol{Deriv}(y, x) \rightarrow \boldsymbol{Deriv}(y * \langle \boldsymbol{Sub}(x, i, z) \rangle, \boldsymbol{Sub}(x, i, z))$$

Proof. We add $\boldsymbol{Term}(z) = 0$, $\boldsymbol{Sub}(x, i, z) > 0$, and $\boldsymbol{Deriv}(y, x)$. That

$$\vdash \boldsymbol{Deriv}(y * \langle \boldsymbol{Sub}(x, i, z) \rangle, \boldsymbol{Sub}(x, i, z)) \tag{12}$$

will then follow at once from II.6.15 once we prove

$$\vdash \boldsymbol{SR}(x, \boldsymbol{Sub}(x, i, z)) \tag{13}$$

We look first at the interesting case: That is, we add $\boldsymbol{Free}(i, x) = 0$.

To prove (13) we need (see definition of $\boldsymbol{SR}$, p. 270) to prove

$$\vdash \boldsymbol{WFF}(x) = 0 \wedge \boldsymbol{WFF}(\boldsymbol{Sub}(x, i, z)) = 0 \wedge$$
$$(\exists i)_{\leq x}(\exists w)_{\leq \boldsymbol{Sub}(x, i, z)}(\boldsymbol{Term}(w) = 0 \wedge \tag{14}$$
$$\boldsymbol{Sub}(x, i, z) = \boldsymbol{Sub}(x, i, w))$$

or, simply,

$$\vdash \boldsymbol{WFF}(x) = 0 \wedge \boldsymbol{WFF}(\boldsymbol{Sub}(x, i, z)) = 0 \wedge$$
$$i \leq x \wedge z \leq \boldsymbol{Sub}(x, i, z)$$

because then the third conjunct of (14) follows by MP and **Ax2** from

$$\vdash i \leq x \wedge z \leq \boldsymbol{Sub}(x, i, z) \wedge \boldsymbol{Term}(z) = 0 \wedge$$
$$\boldsymbol{Sub}(x, i, z) = \boldsymbol{Sub}(x, i, z)$$

That is, we need

$$\vdash \boldsymbol{WFF}(x) = 0 \tag{i}$$

$$\vdash \boldsymbol{WFF}(\boldsymbol{Sub}(x, i, z)) = 0 \tag{ii}$$

$$\vdash i \leq x \tag{iii}$$

$$\vdash z \leq \boldsymbol{Sub}(x, i, z) \tag{iv}$$

We get (i) by $\vdash \boldsymbol{Deriv}(y,x)$ (II.6.16), while (ii) follows from II.6.11, (i), and the assumptions. We get (iii) from the definition of $\boldsymbol{Free}$. Finally, (iv) is by II.6.10.

The other case is to add $\boldsymbol{Free}(i,x) > 0$. Then $\vdash \boldsymbol{Sub}(x,i,z) = x$ (see Exercise II.19); hence we need only show that

$$\vdash \boldsymbol{Deriv}(y,x) \to \boldsymbol{Deriv}(y * \langle x \rangle, x)$$

We leave this as an easy exercise (Exercise II.20). □

 We can also state a special case: *For any formula $\mathscr{A}$, variable v_i and term t substitutable for v_i in $\mathscr{A}$,*

$$\vdash \boldsymbol{Deriv}(y, \ulcorner \mathscr{A} \urcorner) \to \boldsymbol{Deriv}(y * \langle sub_i(\ulcorner \mathscr{A} \urcorner, \ulcorner t \urcorner) \rangle, sub_i(\ulcorner \mathscr{A} \urcorner, \ulcorner t \urcorner))$$

Translating the last two corollaries in terms of Θ we obtain

II.6.19 Corollary.

$$\vdash \Theta(u) \wedge \Theta(\langle \ulcorner \to \urcorner, u, x \rangle) \to \Theta(x)$$

II.6.20 Corollary.

$$\vdash \boldsymbol{Term}(z) = 0 \wedge \boldsymbol{Sub}(x,i,z) > 0 \to \Theta(x) \to \Theta(\boldsymbol{Sub}(x,i,z))$$

 II.6.21 Remark. (1) A special case of Corollary II.6.19 is worth stating: *For any formulas $\mathscr{A}$ and $\mathscr{B}$ over $L_{\mathfrak{A}}$,*

$$\vdash \Theta(\ulcorner \mathscr{A} \urcorner) \wedge \Theta(\ulcorner \mathscr{A} \to \mathscr{B} \urcorner) \to \Theta(\ulcorner \mathscr{B} \urcorner)$$

(2) We omit the rather trivial ("trivial" given the tools we already have developed, that is) proofs of II.6.19–II.6.20. Suffice it to observe that, for example, using **Ax2** and $\models_{\mathbf{Taut}}$, we get from II.6.17

$$\vdash \boldsymbol{Deriv}(y,u) \wedge \boldsymbol{Deriv}(z, \langle \ulcorner \to \urcorner, u, x \rangle) \to \Theta(x)$$

Using ∃-introduction, we can now get rid of y and z. □

In order to focus the mind, we now state the *derivability conditions* that hold for Θ (*Ableitbarkeitsforderungen* of Hilbert and Bernays (1968)). We will need to establish that they indeed do hold in order to meet our goal of this section.

II.6.22 Definition (Derivability Conditions). The following statements are the *derivability conditions* for the derivability predicate Θ appropriate for Γ. For any formulas $\mathscr{A}$ and $\mathscr{B}$ over $L_{\mathfrak{A}}$,[†]

DC 1. If $\vdash_\Gamma \mathscr{A}$, then $\vdash_C \Theta(\ulcorner \mathscr{A} \urcorner)$.

DC 2. $\vdash_C \Theta(\ulcorner \mathscr{A} \urcorner) \wedge \Theta(\ulcorner \mathscr{A} \to \mathscr{B} \urcorner) \to \Theta(\ulcorner \mathscr{B} \urcorner)$.

DC 3. For any primitive recursive term t over $L_{\mathfrak{N}_C}$ and any numbers $a_1, \ldots, a_n$ in $\mathbb{N}$, $\vdash_C t(\widetilde{a_1}, \ldots, \widetilde{a_n}) = 0 \to \Theta(\ulcorner t(\widetilde{a_1}, \ldots, \widetilde{a_n}) = 0 \urcorner)$. $\qquad\square$

II.6.23 Remark. DC 3 above is pretty much the Hilbert-Bernays (1968) third derivability condition. **DC1–DC2** are actually Löb's versions. Löb also has a different third derivability condition – a formalized version of **DC 1** – that we prove later (see II.6.38). $\qquad\square$

Now, **DC 1** was settled in II.6.3(4), while **DC 2** is II.6.21(1). Thus we focus our effort on proving **DC 3**. To this end, we will find it convenient to prove a bit more: namely, that **DC 3** is true even if instead of the $\widetilde{a_1}, \ldots, \widetilde{a_n}$ we use *free variables* $x_1, \ldots, x_n$. To formulate such a version we will employ, for any term *or* formula $E(x_1, \ldots, x_n)$,[‡] a *primitive recursive term* $g_E(x_1, \ldots, x_n)$ such that, for all $a_1, \ldots, a_n$ in $\mathbb{N}$,

$$\vdash_C \ulcorner E(\widetilde{a_1}, \ldots, \widetilde{a_n}) \urcorner = g_E(\widetilde{a_1}, \ldots, \widetilde{a_n}) \tag{1}$$

Assume for a moment that we have obtained such a family of g-terms, and let $g_{t=0}(x_1, \ldots, x_n)$ be appropriate for the formula $t(x_1, \ldots, x_n) = 0$. If now we manage to prove

$$\vdash t(x_1, \ldots, x_n) = 0 \to \Theta(g_{t=0}(x_1, \ldots, x_n))$$

then **DC 3** follows by substitution and **Ax4** from (1) above.

To this end, we first address an important special case: We introduce a function symbol ***Num*** such that

$$\vdash \textbf{\textit{Num}}(\widetilde{n}) = \ulcorner \widetilde{n} \urcorner \qquad \text{for all } n \in \mathbb{N} \tag{2}$$

Thus, we are saying that $\textbf{\textit{Num}} = g_x$, that is, a g-term appropriate for the term x.

The symbol ***Num*** – and its defining axioms below – are in $L_{\mathfrak{N}_C}$ and C respectively, since the introducing axioms make it clear that ***Num*** is primitive recursive.

[†] In the interest of clarity – and emphasis – we have retained the subscript C of $\vdash$, wherever it was applicable throughout the definition.

[‡] Recall the convention on round brackets, p. 18.

We define

$$Num(0) = \ulcorner 0 \urcorner$$
$$Num(Sx) = \langle \ulcorner S \urcorner, Num(x) \rangle \qquad (Num)$$

To see that the above definition behaves as required by (2), we do *meta-mathematical* induction on $n \in \mathbb{N}$: For $n = 0$, the claim is settled by the first equation in the group *(Num)* and by $\vdash \tilde{0} = 0$ (definition of "$\tilde{n}$", p. 171).

Assume the claim for a fixed n. Now (definition), $\vdash \widetilde{n+1} = S\tilde{n}$; thus (**Ax4** and second equation in *(Num)*)

$$\vdash Num(\widetilde{n+1}) = \langle \ulcorner S \urcorner, Num(\tilde{n}) \rangle$$
$$\vdash \langle \ulcorner S \urcorner, Num(\tilde{n}) \rangle = \langle \ulcorner S \urcorner, \ulcorner \tilde{n} \urcorner \rangle \qquad \text{by I.H. and } \mathbf{Ax4}$$
$$\vdash \langle \ulcorner S \urcorner, \ulcorner \tilde{n} \urcorner \rangle = \ulcorner S\tilde{n} \urcorner \qquad \text{by definition of } \text{``}\ulcorner St \urcorner\text{''}$$
$$\vdash \ulcorner S\tilde{n} \urcorner = \ulcorner \widetilde{n+1} \urcorner \qquad \text{by definition of } \tilde{n}$$

Intuitively, $Num(x)$ is the Gödel number of

$$\underbrace{S \ldots S}_{x \text{ copies}} 0 \qquad (3)$$

where x is a *formal variable* (not just a *closed* term $\tilde{n}$). If we are right in this assessment, then it must be case that $\vdash Term(Num(x)) = 0$. Moreover, even though the expression in (3),[†] *intuitively,* "depends" on x, it still contains no variables (it is a sequence of copies of the symbol "S" followed by "0"), so that we expect $\vdash Free(y, Num(x)) > 0$.

Both expectations are well founded.

II.6.24 Lemma. $\vdash Term(Num(x)) = 0$.

Proof. Formal induction on x. For $x = 0$, we want $\vdash Term(\ulcorner 0 \urcorner) = 0$. This is the case, by II.3.8, since $Term(gn(0)) = 0$ is true.

We now prove

$$\vdash Term(Num(Sx)) = 0 \qquad (4)$$

based on the obvious I.H. It suffices (by *(Num)* above) to prove

$$\vdash Term(\langle \ulcorner S \urcorner, Num(x) \rangle) = 0$$

[†] Of course, such an expression is *not* well formed in our formalism, because it is a variable length term. This is why we emphasized the word "intuitively" twice in this comment.

Turning to the definition of **Term** (p. 267), we find that the third disjunct is provable, since

$$\vdash Seq\left(\langle \ulcorner S \urcorner, Num(x)\rangle\right) \wedge lh(\langle \ulcorner S \urcorner, Num(x)\rangle) = SS0$$
$$\wedge func((\langle \ulcorner S \urcorner, Num(x)\rangle)_0, S0)$$
$$\wedge Term((\langle \ulcorner S \urcorner, Num(x)\rangle)_{S0}) = 0\Big), \qquad \text{the last conjunct by I.H.}$$

We are done, by definition by cases ((15), p. 221).   □

II.6.24 says that **Num(x)** "behaves" like the Gödel number of a term. It does *not* say that there is a term t such that $Num(x) = \ulcorner t \urcorner$. (See also the previous footnote, and also recall that Gödel numbers of specific expressions are *closed* terms.)

II.6.25 Lemma. $\vdash Free(y, Num(x)) > 0$.

Proof. We do (formal) induction on x. For $x = 0$ we want $\vdash Free(y, \ulcorner 0 \urcorner) > 0$, which is correct (if we examine the cases in the definition of **Free**, p. 267, only the "otherwise" applies).

To conclude, we examine $Free(y, Num(Sx))$ (with frozen free variables x and y). By II.6.24,

$$\vdash Term(Num(Sx)) = 0$$

so we need only examine (see (Fr), p. 267)

$$(\exists z)_{<SS0}(0 < z \wedge Free(y, (\langle \ulcorner S \urcorner, Num(x)\rangle)_z) = 0) \qquad (5)$$

We want to *refute* (5), so we add it as an assumption (with frozen free variables x and y – proof by contradiction, I.4.21).

We take a new constant, a, and assume

$$a < SS0 \wedge 0 < a \wedge Free(y, (\langle \ulcorner S \urcorner, Num(x)\rangle)_a) = 0$$

One now gets $\vdash a = S0$ from the first two conjuncts;[†] hence

$$\vdash Free(y, Num(x)) = 0$$

from the last conjunct.

We have just contradicted the I.H. $Free(y, Num(x)) > 0$. Thus, the negation of (5) is provable. It now follows that the "otherwise" case (conjunction of the negations of the explicit cases) is provable in the definition of

---

[†] The first yields $a \leq S0$, and the second yields $S0 \leq a$.

$Free(y, Num(Sx))$. Therefore, $\vdash Free(y, Num(Sx)) = S0$ (see also (15), p. 221). □

It is now clear what the g_E-terms must be:

Let $E(v_{i_1}, \ldots, v_{i_n})$ be a term *or* formula where the v_{i_j} are (names for) the *formal object variables*, as these were constructed on p. 166 from the symbols of a finite alphabet. Let us introduce the *informal abbreviation*

$$s_i(x, y) = sub_i(x, Num(y)) \qquad (s)$$

 Informal, because we do not need to introduce a formal symbol. All we need is convenience. Were we to introduce s_i formally, it would then be a primitive recursive function symbol of $L_{\mathfrak{N}_C}$ and (s) would be the defining axiom (essentially, composition). As it stands now, it is just a metatheoretical symbol.

Then $g_E(x_1, \ldots, x_n)$ – where the x_i are arbitrary metavariables, possibly overlapping with the v_{i_j} – is an informal abbreviation for the term[†]

$$s_{i_n}(\ldots s_{i_2}(s_{i_1}(\ulcorner E \urcorner, x_1), x_2), \ldots, x_n) \qquad (6)$$

Indeed **Ax4**, (2) (p. 296), and (s) above imply, for any $n \in \mathbb{N}$,

$$\vdash_C s_{i_j}(x, \widetilde{n}) = sub_{i_j}(x, \ulcorner \widetilde{n} \urcorner)$$

Thus (p. 279)

$$\vdash_C s_{i_j}(\ulcorner E[v_{i_j}] \urcorner, \widetilde{n}) = \ulcorner E[\widetilde{n}] \urcorner$$

Repeating the above for each free variable of E, we obtain

$$\vdash_C g_E(\widetilde{m}_1, \ldots, \widetilde{m}_n) = \ulcorner E(\widetilde{m}_1, \ldots, \widetilde{m}_n) \urcorner$$

It is clear that g_E is a primitive recursive term (each $sub_i(x, y)$ is).

 We abbreviate (6) – i.e., $g_E(x_1, \ldots, x_n)$ – by yet another informal notation as in Hilbert and Bernays (1968):

$$\{E(x_1, \ldots, x_n)\} \qquad (7)$$

The reader will exercise care not to confuse the *syntactic* (meta)variables x_j introduced in (6) and (7) with the formal variables that occur free in the expression E. The latter are the v_{i_j}, as already noted when introducing (6). In particular, the "i" in sub_i is that of the formal v_i, not that of the "syntactic" x_i.[‡]

[†] By II.6.25, the order of substitutions is irrelevant. "E" and "$E(v_{i_1}, \ldots, v_{i_n})$" are, of course, interchangeable. The latter simply indicates explicitly the list of all free variables of E.

[‡] Unless, *totally coincidentally*, x_i denotes v_i in some context. See the informal definition (s) above, and also the definition (sub_i), p. 268.

We may write simply $\{E\}$ if the variables $x_1, \ldots, x_n$ are implied by the context.

We note that while $\ulcorner E(x_1, \ldots, x_n) \urcorner$ is a *closed term*, $\{E(x_1, \ldots, x_n)\}$ is a *term* with precisely $x_1, \ldots, x_n$ as its list of free variables.

II.6.26 Example. $\vdash \{x\} = Num(x)$. Indeed,

$$\vdash \{x\} = sub_m(\ulcorner v_m \urcorner, Num(x))$$

But $\vdash sub_m(\ulcorner v_m \urcorner, Num(x)) = Num(x)$ (by the definition of **Sub**, first case, p. 268). For any terms

$$t(v_{j_1}, \ldots, v_{j_h}, v_{m_1}, \ldots, v_{m_n})$$

and

$$s(v_{j_1}, \ldots, v_{j_h}, v_{k_1}, \ldots, v_{k_r})$$

and any choice of (meta)variables

$$x_1, \ldots, x_h, y_1, \ldots, y_n, z_1, \ldots, z_r \tag{1}$$

We have

$$\vdash \{(t = s)(x_1, \ldots, x_h, y_1, \ldots, y_n, z_1, \ldots, z_r)\} =$$
$$\langle \ulcorner = \urcorner, \{t(x_1, \ldots, x_h, z_1, \ldots, z_r)\}, \{s(x_1, \ldots, x_h, y_1, \ldots, y_n)\} \rangle \tag{2}$$

Now that the list (1) has been recorded, we will feel free to abbreviate (2) as

$$\vdash \{t = s\} = \langle \ulcorner = \urcorner, \{t\}, \{s\} \rangle \tag{3}$$

We proceed to establish (2) (or (3)) recalling that the order of substitution is irrelevant, and using "=" *conjunctionally* below (i.e., $\vdash t = s = r$ means $\vdash t = s$ *and* $\vdash s = r$):

$$\vdash \{(t = s)(x_1, \ldots, x_h, y_1, \ldots, y_n, z_1, \ldots, z_r)\} =$$
$$s_{k_r}(\ldots s_{m_n}(\ldots s_{j_h}(\ldots s_{j_1}(\ulcorner t = s \urcorner, x_1), \ldots, x_h), \ldots, y_n), \ldots, z_r) =$$
$$\langle \ulcorner = \urcorner, s_{k_r}(\ldots s_{m_n}(\ldots s_{j_h}(\ldots s_{j_1}(\ulcorner t \urcorner, x_1), \ldots, x_h), \ldots, y_n), \ldots, z_r),$$
$$s_{k_r}(\ldots s_{m_n}(\ldots s_{j_h}(\ldots s_{j_1}(\ulcorner s \urcorner, x_1), \ldots, x_h), \ldots, y_n), \ldots, z_r) \rangle =$$
$$\langle \ulcorner = \urcorner, \{t(x_1, \ldots, x_h, z_1, \ldots, z_r)\}, \{s(x_1, \ldots, x_h, y_1, \ldots, y_n)\} \rangle$$

The computations above iterate case 4 of the definition of **Sub** (subcase for **AF**, p. 268), keeping in mind informal definition (s) (p. 299). This iterative calculation relies on II.6.4 and II.6.24 to ensure that

$$\vdash Term(Sub(\ulcorner E \urcorner, \widetilde{i}, Num(w))) = 0$$

at every step (E is t or s).

We have abbreviated

$$\ulcorner (t = s)(v_{j_1}, \ldots, v_{j_h}, v_{m_1}, \ldots, v_{m_n}, v_{k_1}, \ldots, v_{k_r}) \urcorner \quad \text{by} \quad \ulcorner t = s \urcorner$$

$$\ulcorner t(v_{j_1}, \ldots, v_{j_h}, v_{m_1}, \ldots, v_{m_n}) \urcorner \quad \text{by} \quad \ulcorner t \urcorner$$

and

$$\ulcorner s(v_{j_1}, \ldots, v_{j_h}, v_{k_1}, \ldots, v_{k_r}) \urcorner \quad \text{by} \quad \ulcorner s \urcorner$$

We have also freely used during our computation the fact that

$$\vdash Free(i, x) > 0 \wedge Term(x) = 0 \rightarrow Sub(x, i, z) = x$$

(see Exercise II.19).

The same type of computation – employing II.6.11, II.6.12, II.6.24, and II.6.25 – establishes that

$$\vdash \{\mathscr{A} \rightarrow \mathscr{B}\} = \langle \ulcorner \rightarrow \urcorner, \{\mathscr{A}\}, \{\mathscr{B}\} \rangle$$

We have used the short { }-notation above.

The behaviour of the symbol { } is not very stable with respect to substitution. For example, we have found that $\vdash \{x\} = Num(x)$ above. However, it is not true that for arbitrary unary f

$$\vdash \{fx\} = Num(fx) \tag{4}$$

(see Exercise II.21). On the other hand, (4) does hold if $f \equiv S$. Indeed (using, once again, "=" conjunctionally), we find that

$$\begin{aligned}
\vdash \{Sx\} &= sub_1(\ulcorner Sv_1 \urcorner, Num(x)) \\
&= \langle \ulcorner S \urcorner, sub_1(\ulcorner v_1 \urcorner, Num(x)) \rangle \\
&= \langle \ulcorner S \urcorner, Num(x) \rangle \\
&= Num(Sx)
\end{aligned}$$

$\square$

II.6.27 Lemma. $\vdash_C \Theta(x) \rightarrow \Theta(Sub(x, i, Num(y)))$.

Proof. By modus ponens from II.6.20, since

$$\vdash Term(Num(y)) = 0$$

by II.6.24, and

$$\vdash \Theta(x) \rightarrow Sub(x, i, Num(y)) > 0$$

– the latter from $\vdash \Theta(x) \rightarrow WFF(x) = 0$ (cf. II.6.16), II.6.12, and II.6.25. $\square$

II.6.28 Corollary. *For any formula $\mathcal{A}$ over $L_{\mathfrak{A}}$ and variable v_i,*

$$\vdash_C \Theta(\ulcorner \mathcal{A}[v_i] \urcorner) \to \Theta\Big(Sub\big(\ulcorner \mathcal{A}[v_i] \urcorner, \tilde{i}, Num(x)\big)\Big)$$

Iterating application of the above for all the free variables in $\mathcal{A}$, we obtain at once:

II.6.29 Corollary. *For any formula $\mathcal{A}$ over $L_{\mathfrak{A}}$,*

$$\vdash_C \Theta(\ulcorner \mathcal{A}(v_{i_1}, \ldots, v_{i_n}) \urcorner) \to \Theta(\{\mathcal{A}(x_1, \ldots, x_n)\})$$

Of course, we also obtain a useful special case – where x_j denotes v_{i_j} – as follows: *For any formula $\mathcal{A}$ over $L_{\mathfrak{A}}$,*

$$\vdash_C \Theta(\ulcorner \mathcal{A}(x_1, \ldots, x_n) \urcorner) \to \Theta(\{\mathcal{A}(x_1, \ldots, x_n)\})$$

The next corollary follows at once from the above remark and **DC 1**.

II.6.30 Corollary (The Free Variables Version of DC 1). *For any formula $\mathcal{A}$ over $L_{\mathfrak{A}}$, if $\vdash_\Gamma \mathcal{A}(x_1, \ldots, x_n)$, then $\vdash_C \Theta(\{\mathcal{A}(x_1, \ldots, x_n)\})$.*

II.6.31 Lemma (The Free Variables Version of DC 2). *For any formulas $\mathcal{A}$ and $\mathcal{B}$ over $L_{\mathfrak{A}}$,*

$$\vdash_C \Theta(\{\mathcal{A} \to \mathcal{B}\}) \to \Theta(\{\mathcal{A}\}) \to \Theta(\{\mathcal{B}\})$$

Proof. $\{\mathcal{A} \to \mathcal{B}\}$ abbreviates

$$\langle \ulcorner \to \urcorner, sub_{i_n}(\ldots sub_{i_1}(\ulcorner \mathcal{A} \urcorner, Num(x_1)), \ldots, Num(x_n)),$$
$$sub_{i_n}(\ldots sub_{i_1}(\ulcorner \mathcal{B} \urcorner, Num(x_1)), \ldots, Num(x_n)) \rangle$$

while $\{\mathcal{A}\}$ and $\{\mathcal{B}\}$ abbreviate

$$sub_{i_n}(\ldots sub_{i_1}(\ulcorner \mathcal{A} \urcorner, Num(x_1)), \ldots, Num(x_n))$$

and

$$sub_{i_n}(\ldots sub_{i_1}(\ulcorner \mathcal{B} \urcorner, Num(x_1)), \ldots, Num(x_n))$$

respectively. We are done, by II.6.19.

Pause. Why is

$$\vdash WFF\big(sub_{i_n}(\ldots sub_{i_1}(\ulcorner \mathcal{B} \urcorner, Num(x_1)), \ldots, Num(x_n))\big) = 0?$$ □

II.6.32 Remark. In what follows, all the way to the end of the section, we assume – without loss of generality – that our unspecified consistent Σ_1 extension, Γ, of **PA** also extends C, the theory of Gödel coding.

Here is why generality is not lost in the case where we were unfortunate enough to start with a Γ that did *not* satisfy the assumption:

(1) We start with **PA** $\leq \Gamma$,[†] where Γ is $\Sigma_1(L_{\mathfrak{N}_C})$.
(2) We extend Γ – to Γ' – by adding the axioms for all the defined symbols that were added to **PA** in order to form C and $L_{\mathfrak{N}_C}$. This results to the language $L_{\mathfrak{N}'}$. So, not only do we have **PA** $\leq C$ conservatively, but also $\Gamma \leq \Gamma'$ conservatively.
(3) Since Γ is consistent, the "conservatively" part above ensures that so is Γ'. Trivially, **PA** $\leq \Gamma'$ and $C \leq \Gamma'$.
(4) If Q is a formula that semantically defines the nonlogical axioms of C, that is, $Q(\ulcorner \mathscr{A} \urcorner)$ means that $\mathscr{A}$ is a C-axiom, then $\Gamma'(x)$ can be taken to abbreviate

$$\Gamma(x) \vee Q(x) \qquad (*)$$

Given that $Q(x)$ is primitive recursive (Exercise II.23[‡]), the formula $(*)$ is $\Sigma_1(L_{\mathfrak{N}_C})$ (Exercise II.24). □

II.6.33 Lemma. *For any terms t, s and variable z, if $\vdash_C t = s$ and*

$$\vdash_C t = z \rightarrow \Theta(\{t = z\})$$

then also

$$\vdash_C s = z \rightarrow \Theta(\{s = z\})$$

where we have written "Θ" for "Θ_Γ".

Proof. "$\vdash$" means "$\vdash_C$". By **Ax4**

$$\vdash t = z \rightarrow s = z \qquad (1)$$

and

$$\vdash s = z \rightarrow t = z \qquad (2)$$

[†] The relation "$\leq$" that compares two theories was introduced on p. 46.

[‡] C contains the finitely many defining axioms we have introduced towards the Gödel coding. Moreover it contains the infinitely many axioms introducing the primitive recursive symbols. A defining predicate for this part can be introduced by a course-of-values recursion, as can be easily deduced from the specific allocation scheme chosen for primitive recursive function symbols. See Remark II.3.7, p. 254.

By **DC 1** and **DC 2** (free variables versions, II.6.30 and II.6.31) in that order, (1) yields

$$\vdash \Theta(\{t = z\}) \rightarrow \Theta(\{s = z\}) \tag{3}$$

where we have used the blanket assumption "$C \leq \Gamma$" (II.6.32) in the application of **DC 1** (**DC 1** assumes "$\vdash_\Gamma \ldots$"). Our unused assumption, along with (2) and (3), yields what we want by tautological implication.   □

We now have all the tools we need towards proving **DC 3**.

II.6.34 Main Lemma. *For any primitive recursive term* t *over* $L_{\mathfrak{M}_C}$ *and any variable* z *not free in* t,

$$\vdash_C t = z \rightarrow \Theta(\{t = z\}) \tag{1}$$

where we have written "Θ" for "Θ_Γ".

We continue the practice of omitting the subscript C from $\vdash$ (a subscript will be used if other than C, or as a periodic reminder (or emphasis) if it is C). The implicit assumption "$C \leq \Gamma$" (II.6.32) enables II.6.30 (**DC 1**, free variables version) exactly as in the case of II.6.33.

Here is how *not* to prove the lemma: "By **Ax4** it suffices to prove

$$\vdash \Theta(\{t = t\})$$

This follows from $\vdash t = t$ and II.6.30."

Such an attempt does not heed the warning in Example II.6.26. Recall that $\vdash \{t = z\} = \langle \ulcorner = \urcorner, \{t\}, Num(z)\rangle$. However, we have already noted that it is *not* true in general that $\vdash Num(t) = \{t\}$. Thus, *in general*,

$$\nvdash \{t = z\}[z \leftarrow t] = \{t = t\}$$

Proof. The proof is by (metamathematical) induction on the formation of t and follows the one given by Hilbert and Bernays (1968) (however, unlike them, we do not restrict primitive recursive terms to "normalized" forms).

We have three *basis* cases:

Case 1. $t \equiv Zv_0$. Since $\vdash Zv_0 = 0$, it suffices, by II.6.33, to prove the following version of (1):

$$\vdash 0 = z \rightarrow \Theta(\{0 = z\})$$

By **Ax4** it suffices to prove

$$\vdash \Theta(\{0 = 0\})$$

This follows from $\vdash 0 = 0$ and II.6.30.

Pause. Wait a minute! How does this differ from what I said above that we should *not* do?

Case 2. $t \equiv U_i^n(v_0, \ldots, v_{n-1})$.

Again, by II.6.33 and $\vdash U_i^n(v_0, \ldots, v_{n-1}) = v_{i-1}$, (1) now becomes (where we have simplified the metanotation: x rather than the "actual" v_{i-1})

$$\vdash x = z \rightarrow \Theta(\{x = z\})$$

By **Ax4** it suffices to prove

$$\vdash \Theta(\{x = x\})$$

This follows from $\vdash x = x$ and II.6.30.

Pause. Was this O.K.?

Case 3. $t \equiv Sx$. (1) now is

$$\vdash Sx = z \rightarrow \Theta(\langle \ulcorner = \urcorner, \{Sx\}, Num(z)\rangle)$$

By **Ax4** it suffices to prove

$$\vdash \Theta(\langle \ulcorner = \urcorner, \{Sx\}, Num(Sx)\rangle)$$

– that is (II.6.26), $\vdash \Theta(\langle \ulcorner = \urcorner, \{Sx\}, \{Sx\}\rangle)$, or (II.6.26 again)

$$\vdash \Theta(\{Sx = Sx\})$$

This follows from $\vdash Sx = Sx$ and II.6.30.

We now embark on the two induction steps:

Composition. Suppose that $t \equiv fs_1 \ldots s_n$, where the function f and the terms s_i are primitive recursive.

Let (I.H.)

$$\vdash fx_1 \ldots x_n = z \rightarrow \Theta(\{fx_1 \ldots x_n = z\}) \tag{2}$$

and

$$\vdash s_i = x_i \rightarrow \Theta(\{s_i = x_i\}) \qquad \text{for } i = 1, \ldots, n \tag{3}$$

where none of the x_i are free in any of the s_j, and, moreover, z is not one of the x_i. By **Ax4**

$$\vdash s_1 = x_1 \rightarrow s_2 = x_2 \rightarrow \cdots \rightarrow s_n = x_n \rightarrow$$
$$f s_1 \ldots s_n = z \rightarrow f x_1 \ldots x_n = z \tag{4}$$

and

$$\vdash s_1 = x_1 \rightarrow s_2 = x_2 \rightarrow \cdots \rightarrow s_n = x_n \rightarrow$$
$$f x_1 \ldots x_n = z \rightarrow f s_1 \ldots s_n = z \tag{5}$$

By (2) and (4) (and tautological implication),

$$\vdash s_1 = x_1 \rightarrow s_2 = x_2 \rightarrow \cdots \rightarrow s_n = x_n \rightarrow$$
$$f s_1 \ldots s_n = z \rightarrow \Theta(\{f x_1 \ldots x_n = z\}) \tag{6}$$

By II.6.30 and (5),

$$\vdash \Theta(\{s_1 = x_1 \rightarrow s_2 = x_2 \rightarrow \cdots \rightarrow s_n = x_n \rightarrow$$
$$f x_1 \ldots x_n = z \rightarrow f s_1 \ldots s_n = z\})$$

Hence, by II.6.31,

$$\vdash \Theta(\{s_1 = x_1\}) \rightarrow \Theta(\{s_2 = x_2\}) \rightarrow \cdots \rightarrow \Theta(\{s_n = x_n\}) \rightarrow$$
$$\Theta(\{f x_1 \ldots x_n = z\}) \rightarrow \Theta(\{f s_1 \ldots s_n = z\})$$

The above and (3) tautologically imply

$$\vdash s_1 = x_1 \rightarrow s_2 = x_2 \rightarrow \cdots \rightarrow s_n = x_n \rightarrow$$
$$\Theta(\{f x_1 \ldots x_n = z\}) \rightarrow \Theta(\{f s_1 \ldots s_n = z\})$$

The above and (6) tautologically imply

$$\vdash s_1 = x_1 \rightarrow s_2 = x_2 \rightarrow \cdots \rightarrow s_n = x_n \rightarrow$$
$$f s_1 \ldots s_n = z \rightarrow \Theta(\{f s_1 \ldots s_n = z\}) \tag{7}$$

Finally, since the x_i are not free in any of the s_j and are distinct from z, the substitutions $[x_i \leftarrow s_i]$ into (7), and tautological implication, imply

$$\vdash f s_1 \ldots s_n = z \rightarrow \Theta(\{f s_1 \ldots s_n = z\})$$

Primitive recursion. We are given that $h(\vec{y})$ and $g(x, \vec{y}, w)$ are primitive recursive terms, and that f has been introduced to satisfy

$$\vdash f(0, \vec{y}) = h(\vec{y})$$
$$\vdash f(Sx, \vec{y}) = g(x, \vec{y}, f(x, \vec{y})) \tag{8}$$

Supposing that z is distinct from $x, \vec{y}$, we want to show

$$\vdash f(x, \vec{y}) = z \rightarrow \Theta(\{f(x, \vec{y}) = z\})$$

To allow the flexibility of splitting the induction step into an I.H. and a conclusion (as usual practice dictates, using the deduction theorem), we prove the following (provably) equivalent form instead:

$$\vdash (\forall z)\Big(f(x, \vec{y}) = z \rightarrow \Theta(\{f(x, \vec{y}) = z\})\Big) \tag{9}$$

under the same restriction, that the bound variable z is not among $x, \vec{y}$.

Now, our *metamathematical* I.H. (on the formation of primitive recursive terms t) – under the assumption that z is not among $x, \vec{y}, w$ – is

$$\vdash h(\vec{y}) = z \rightarrow \Theta(\{h(\vec{y}) = z\})$$
$$\vdash g(x, \vec{y}, w) = z \rightarrow \Theta(\{g(x, \vec{y}, w) = z\}) \tag{10}$$

(9) is proved by *formal* induction on x.

For the *formal basis of* (9), let us set $x \leftarrow 0$. By the first of (8), the first of (10), and II.6.33 (followed by generalization) we deduce the basis, namely,

$$\vdash (\forall z)\Big(f(0, \vec{y}) = z \rightarrow \Theta(\{f(0, \vec{y}) = z\})\Big)$$

Now assume (9) (formal I.H.) for frozen $x, \vec{y}$, and prove

$$\vdash (\forall z)\Big(f(Sx, \vec{y}) = z \rightarrow \Theta(\{f(Sx, \vec{y}) = z\})\Big) \tag{11}$$

By generalization, it suffices to prove

$$\vdash f(Sx, \vec{y}) = z \rightarrow \Theta(\{f(Sx, \vec{y}) = z\}) \tag{11'}$$

(where z is not among $x, \vec{y}$). We choose w, *not* among $x, \vec{y}, z$. By **Ax4**,

$$\vdash f(x, \vec{y}) = w \rightarrow g(x, \vec{y}, f(x, \vec{y})) = z \rightarrow g(x, \vec{y}, w) = z$$
$$\vdash f(x, \vec{y}) = w \rightarrow g(x, \vec{y}, w) = z \rightarrow g(x, \vec{y}, f(x, \vec{y})) = z$$

which by the second part of (8) and **Ax4** translate into

$$\vdash f(x, \vec{y}) = w \rightarrow f(Sx, \vec{y}) = z \rightarrow g(x, \vec{y}, w) = z$$
$$\vdash f(x, \vec{y}) = w \rightarrow g(x, \vec{y}, w) = z \rightarrow f(Sx, \vec{y}) = z \tag{12}$$

Tautological implication using the second part of (10) and the first of (12) yields

$$\vdash f(x, \vec{y}) = w \rightarrow f(Sx, \vec{y}) = z \rightarrow \Theta(\{g(x, \vec{y}, w) = z\}) \tag{13}$$

II.6.30–II.6.31 and the second part of (12) yield

$$\vdash \Theta(\{f(x, \vec{y}) = w\}) \rightarrow \Theta(\{g(x, \vec{y}, w) = z\}) \rightarrow \Theta(\{f(Sx, \vec{y}) = z\})$$

Using the I.H. (9), followed by specialization (and with the help of $\models_{Taut}$), this yields

$$\vdash f(x,\vec{y}) = w \rightarrow \Theta(\{g(x,\vec{y},w) = z\}) \rightarrow \Theta(\{f(Sx,\vec{y}) = z\})$$

Here is where the (universally quantified) form of (9) helped formally. We have been manipulating the z-variable by substituting w. This would be enough to unfreeze variables frozen between I.H. and conclusion, thus invalidating the deduction theorem step. However, that is not the case here, because this variable manipulation is hidden from the deduction theorem. z is a bound variable in the assumption (9).

Tautological implication, using the above and (13), furnishes

$$\vdash f(x,\vec{y}) = w \rightarrow f(Sx,\vec{y}) = z \rightarrow \Theta(\{f(Sx,\vec{y}) = z\})$$

Finally, since $x, w, z, \vec{y}$ are all distinct, the substitution $[w \leftarrow f(x,\vec{y})]$ in the above yields (11′) by tautological implication, and we thus have (11) by generalization.   □

At this point the reader may take a well-deserved break.

Next, we prove

II.6.35 Corollary. *For any primitive recursive predicate P over $L_{\mathfrak{N}_C}$,*

$$\vdash_C P(x_1,\ldots,x_n) \rightarrow \Theta(\{P(x_1,\ldots,x_n)\})$$

Proof. Indeed,

$$\vdash P(x_1,\ldots,x_n) \rightarrow \chi_P(x_1,\ldots,x_n) = 0 \tag{1}$$

and

$$\vdash \chi_P(x_1,\ldots,x_n) = 0 \rightarrow P(x_1,\ldots,x_n) \tag{2}$$

where χ_P is "the" (primitive recursive) characteristic function (see II.3.6, p. 253, and II.1.25, p. 220) for $P(x_1,\ldots,x_n)$. By the main lemma (II.6.34),

$$\vdash \chi_P(x_1,\ldots,x_n) = 0 \rightarrow \Theta(\{\chi_P(x_1,\ldots,x_n) = 0\})$$

Hence (by (1))

$$\vdash P(x_1,\ldots,x_n) \rightarrow \Theta(\{\chi_P(x_1,\ldots,x_n) = 0\}) \tag{3}$$

By (2) and II.6.30–II.6.31,

$$\vdash \Theta(\{\chi_P(x_1,\ldots,x_n)=0\}) \to \Theta(\{P(x_1,\ldots,x_n)\})$$

This and (3) yield what we want, by tautological implication. $\qquad\square$

We can prove a bit more:

II.6.36 Corollary. *For any* $\Sigma_1(L_{\mathfrak{N}_c})$ *formula* $\mathscr{A}$,

$$\vdash_C \mathscr{A}(x_1,\ldots,x_n) \to \Theta(\{\mathscr{A}(x_1,\ldots,x_n)\})$$

Proof. Let $\mathscr{A}(x_1,\ldots,x_n)$ be $\Sigma_1(L_{\mathfrak{N}_c})$.

Then, for some $\Delta_0(L_{\mathfrak{N}_c})$ formula $\mathscr{Q}(y,x_1,\ldots,x_n)$, where y is distinct from $x_1,\ldots,x_n$,

$$\vdash \mathscr{A}(x_1,\ldots,x_n) \leftrightarrow (\exists y)\mathscr{Q}(y,x_1,\ldots,x_n)$$

We introduce a predicate R by the definition

$$R(y,\vec{x}_n) \leftrightarrow \mathscr{Q}(y,\vec{x}_n) \tag{4}$$

Then R is primitive recursive (Exercise II.22).

By II.6.35,

$$\vdash R(y,x_1,\ldots,x_n) \to \Theta(\{R(y,x_1,\ldots,x_n)\})$$

Hence, by (4) and tautological implication,

$$\vdash \mathscr{Q}(y,x_1,\ldots,x_n) \to \Theta(\{R(y,x_1,\ldots,x_n)\}) \tag{5}$$

By II.6.30–II.6.31 and the $\to$-part of (4),

$$\vdash \Theta(\{R(y,x_1,\ldots,x_n)\}) \to \Theta(\{\mathscr{Q}(y,x_1,\ldots,x_n)\})$$

Hence (by (5))

$$\vdash \mathscr{Q}(y,x_1,\ldots,x_n) \to \Theta(\{\mathscr{Q}(y,x_1,\ldots,x_n)\}) \tag{6}$$

Now,

$$\vdash \mathscr{Q}(y,x_1,\ldots,x_n) \to (\exists y)\mathscr{Q}(y,x_1,\ldots,x_n)$$

Hence, by II.6.30–II.6.31,

$$\vdash \Theta(\{\mathscr{Q}(y,x_1,\ldots,x_n)\}) \to \Theta(\{(\exists y)\mathscr{Q}(y,x_1,\ldots,x_n)\})$$

310   *II. The Second Incompleteness Theorem*

Combining the above with (6) via $\models_{\text{Taut}}$, we get

$$\vdash \mathcal{Q}(y, x_1, \ldots, x_n) \rightarrow \Theta\big(\{(\exists y)\mathcal{Q}(y, x_1, \ldots, x_n)\}\big)$$

which yields what we want via $\exists$-introduction. $\qquad\qquad\square$

The following corollary is a step backward in terms of degree of generality, but it is all we really need. It is a formalized **DC 1**, for it says "if it is true that $\mathcal{A}$ is provable, then it is also true that $\Theta(\ulcorner\mathcal{A}\urcorner)$ is provable".

II.6.37 Corollary (Löb). *For any formula $\mathcal{A}$ over $L_{\mathfrak{N}_c}$,*

$$\vdash_C \Theta(\ulcorner\mathcal{A}\urcorner) \rightarrow \Theta\big(\ulcorner\Theta(\ulcorner\mathcal{A}\urcorner)\urcorner\big)$$

Proof. For any fixed $\mathcal{A}$, $\Theta(\ulcorner\mathcal{A}\urcorner)$ is a $\Sigma_1(L_{\mathfrak{N}_c})$ sentence. $\qquad\square$

 II.6.38 Remark. The above corollary is Löb's *third derivability condition*, **DC 3**. $\qquad\qquad\square$

II.6.39 Theorem (The Fixpoint Theorem, or Diagonalization Lemma). *For any formula $\mathcal{A}(x)$ over $L_{\mathfrak{N}_c}$, there is a sentence $\mathcal{B}$ over the basic language of* **PA**, $L_{\mathfrak{N}}$, *such that*

$$\vdash_C \mathcal{B} \leftrightarrow \mathcal{A}(\ulcorner\mathcal{B}\urcorner)$$

$\mathcal{B}$ is called a *fixed point or fixpoint of $\mathcal{A}$ in C*.
The assumption $C \leq \Gamma$ of II.6.32 is not used here.

Proof. Let $\mathcal{C}(x)$ be the formula obtained from $\mathcal{A}(s_1(x, x))$ after removing all defined symbols (following I.7.1 and I.7.3). Then $\mathcal{C}$ is over $L_{\mathfrak{N}}$, and

$$\vdash_C \mathcal{C}(x) \leftrightarrow \mathcal{A}(s_1(x, x)) \qquad\qquad (1)$$

Let next $n \in \mathbb{N}$ be such that (cf. (1), p. 274)

$$\vdash_C \tilde{n} = \ulcorner\mathcal{C}(x)\urcorner$$

Hence (p. 279, 297, and 299)

$$\vdash_C s_1(\tilde{n}, \tilde{n}) = \ulcorner\mathcal{C}(\tilde{n})\urcorner \qquad\qquad (2)$$

By (1) and substitution,

$$\vdash_C \mathcal{C}(\tilde{n}) \leftrightarrow \mathcal{A}(s_1(\tilde{n}, \tilde{n}))$$

By **Ax4** and (2), the above yields

$$\vdash_C \mathscr{C}(\tilde{n}) \leftrightarrow \mathscr{A}(\ulcorner \mathscr{C}(\tilde{n})\urcorner)$$

Thus "$\mathscr{C}(\tilde{n})$" is the sentence "$\mathscr{B}$" we want. □

Applying the above to $\neg\Theta_\Gamma$, we obtain a sentence that semantically says "I am not a theorem of Γ".

II.6.40 Corollary (Gödel). *There is a sentence $\mathscr{G}$ over the basic language $L_\mathfrak{N}$ such that*

$$\vdash_C \mathscr{G} \leftrightarrow \neg\Theta_\Gamma(\ulcorner\mathscr{G}\urcorner) \tag{1}$$

We now need to revisit (half of) Gödel's first incompleteness theorem. We will show that the sentence $\mathscr{G}$ above is not provable in Γ.[†] For the balance of the section we will be careful to subscript Θ with the name, say Γ, of the theory for which it is a *provability predicate*.

II.6.41 Lemma (Gödel). *Let Γ over $L_\mathfrak{A}$ be a* consistent *extension of* **PA** *such that $\Gamma(x)$ is $\Sigma_1(L_{\mathfrak{N}_C})$. Let* **the sentence** *$\mathscr{G}$ over $L_\mathfrak{N}$ be a fixed point of $\neg\Theta_\Gamma$ in C. Then $\nvdash_\Gamma \mathscr{G}$.*

 The assumption $C \leq \Gamma$ of II.6.32 is not used here.

Proof. Assume that

$$\vdash_\Gamma \mathscr{G} \tag{2}$$

By the assumption on $\Gamma(x)$, **DC 1** is applicable; hence

$$\vdash_C \Theta_\Gamma(\ulcorner\mathscr{G}\urcorner)$$

(1) of II.6.40 now yields

$$\vdash_C \neg\mathscr{G}$$

Since **PA** $\leq C$ conservatively, and $\mathscr{G}$ is over $L_\mathfrak{N}$, we have $\vdash_{\textbf{PA}} \neg\mathscr{G}$, and hence, by **PA** $\leq \Gamma$,

$$\vdash_\Gamma \neg\mathscr{G}$$

The above and (2) contradict the consistency of Γ. □

[†] The other half, which we do not need towards the proof of the second incompleteness theorem, uses a stronger consistency assumption – ω-consistency (p. 189) – and states that $\mathscr{G}$ is not refutable either.

We continue honouring the assumption of II.6.32; however, we make it explicit here. That is, we have the following situation:

(1) **PA** $\leq \Gamma$ consistently (Γ in $\Sigma_1(L_{\mathfrak{N}_c})$).
(2) $\Gamma \leq \Gamma'$ conservatively, by adding the symbols (and their definitions) of C. Γ' is in $\Sigma_1(L_{\mathfrak{N}_c})$

We let the *metasymbol* "Con_Γ" stand for "Γ is consistent". With the help of the arithmetization tools this can be implemented as a sentence whose semantics is that some convenient refutable formula is not a Γ-theorem.

"$0 = S0$" is our choice of a refutable formula (cf. *S1*). We may now define

$$Con_\Gamma \quad \text{abbreviates} \quad \neg\Theta_{\Gamma'}(\ulcorner 0 = S0 \urcorner)$$

since $0 = S0$ is unprovable in Γ iff it is so in Γ'; therefore $\neg\Theta_{\Gamma'}(\ulcorner 0 = S0 \urcorner)$ – that is, Con_Γ – says both "Γ' is consistent" and "Γ is consistent".

We will next want to show that, under some reasonable assumptions, Γ cannot prove Con_Γ, that is, *it cannot prove its own consistency.*

For this task to be meaningful (and "fair" to Γ), Con_Γ (that is, the actual formula it stands for) must be in the language of Γ, $L_{\mathfrak{A}}$. We can guarantee this if we define Con_Γ to instead abbreviate the sentence $\mathscr{S}$ obtained from $\neg\Theta_{\Gamma'}(\ulcorner 0 = S0 \urcorner)$ *by elimination of defined symbols.* Thus, we finalize the definition as

$$Con_\Gamma \quad \text{abbreviates} \quad \mathscr{S} \qquad\qquad (*)$$

where $\mathscr{S}$ is over the basic language $L_{\mathfrak{N}}$ of **PA** and satisfies

$$\vdash_C \mathscr{S} \leftrightarrow \neg\Theta_{\Gamma'}(\ulcorner 0 = S0 \urcorner) \qquad\qquad (**)$$

II.6.42 Theorem (Gödel's Second Incompleteness Theorem). *Let Γ be a consistent extension of* **PA** *such that $\Gamma(x)$ is $\Sigma_1(L_{\mathfrak{N}_c})$. Then $\nvdash_\Gamma Con_\Gamma$.*

Proof. Let $\Gamma \leq \Gamma'$ as in the discussion above (we continue being explicit here about our blanket assumption II.6.32; thus $C \leq \Gamma'$, while $C \leq \Gamma$ *might* fail).

The proof utilizes the fixed point $\mathscr{S}$ of $\neg\Theta_{\Gamma'}$ that is obtained in the manner of II.6.40–II.6.41 above (note however the prime). Our aim is to show that[†]

$$\vdash_{\Gamma'} Con_\Gamma \rightarrow \mathscr{S} \qquad\qquad (3)$$

We start with the observation

$$\models_{\text{Taut}} \neg\mathscr{S} \rightarrow (\mathscr{S} \rightarrow 0 = S0)$$

[†] Cf. the informal discussion at the beginning of this chapter, p. 205.

Hence (absolutely)

$$\vdash \neg\mathscr{S} \to (\mathscr{S} \to 0 = S0)$$

which by (1) of II.6.40 (but using the theory Γ' rather than Γ) and the Leibniz rule implies

$$\vdash_C \Theta_{\Gamma'}(\ulcorner\mathscr{S}\urcorner) \to (\mathscr{S} \to 0 = S0)$$

DC 1 now yields (this uses $C \leq \Gamma'$)

$$\vdash_C \Theta_{\Gamma'}\Big(\ulcorner\Theta_{\Gamma'}(\ulcorner\mathscr{S}\urcorner) \to (\mathscr{S} \to 0 = S0)\urcorner\Big)$$

which further yields, via **DC 2**,

$$\vdash_C \Theta_{\Gamma'}\Big(\ulcorner\Theta_{\Gamma'}(\ulcorner\mathscr{S}\urcorner)\urcorner\Big) \to \Theta_{\Gamma'}(\ulcorner\mathscr{S}\urcorner) \to \Theta_{\Gamma'}(\ulcorner 0 = S0\urcorner)$$

Löb's **DC 3** (II.6.37, which also uses $C \leq \Gamma'$) and the above yield, by tautological implication,

$$\vdash_C \Theta_{\Gamma'}(\ulcorner\mathscr{S}\urcorner) \to \Theta_{\Gamma'}(\ulcorner 0 = S0\urcorner)$$

By contraposition, and using (1) of II.6.40 (for Γ') and $(**)$,

$$\vdash_C \mathscr{S} \to \mathscr{G} \tag{4}$$

Since $C \leq \Gamma'$, (4) implies (3) (via $(*)$).

Now if $\vdash_\Gamma Con_\Gamma$, then also $\vdash_{\Gamma'} Con_\Gamma$ by $\Gamma \leq \Gamma'$. Thus (3) would yield $\vdash_{\Gamma'} \mathscr{S}$, contradicting II.6.41 (for Γ'). □

(1) The above proof is clearly also valid for the trivial extension $\Gamma = \textbf{PA}$. In this case, $\Gamma' = C$. Moreover, we note that, since $\textbf{PA} \leq C$ conservatively and $\mathscr{S} \to \mathscr{G}$ is over $L_\mathfrak{N}$, (4) implies

$$\vdash_{\textbf{PA}} \mathscr{S} \to \mathscr{G}$$

as well, from which (via $\textbf{PA} \leq \Gamma$ – no prime) $\vdash_\Gamma \mathscr{S} \to \mathscr{G}$.

(2) It is also true that $\vdash_\Gamma \mathscr{G} \to Con_\Gamma$; thus $\vdash_\Gamma \mathscr{G} \leftrightarrow Con_\Gamma$ (Exercise II.25).

II.6.41 and II.6.42 provide two examples of sentences unprovable by Γ, a consistent extension of **PA** with a Σ_1 set of axioms. As remarked above, the two sentences are provably equivalent (in Γ). The former says "I am not a theorem",[†] and thus its unprovability shows that it is true in $\mathfrak{A}$, the natural structure appropriate for $L_\mathfrak{A}$.

[†] Of Γ' – being a fixed point of $\neg\Theta_{\Gamma'}$ – and hence not of Γ either.

Let us put II.6.32 back into (implicit) force to avoid verbosity in the discussion and results that follow. So we have $C \leq \Gamma$.

What about a sentence that says "I *am* a theorem of Γ"?[†] Such a sentence, $\mathscr{H}$, is a fixed point of Θ_Γ, since $\Theta_\Gamma(\ulcorner \mathscr{H} \urcorner)$ "says" that (i.e., is true in $\mathfrak{N}_c$ iff) $\Gamma \vdash \mathscr{H}$. A theorem of Löb (1955) shows that such sentences are provable, and hence true as well, since then $\Theta_\Gamma(\ulcorner \mathscr{H} \urcorner)$ is true, and we also know that $\vdash_\Gamma \mathscr{H} \leftrightarrow \Theta_\Gamma(\ulcorner \mathscr{H} \urcorner)$.

II.6.43 Theorem (Löb's Theorem (1955)). *Let* **PA** $\leq \Gamma$ *consistently, and assume that the formula* Γ *is* $\Sigma_1(L_{\mathfrak{N}_c})$. *Then, for any sentence* $\mathscr{A}$,

$$\vdash_\Gamma \Theta_\Gamma(\ulcorner \mathscr{A} \urcorner) \to \mathscr{A} \quad \text{implies that} \quad \vdash_\Gamma \mathscr{A}.$$

Proof. Equivalently, let us prove that

$$\nvdash_\Gamma \mathscr{A} \quad \text{implies that} \quad \nvdash_\Gamma \Theta_\Gamma(\ulcorner \mathscr{A} \urcorner) \to \mathscr{A}$$

So assume that $\nvdash_\Gamma \mathscr{A}$. By I.4.21, $\Gamma + \neg\mathscr{A}$ is a consistent $\Sigma_1(L_{\mathfrak{N}_c})$ extension of C (hence of **PA**), that is, it is semantically defined by a $\Sigma_1(L_{\mathfrak{N}_c})$ formula.

Pause. Why is $\Gamma + \neg\mathscr{A}$ $\Sigma_1(L_{\mathfrak{N}_c})$?

By II.6.42,

$$\nvdash_{\Gamma+\neg\mathscr{A}} Con_{\Gamma+\neg\mathscr{A}} \tag{1}$$

Now, we choose as an "implementation" of the sentence "$Con_{\Gamma+\neg\mathscr{A}}$" the sentence "$\neg\Theta_\Gamma(\ulcorner \mathscr{A} \urcorner)$", because of I.4.21. Thus, (1) can be written as

$$\nvdash_{\Gamma+\neg\mathscr{A}} \neg\Theta_\Gamma(\ulcorner \mathscr{A} \urcorner)$$

Hence, by modus ponens,

$$\nvdash_\Gamma \neg\mathscr{A} \to \neg\Theta_\Gamma(\ulcorner \mathscr{A} \urcorner)$$

The contrapositive is what we want: $\nvdash_\Gamma \Theta_\Gamma(\ulcorner \mathscr{A} \urcorner) \to \mathscr{A}$. □

The statement in II.6.43 is actually an equivalence. The other direction is trivially true by tautological implication.

[†] This question was posed by Henkin (1952).

Löb's theorem can be proved from first principles, i.e., without reliance on Gödel's second incompleteness theorem. One starts by getting a fixed point $\mathscr{B}$ of $\Theta_\Gamma(x) \to \mathscr{A}$ in C and then judiciously applying the derivability conditions (Exercise II.26).

Conversely, Löb's theorem implies Gödel's (second) theorem: Indeed, the tautology

$$\neg\Theta_\Gamma(\ulcorner 0 = S0\urcorner) \to \Theta_\Gamma(\ulcorner 0 = S0\urcorner) \to 0 = S0$$

yields[†]

$$\vdash_\Gamma \neg\Theta_\Gamma(\ulcorner 0 = S0\urcorner) \to \Theta_\Gamma(\ulcorner 0 = S0\urcorner) \to 0 = S0$$

Thus

$$\vdash_\Gamma Con_\Gamma \to \Theta_\Gamma(\ulcorner 0 = S0\urcorner) \to 0 = S0$$

Suppose now that $\vdash_\Gamma Con_\Gamma$. Then, by the above,

$$\vdash_\Gamma \Theta_\Gamma(\ulcorner 0 = S0\urcorner) \to 0 = S0$$

Hence, by Löb's theorem, $\vdash_\Gamma 0 = S0$, contradicting the consistency of Γ.

We have simplified the hypotheses in Gödel's two theorems in allowing Γ to *extend* **PA** (or, essentially equivalently, extend C in the proof of the second). The theorems actually hold under a more general hypothesis that Γ *contains* **PA** (or C). This containment can be understood intuitively, at least in the case of formal set theory (formalized as ZFC for example):[‡] There is enough machinery inside ZFC to *construct* the set of natural numbers (ω) and show that this set satisfies the axioms **PA**.[§] One can then carry out the arithmetization of Section II.5 for terms, formulas, proofs, and theorems, *of ZFC this time, in ZFC* exactly as we did for Peano arithmetic, and prove that both incompleteness theorems hold for ZFC.[¶]

[†] Under the assumption of II.6.32 – $C \leq \Gamma - \Theta_\Gamma$ is in the language of Γ.
[‡] The formal treatment requires the concept of interpreting one theory inside another. This is one of many interesting topics that we must leave out in this brief course in logic. See however our volume 2 for a complete discussion of this topic and its bearing on this comment here.
[§] Thus, one constructs a model of **PA** inside ZFC. We show how this is done in volume 2.
[¶] In other words, the elements of the set ω will furnish us with Gödel numbers for the formulas and terms of ZFC, while certain terms and formulas over ω will allow ZFC to "argue" about these Gödel numbers.

II.7. Exercises

II.1. Prove that **PA** can actually prove axiom $<$**3**, and therefore this axiom is dependent (redundant).

II.2. Prove in **PA**: $\vdash x \leq y \rightarrow \neg y < x$.

II.3. Prove in **PA**: $\vdash x + y = y + x$.

II.4. Prove in **PA**: $\vdash x + (y + z) = (x + y) + z$.

II.5. Prove in **PA**: $\vdash x \times y = y \times x$.

II.6. Prove in **PA**: $\vdash x \times (y \times z) = (x \times y) \times z$.

II.7. Prove in **PA**: $\vdash x \times (y + z) = (x \times y) + (x \times z)$.

II.8. Settle the Pause regarding the displayed formula (f) on p. 218.

II.9. Prove the formula (LCM') on p. 230.

II.10. Prove in **PA**: $\vdash x > 0 \rightarrow x = S\delta(x, S0)$.

II.11. Prove Lemma II.1.36.

II.12. Prove in **PA**: $\vdash x \leq J(x, y)$, and $\vdash y \leq J(x, y)$, where J is that of p. 234.

II.13. Prove the concluding claim in II.3.7.

II.14. Conclude the proof of Lemma II.4.6.

II.15. Prove that there is a unary formal primitive recursive predicate Γ such that, for any formula $\mathscr{A}$, $\Gamma(\ulcorner \mathscr{A} \urcorner)$ means that $\mathscr{A}$ is (an instance of) a Peano axiom.

II.16. Prove that if the formula Γ that semantically defines the nonlogical axioms of an extension of **PA** is $\Sigma_1(L_{\mathfrak{N}'})$ then so are **Proof**, **Deriv**, and Θ.

II.17. Settle the Pause on p. 273.

II.18. Settle the Pause on p. 289.

II.19. Prove that

$$\vdash \textbf{\textit{Free}}(i,x) > 0 \wedge (\textbf{\textit{Term}}(x) = 0 \vee WFF(x) = 0)$$
$$\rightarrow \textbf{\textit{Sub}}(x, i, z) = x$$

II.20. Prove that

$$\vdash \textbf{\textit{Deriv}}(y, x) \rightarrow \textbf{\textit{Deriv}}(y * \langle x \rangle, x)$$

II.21. Show that it is not true that for arbitrary unary f

$$\vdash \{fx\} = Num(fx)$$

II.22. Prove that if $\mathcal{Q}$ is $\Delta_0(L_{\mathfrak{N}_c})$ and P is introduced by $P\vec{x} \leftrightarrow \mathcal{Q}(\vec{x})$, then P is primitive recursive.

II.23. Prove that the characterizing formula for the nonlogical axioms of the theory C – that is, Γ such that, for any formula $\mathscr{A}$, $\Gamma(\ulcorner \mathscr{A} \urcorner)$ means that $\mathscr{A}$ is a nonlogical axiom – is primitive recursive.

II.24. Prove that if the one-variable formulas Q and Γ are primitive recursive and $\Sigma_1(L_{\mathfrak{A}})$ respectively, then the formula $Q(x) \vee \Gamma(x)$ is $\Sigma_1(L_{\mathfrak{A}})$.

II.25. Refer to II.6.42. Prove that it is also true that $\vdash_\Gamma \mathscr{G} \to Con_\Gamma$, and thus

$$\vdash_\Gamma \mathscr{G} \leftrightarrow Con_\Gamma$$

II.26. Prove Löb's theorem without the help of Gödel's second incompleteness theorem.

II.27. Prove Tarski's theorem (see I.9.31, p. 174) on the (semantic) undefinability of truth using the fixpoint theorem (II.6.39) to find a sentence that says "I am false".

II.28. Refer to II.6.41. Let Γ over $L_{\mathfrak{A}}$ be an ω-consistent extension of **PA** such that $\Gamma(x)$ is $\Sigma_1(L_{\mathfrak{N}_c})$. Let the sentence $\mathscr{G}$ over $L_{\mathfrak{N}}$ be a fixed point of $\neg \Theta_\Gamma$ in C. Then $\nvdash_\Gamma \neg\mathscr{G}$.

II.29. Let $\mathscr{A}$ be any sentence in the language of a Γ – where $\Gamma(x)$ is $\Sigma_1(L_{\mathfrak{N}_c})$ – that extends **PA** consistently. Establish that $\vdash_C \neg\Theta(\ulcorner\mathscr{A}\urcorner) \to Con_\Gamma$.

II.30. With the usual assumptions on Γ and C (II.6.32), prove that

$$\vdash_C Con_\Gamma \to Con_{\Gamma + \neg Con_\Gamma}$$

Bibliography

Barwise, Jon, editor (1978). *Handbook of Mathematical Logic.* North-Holland, Amsterdam.

Bennett, J. (1962). *On Spectra.* PhD thesis, Princeton University.

Bourbaki, N. (1966a). *Éléments de mathématique.* Hermann, Paris.

———(1966b). *Éléments de mathématique; théorie des ensembles.* Hermann, Paris.

Chang, C.C., and H. Jerome, Keisler (1973). *Model Theory.* North-Holland, Amsterdam.

Church, Alonzo (1936). A note on the Entscheidungsproblem, *J. Symbolic Logic*, 1:40–41, 101–102.

Cohen, P. J. (1963). The independence of the continuum hypothesis, part I, *Proc. Nat. Acad. Sci. U.S.A.*, 50: 1143–1148; part II 51:105–110 (1964).

Davis, M. (1965). *The Undecidable.* Raven Press, Hewlett, N. Y.

Dekker, James C. E. (1955). Productive sets, *Trans. Amer. Math. Soc.*, 78:129–149.

Dijkstra, Edsger W., and Carel S. Scholten (1990). *Predicate Calculus and Program Semantics.* Springer-Verlag, New York.

Enderton, Herbert B. (1972). *A Mathematical Introduction to Logic.* Academic Press, New York.

Gödel, K. (1931). Über formale unentscheidbare Sätze der Principia Mathematica und verwandter Systeme I, *Monatsh. Math. u. Phys.* 38:173–198. (English transl. in Davis (1965, pp. 5–38).

———(1938). The consistency of the axiom of choice and of the generalized continuum hypothesis, *Proc. Nat. Acad. Sci. U.S.A.*, 24:556–557.

Grzegorczyk, A. (1953). Some classes of recursive functions, *Rozprawy Mat.*, 4:1–45.

Gries, David, and Fred B. Schneider (1994). *A Logical Approach to Discrete Math.* Springer-Verlag, New York.

———(1995). Equational propositional logic, *Inf. Process. Lett.*, 53:145–152.

Henkin, Leon (1952). A problem concerning provability, *J. Symbolic Logic*, 17:160.

Henle, James M., and Eugene M. Kleinberg (1980). *Infinitesimal Calculus.* The MIT Press, Cambridge, Mass.

Hermes, H. (1973). *Introduction to Mathematical Logic.* Springer-Verlag, New York.

Hilbert, D., and P. Bernays (1968). *Grundlagen der Mathematik I, II.* Springer-Verlag, New York.

Hinman, P. G. (1978). *Recursion-Theoretic Hierarchies.* Springer-Verlag, New York.

Kalmár, L. (1957). An argument against the plausibility of Church's thesis. In *Constructivity in Mathematics*, Proc. of the Colloquium held at Amsterdam, pp. 72–80.

319

Keisler, H. Jerome (1976). *Foundations of Infinitesimal Calculus*. PWS Publishers, Boston.

————Fundamentals of model theory. In Barwise (1978), Chapter A.2, pp. 47–104.

————(1982). *Elementary Calculus; an Infinitesimal Approach*. PWS Publishers, Boston.

Kleene, S. C. (1943). Recursive predicates and quantifiers, *Trans. Amer. Math. Soc.*, 53:41–73, 1943. In Davis (1965, pp. 255–287).

LeVeque, William J. (1956). *Topics in Number Theory*, volume I. Addison-Wesley, Reading, Mass.

Löb, Martin H. (1955). Solution to a problem of Leon Henkin, *J. Symbolic Logic*, 20:115–118.

Manin, Yu. I. (1977). *A Course in Mathematical Logic*. Springer-Verlag, New York.

Mendelson, Elliott (1987). *Introduction to Mathematical Logic*, 3rd edition. Wadsworth & Brooks, Monterey, Calif.

Mostowski, A. (1947). On definable sets of positive integers, *Fund. Math.* 34:81–112.

Rasiowa, H., and R. Sikorski (1963). *The Mathematics of Metamathematics*. Państwowe Wydawnictwo Naukowe, Warszawa.

Robinson, A. (1961) Non-standard analysis, *Proc. Roy. Acad. Amsterdam Ser. A*, 64:432–440.

————(1966). *Non-standard Analysis*. North-Holland, Amsterdam. Revised ed., 1974.

Robinson, Raphael M. (1950). An essentially undecidable axiom system (Abstract). In *Proc. International Congress of Mathematicians*, volume 1, pp. 729–730.

Rogers, H. (1967). *Theory of Recursive Functions and Effective Computability*. McGraw-Hill, New York.

Rosser, J. Barkley (1936). Extensions of some theorems of Gödel and Church, *J. Symbolic Logic*, 1:87–91.

Schütte, K. (1977). *Proof Theory*. Springer-Verlag, New York.

Schwichtenberg, Helmut (1978). Proof theory: some applications of cut-elimination. In Barwise (1978), Chapter D.2, pp. 867–895.

Shoenfield, Joseph R. (1967). *Mathematical Logic*. Addison-Wesley, Reading, Mass.

Smoryński, C. (1978). The incompleteness theorems. In Barwise (1978), Chapter D.1, pp. 821–865.

————(1985). *Self-Reference and Modal Logic*. Springer-Verlag, New York.

Smullyan, Raymond M. (1961). *Theory of Formal Systems*. Ann. Math. Stud. 47. Princeton University Press, Princeton.

————(1992). *Gödel's Incompleteness Theorems*. Oxford University Press, Oxford.

Tourlakis, G. (1984). *Computability*. Reston Publishing Company, Reston, Virginia.

————(1986). Some reflections on the foundations of ordinary recursion theory, and a new proposal, *Z. math. Logik*, 32(6):503–515.

————(1996). Recursion in partial type-1 objects with well-behaved oracles, *Math. Logic Quart. (MLQ)*, 42:449–460.

————(2000a). A basic formal equational predicate logic – part I, *BSL*, 29(1–2):43–56.

————(2000b). A basic formal equational predicate logic–part II, *BSL*, 29(3):75–88.

————(2001a). Computability in type-2 objects with well-behaved type-1 oracles is *p*-normal, *Fund. Inform.*, 48(1):83–91.

————(2001b). On the soundness and completeness of equational predicate logics, *J. Comput. and Logic*, 11(4):623–653.

Veblen, Oswald, and John Wesley Young (1916). *Projective Geometry*, volume I. Ginn, Boston.

Wilder, R. L. (1963). *Introduction to the Foundations of Mathematics*. Wiley, New York.

Whitehead, A. N., and B. Russell (1912). *Principia Mathematica*, volume 2. Cambridge University Press, Cambridge.

List of Symbols

Index